Enhancing Learning Opportunities Through Student, Scientist, and Teacher Partnerships

Donna Farland-Smith
The Ohio State University, USA

IGI Global
PUBLISHER of TIMELY KNOWLEDGE

A volume in the Advances in Educational Technologies and Instructional Design (AETID) Book Series

Published in the United States of America by
IGI Global
Information Science Reference (an imprint of IGI Global)
701 E. Chocolate Avenue
Hershey PA, USA 17033
Tel: 717-533-8845
Fax: 717-533-8661
E-mail: cust@igi-global.com
Web site: http://www.igi-global.com

Library of Congress Cataloging-in-Publication Data

Names: Farland-Smith, Donna, editor.
Title: Enhancing learning opportunities through student, scientist, and teacher partnerships / Donna Farland-Smith, editor.
Description: Hershey : Information Science Reference, 2021. | Includes bibliographical references and index. | Summary: "This book will serve as a resource for those in STEM education, both in formal settings and informal settings and bring together all different kinds of student-scientist- teacher partnerships programs with research documentation in one source"-- Provided by publisher.
Identifiers: LCCN 2020038795 (print) | LCCN 2020038796 (ebook) | ISBN 9781799849667 (Hardcover) | ISBN 9781799876182 (Softcover) | ISBN 9781799849674 (eBook)
Subjects: LCSH: Science--Study and teaching--Research. | Technology--Study and teaching--Research. | Engineering--Study and teaching--Research. | Mathematics--Study and teaching--Research.
Classification: LCC Q181 .E632 2021 (print) | LCC Q181 (ebook) | DDC 507.1--dc23
LC record available at https://lccn.loc.gov/2020038795
LC ebook record available at https://lccn.loc.gov/2020038796

This book is published in the IGI Global book series Advances in Educational Technologies and Instructional Design (AE-TID) (ISSN: 2326-8905; eISSN: 2326-8913)

British Cataloguing in Publication Data
A Cataloguing in Publication record for this book is available from the British Library.

For electronic access to this publication, please contact: eresources@igi-global.com.

Advances in Educational Technologies and Instructional Design (AETID) Book Series

Lawrence A. Tomei
Robert Morris University, USA

ISSN:2326-8905
EISSN:2326-8913

Mission

Education has undergone, and continues to undergo, immense changes in the way it is enacted and distributed to both child and adult learners. In modern education, the traditional classroom learning experience has evolved to include technological resources and to provide online classroom opportunities to students of all ages regardless of their geographical locations. From distance education, Massive-Open-Online-Courses (MOOCs), and electronic tablets in the classroom, technology is now an integral part of learning and is also affecting the way educators communicate information to students.

The **Advances in Educational Technologies & Instructional Design (AETID) Book Series** explores new research and theories for facilitating learning and improving educational performance utilizing technological processes and resources. The series examines technologies that can be integrated into K-12 classrooms to improve skills and learning abilities in all subjects including STEM education and language learning. Additionally, it studies the emergence of fully online classrooms for young and adult learners alike, and the communication and accountability challenges that can arise. Trending topics that are covered include adaptive learning, game-based learning, virtual school environments, and social media effects. School administrators, educators, academicians, researchers, and students will find this series to be an excellent resource for the effective design and implementation of learning technologies in their classes.

Coverage

- Instructional Design
- Social Media Effects on Education
- Web 2.0 and Education
- Instructional Design Models
- Curriculum Development
- Game-Based Learning
- E-Learning
- Digital Divide in Education
- Higher Education Technologies
- Virtual School Environments

IGI Global is currently accepting manuscripts for publication within this series. To submit a proposal for a volume in this series, please contact our Acquisition Editors at Acquisitions@igi-global.com or visit: http://www.igi-global.com/publish/.

The Advances in Educational Technologies and Instructional Design (AETID) Book Series (ISSN 2326-8905) is published by IGI Global, 701 E. Chocolate Avenue, Hershey, PA 17033-1240, USA, www.igi-global.com. This series is composed of titles available for purchase individually; each title is edited to be contextually exclusive from any other title within the series. For pricing and ordering information please visit http://www.igi-global.com/book-series/advances-educational-technologies-instructional-design/73678. Postmaster: Send all address changes to above address.

Titles in this Series

For a list of additional titles in this series, please visit:
http://www.igi-global.com/book-series/advances-educational-technologies-instructional-design/73678

Applying Universal Design for Learning Across Disciplines Case Studies on Implementation
Frederic Fovet (Royal Roads University, Canada)
Information Science Reference • © 2021 • 335pp • H/C (ISBN: 9781799871064) • US $195.00

Advancing Online Course Design and Pedagogy for the 21st Century Learning Environment
Daniel Chatham (Middlebury Institute of International Studies at Monterey, USA)
Information Science Reference • © 2021 • 382pp • H/C (ISBN: 9781799855989) • US $195.00

Affordances and Constraints of Mobile Phone Use in English Language Arts Classrooms
Clarice M. Moran (Appalachian State University, USA)
Information Science Reference • © 2021 • 229pp • H/C (ISBN: 9781799858058) • US $175.00

Examining an Operational Approach to Teaching Probability
Alessio Drivet (Geogebra Institute of Turin, Italy)
Information Science Reference • © 2021 • 369pp • H/C (ISBN: 9781799838715) • US $195.00

4C-ID Model and Cognitive Approaches to Instructional Design and Technology Emerging Research and Opportunities
Guilhermina Maria Lobato Ferreira de Miranda (Instituto de Educação, Universidade de Lisboa, Portugal) Manuel Joaquim Henriques Rafael (Universidade de Lisboa, Portugal) Mário Marcelino Luis de Melo (UI, DEF, Instituto de Educação, Universidade de Lisboa, Portugal) Joana Martinho de Almeida Costa Pardal (Instituto Universitário de Lisboa (ISCTE-IUL), ISTAR, Portugal) and Thiago Bessa Pontes (Universidade Federal do Cariri, Brazil)
Information Science Reference • © 2021 • 243pp • H/C (ISBN: 9781799840961) • US $175.00

Implementing Augmented Reality Into Immersive Virtual Learning Environments
Donna Russell (Walden University, USA)
Information Science Reference • © 2021 • 266pp • H/C (ISBN: 9781799842224) • US $195.00

Handbook of Research on Innovations in Non-Traditional Educational Practices
Jared Keengwe (University of North Dakota, USA)
Information Science Reference • © 2021 • 462pp • H/C (ISBN: 9781799843603) • US $245.00

Handbook of Research on Using Educational Robotics to Facilitate Student Learning
Stamatios Papadakis (University of Crete, Greece) and Michail Kalogiannakis (University of Crete, Greece)
Information Science Reference • © 2021 • 651pp • H/C (ISBN: 9781799867173) • US $275.00

IGI Global
PUBLISHER of TIMELY KNOWLEDGE

701 East Chocolate Avenue, Hershey, PA 17033, USA
Tel: 717-533-8845 x100 • Fax: 717-533-8661
E-Mail: cust@igi-global.com • www.igi-global.com

Table of Contents

Detailed Table of Contents

The purpose of this chapter is to show the long-lasting impacts of educator-scientist-student partnerships on both educators and students as perceived through the eyes of an educator. The chapter chronicles the educator's experiences and changes in teaching philosophy before, during, and after participation in the PolarTREC program. The goal of the chapter is to inform the reader on the benefits of participation in educator-scientist-student partnerships and encourage the reader to participate in future partnerships.

Partnerships between educators and science faculty can be professionally productive and personally fulfilling, but developing such partnerships can be challenging. This chapter describes the career trajectory of a geoscience faculty member who evolved from a focus on research within his science discipline to a portfolio of activities including disciplinary science research, geoscience education activities, and geoscience education research. The author provides suggestions for how to identify potential science faculty partners and how to develop and maintain successful partnerships.

This chapter focuses on the language and interactions of a visiting scientist and a group of second-grade students (N=18). Guiding this study was the understanding that a student-scientist-teacher partnership supports a social cultural perspective that provides a zone of proximal development in a multifaceted

number of ways to do just this. As a result, teachers and students can be made aware of the professional work of scientists and the ways in which they can participate in and mentor others to engage in authentic research both in and out of the classroom.

Section 3

This chapter describes an interdisciplinary program between a College of Education and a College of Natural Resources and their partnerships with rural high schools, regional colleges, and bioeconomy industries. The overarching goal of the program was to provide engaging professional development and support for teachers and diverse undergraduate students to prepare and promote diverse students to consider STEM majors and careers related to bioproducts and bioenergy. A team of faculty and graduate students from a Forest Biomaterials department and a Science Education department developed online courses, workshops, and laboratory activities and internship placements for undergraduate students and high school science and CTE teachers. This chapter details the need for the program, its context, online course development, and laboratory activities. For each of the key partners in the program—the faculty and graduate students, the high school teachers, and the undergraduate students—key strategies, lessons learned, and recommendations are shared.

University students go to school several times a week, as voluntary supporters, and help teachers not only conduct lessons, but also create self-made teaching materials with newly developed technologies

Preface

"Alone we do so little, together we do so much." Helen Keller

In every partnership collaboration is key. The benefits are mutual. In the case of student-scientist-teacher partnerships a scientist might learn a new way to communicate an idea in a café (Ch 5) or learn the real meaning of science *education* after 20 years (Ch 2) or the rules of engagement in an elementary classroom (Ch 3). A teacher may benefit as a member from such relationships by traveling to Antarctica to see how scientific research is collected and analyzed (Ch 1), or take part in professional development developed and delivered by content experts even though they live in a rural part of the state (Ch 7). Students benefit when a newly developed technology allows them to read (Ch 8), or when they are able to meet and chat with five different scientists over Skype and completely changes their idea about who scientists are, where they work and what they do (Ch 4).

Student-scientist-teacher partnerships are not limited to K-12 classroom settings. Scientists can work with the general public (CH 5; Ch 10; Ch 11; Ch 12) as a member of these partnerships. Scientists can work with higher education (Ch 7; Ch 8; Ch 9; Ch 10; Ch 11; Ch 12) in their fields of scientific disciplines of teacher education programs.

Through a collaborative approach in science education, we are wiser, stronger, more appreciative, and more aware of other's roles and responsibilities and quite simply more knowledgeable. One thing is for certain, scientists are story tellers, and they share bits and pieces of their lives and their careers with educators and younger generations through online interactions (Ch 4), Cafes (Ch5), and through authentic experiences (Ch 6) These partnerships humanize the field of science and attract others to science allowing the public to see people who choose science as a career. The surprise is, they are just like us. Or perhaps we are just like them?

Historically, there have been may different kinds of student- scientist- teacher partnerships, however they have not always been labeled as such. The earliest examples date back to the 1800s when the US Weather Service used volunteers to record and documented rainfall and air temperature (Firehock & West, 1995). This is very similar to the term 'citizen scientist' that refers to participates as field assistants in scientific studies who are volunteers. Citizen scientists represent a partnership between volunteers and scientists to answer real-world problems (Cohen, 2008).

The late 1990's brought two science reform efforts, although generated independently, they both called for science literacy. Project 2061 (AAA, 1989) and the National Science Education Standards (NRC, 1995) both had clear language around students achieving scientific literacy. This is when a re-birth of student/scientist partnerships (SSP) became in vogue. The definition of the student/scientist partnership defined by Lawless and Rock (1998) was an approach to introducing students to hands-on science by involving them with on-going student/scientist investigations. The only problem with this partnership was that it left an extremely valuable member, the classroom teacher.

The emphasis to help students see science as a human endeavor has been consistent in the following documents: *National Science Education Standards* (NRC,1996), *A Framework for K-12 Science Education: Practices, Crosscutting Concepts, and Core Ideas* (NRC, 2012) and *Next Generation Science Standards* (NGSS, 2013). Reform documents highlight science as a human endeavor to include the ways in which men and women of different races and ethnicities engage with science and technology have contributed to science. In an attempt to "see scientists as humans" educators focused on bringing them into their classrooms as visitors.

Visiting scientist programs were designed to improve children's perceptions of scientists. In such studies like Bozdin and Gerhinger (2001) and Flick (1990) these approaches worked to decrease stereotypical beliefs about scientists. However, little research exists to document short-term scientists' effectiveness in classroom programs (Larson, Liston, Thiry, & Graf, 2007). In another study, no change in students' stereotypical images of scientists when three female scientists visited an upper elementary classroom (Buck, Leslie-Pelecky & Kirby, 2002). Therefore, the kinds of visits and relationships between students, scientists, and teachers are infinite and the benefit of having this book as a resource for students-scientist-teacher partnerships is to gather several examples of quality programs/partnerships with research in one place.

Student-teacher scientist partnerships (STSPs) as defined by (Lawless & Rock, 1998, Sadler et al., 2010) include students and teachers partnering with scientists to gather data or analyze data in relation to the work the scientist does. The relationship involves three active members students, scientists, and teachers either in actual research or in some other capacity.

Student-scientist-teacher interactions provide students with several advantages. They provide opportunities to interact with experts and professionals in the field, give students a chance at meeting a role model that may impact students' career choices, and increase awareness of available career options combined with an understanding of how their skills and interests affect their career decisions. Additionally, they have the potential to enhance attitudes and interest toward STEM professions for students and grant opportunities to connect with scientists as human beings and see them as "real people," replacing stereotypical perceptions of scientists. Moreover, there are many advantages for the teacher or informal educator when these partnerships are established. For these reasons and more, numerous studies are often conducted involving the partnerships of students, scientists, and teachers.

It is important to think broadly about who can serve in the role of students. We do not want to limit this role to K-12 students. It can include college students as well. For these reasons it is also important to think broadly in terms of who is a scientist and who is not. A scientist does not have to be a professional scientist. A scientist in this partnership is any individual with the scientific knowledge. The same is true of the role of the teacher, this includes preschool educators, informal educators, and college professors.

ORGANIZATION OF THE BOOK

The book has organized the 12 chapters into four sections based on the age of the students. A brief description of each of the chapters follows:

Section 1 is made up of two chapters each with their own perspective of the student, scientist, teacher partnership. The first chapter is the perspective of a teacher and the second chapter is the perspective of a scientist. We regret that we do not have a students' perceptive to include in this section.

Chapter 1, "A Voice from the Field: One Teacher's Perspective on Educator-Scientist-Student Partnerships," is a teacher's journey through four different out of the classroom experiences working with scientists. Her reflection is included so that others can learn from it those that may not be teachers can understand the extreme sacrifices she went through to participate in these types of programs. Those who are educators will benefit from knowing these kinds of opportunities exist.

Chapter 2, "Meet Them Where They Are (A Perspective That Works Both Ways): Reflections on Scientist-Educator Collaborations," is one scientist's journey in collaborating with teachers nd teacher educators over a thirty-five year career. His reflection is included so that other can learn from it those that may not be scientists can understand how scientists can integrate into education. Those who are scientists will benefit from navigating the academic pressures and working with educators, while knowing these kinds of opportunities exist. He highlights different stages in one's career and how much of his view of science education changed as a result of these partnerships.

Section 2 is composed of four chapters at the K-12 level in which students, scientists and teachers collaborate to enhance learning outcomes in very different ways. The first chapter in this section deals with elementary, or second grade students. The second chapter in this section discusses middle school, or seventh and eighth grade students. The third and fourth chapters in this section

involves high school students.

Chapter 3 is "The Role of the Knowledgeable Other in Student-Scientist-Teacher Partnerships: An Examination of a Second Grade Classroom." When Dr. Z visited, he brought artifacts of animals, such as bones and skulls and other props to support the engagement and experience for Ms. Smith's young students. What began as a way to enrich his daughter's class lasted as a student-scientist-teacher partnership year after year. This research looked at the *role of the "more knowledgeable other"* in student-scientist-teacher partnerships in elementary classroom. At times the scientist served in that role, at times it was the teacher, still in other exchanges it was the classroom teacher. In this partnership, the students are identified as (second grade students), the teacher (second grade teacher), and the scientist (Veterinarian).

Chapter 4, "'The Way I Think About Scientists Has Changed': Measuring the Impact of Virtual Scientists on Middle School Students' Perceptions of Scientists," focuses on the impact a weeklong series of virtual visits between five scientists and middle school students (N=48) had on students' perceptions of scientists and their work. Overall, based on students' self-reflections and results from Enhanced-Drawings of Scientists Test (E-DAST), the week of virtual discussion-based interactions broadened students' perceptions of what scientists do (*activity*), where scientists work (*location*), and what scientists look like (*appearance*). The virtual discussion-based experiences had the most significant impact on middle school students' perceptions of the work scientists do (*activity*). As an important finding demonstrates—through middle school students own words—that students' perceptions of the *appearance* and *location* of scientists may be 'gatekeepers' for the *activity* of scientists. In other words, until students can 'get past' or debunk the stereotype of what scientists look like (*appearance*) and where they work (*location*), students may not truly understand the work scientists do (*activity*). In this partnership, the students are identified as (middle school), the teacher (middle school teacher and college professor), and the scientists (five different scientists selected by the classroom teacher).

Chapter 5, "Enriching the Lives of High School Teenagers With Science Cafés," describes the development of the Teen Science Café Network whose mission is to connect high school teenagers with science and scientists. This chapter not only describes the history of the program, but outlines the roles involved and research collected. Since its inception, teen science cafes have been launched by a myriad of organizations, some with adult leaders and others that are organized completely by teens. They have been implemented in rural areas, and cities to serve over 160,000 teens. In this partnership, the students are identified as (high school students), the teacher (none), and the scientists (scientists selected to speak at the cafe).

Chapter 6, "Learning About the Authentic Nature of Scientific Work From Professional Scientists: Preparation for Research Participation and Mentoring," discusses the student, teacher, scientist, partnership by focusing on authenticity and claims that in order to make students' work in the classroom more authentic, teachers must connect with real working scientists. Science educators then need to work with science teachers to help them translate this knowledge of authenticity into realistic authentic classroom experiences for their students. In this partnership, the students are identified as (high school), the teacher (college professor), and the scientist (Chemist).

Section 3 is composed of three chapters in which student, scientist, teacher partnerships occur at the college level. The first chapter in this section does include high school students but the bulk of the partnership is about the students at the college level and for this reason is included in this section. The second chapter in this section is our only international chapter and special education dealing with university preservice teachers? The third chapter in this section is about preservice teachers serving the role of the students.

Chapter 7, "An Interdisciplinary Model to Diversify STEM Participation: College, High School, and Industry Partnerships," describes an interdiscinplinary program that provided engaging professional development for teachers and diverse undergraduate students to prepare diverse students to consider STEM majors and promote regional careers related to bioproducts and bioenergy. In this partnership, the students are identified as (undergraduate and graduate), the teacher (college professor), and the scientists (Researchers from biotech fields).

Chapter 8 is "University Students-Schoolteachers Partnership With Newly Developed Technologies." The collaborative research works of the university professor with special needs students in Japan, preservice teachers, and the ICT business companies are crucial to produce newly creative and gifted software and tools for students with various reading and speaking challenges. This chapter paper briefly presents newly developed software and tools used for class lessons, and then it describes fruitful collaboration with schoolteachers in creating self-made teaching materials and conducting activities. This chapter follows the development of the state-of-the-art Grid Onput dot-code technology used to in creating original self-made teaching materials. In this partnership, the students are identified as (university students-preservice teachers), the teacher (schoolteachers in Japan), and scientists (ICT business companies) to develop the technology.

Chapter 9, "Connecting Pre-Service Teachers to Astronauts and the International Space Station Through an Educational Downlink: An Out-of-This-World Experience," discusses the steps their institution took to host a student teacher scientist partnership with the focus on preservice elementary teachers connected with scientists in space thus initiating the Year of Education on the Station former classroom teachers and astronauts. Two former teachers on International Space Station. Through the use of a downlink, a way for students on Earth to speak with astronauts in space, they were able to communicate. Sounds simple, but setting up this all up from an institutional perspective had its challenges. The partnership between NASA and UHD and a structure for the partnership was developed. Three goals, strengthening teacher candidates science content knowledge, increasing the quality and quantity of hands-on opportunities for teacher candidates to engage in the inquiry process, and introducing pre-service teachers to and engaging them in NASA's educational resources. In this partnership, the students are identified as (preservice education), the teacher (college professor), and the scientists (NASA astronauts).

Section 4 is composed of three chapters in which the role of scientist is placed in the general public, aka, informal science spaces. The first chapter in this section uses college students in the role of students. The second chapter discusses how one science center implores student, teacher, scientist partnerships. The third chapter looks at zoos and aquariums and their role in student, scientist, teacher partnerships.

Chapter 10, "Training Students to Conduct Outreach in an Informal Science Learning Environment," describes a course that teaches undergraduate and graduate students to engage in accessible public communication about language science topics within a science museum. In this partnership, the students are identified as (undergraduate and graduate), the teacher (college professor), and the scientists (Researchers from the College of Optometry, the School of Education's Exercise Science program, Communications department, College of Pharmacy).

Chapter 11 is "Real-World Connections With Conservation Science Through Zoos and Aquariums." In this partnership, the students are identified as (undergraduate and graduate), the teacher (college professor), and the scientists (Researchers from the College of Optometry, the School of Education's Exercise Science program, Communications department, College of Pharmacy).

Chapter 12 is "The Potential for Science Centers in Student-Teacher-Scientist Partnerships: Explorations From One Museum." In this partnership, the students are identified as (undergraduate and graduate), the teacher (college professor), and the scientists (Researchers from the College of Optometry, the School of Education's Exercise Science program, Communications department, College of Pharmacy).

Enhancing Learning Opportunities Through Student, Scientist, and Teacher Partnerships is an organized collection of student-scientist-teacher partnerships with past and present models. It includes research, justifications, benefits, implementation, and learning outcomes of these interactions. This book presents a variety of different scientist-student-teacher partnerships with research data to support different learning outcomes in settings like schools, after-school programs, museums, science centers, zoos, aquariums, children's museums, space centers, nature centers, and more. This book is ideal for in-service and preservice teachers, administrators, teacher educators, practitioners, stakeholders, researchers, academicians, and students interested in research on beneficial student-scientist-teacher partnerships/models in formal and informal settings. It may be helpful to anyone writing grants or looking to replicate models of beneficial student-scientist-teacher partnerships/programs. This handbook serves as a contribution to the field of science education as a collection of different models of these unique, timely, and interesting partnerships. This is why the idea of students, scientists, and teachers collaborating together are so promising for our future; they each have strengths the other may not.

REFERENCES

American Association for the Advancement of Science (AAAS). (1989). *Project 2061: Science for all Americans*. Oxford Press.

Cohen, J. (2008). Citizen Science: Can volunteers do real research? *Bioscience*, *58*(3), 192–197. doi:10.1641/B580303

Firehock, K., & West, J. (1995). A brief history of volunteer biological water monitoring using macroinvertebrates. *Journal of the North American Benthological Society*, *14*(1), 197–202. doi:10.2307/1467734

Lawless, J., & Rock, B. (1998). Student Scientist Partnerships and Data Quality. *Journal of Science Education and Technology*, *7*(1), 5–13. doi:10.1023/A:1022575914118

National Research Council (NRC). (1995). *National Science Education Standards*. Academy Press.

Section 1

Chapter 1
A Voice From the Field: One Teacher's Perspective on Educator-Scientist-Student Partnerships

Lindsay M. Smith
Mooresville Graded School District, USA

ABSTRACT

The purpose of this chapter is to show the long-lasting impacts of educator-scientist-student partnerships on both educators and students as perceived through the eyes of an educator. The chapter chronicles the educator's experiences and changes in teaching philosophy before, during, and after participation in the PolarTREC program. The goal of the chapter is to inform the reader on the benefits of participation in educator-scientist-student partnerships and encourage the reader to participate in future partnerships.

INTRODUCTION

Teachers become teachers for several reasons, but for me and many of the teachers I have worked with, the reason lies in the proverbial "lightbulb" or "ah-ha" moments we have with our students. We want to inspire the next generation, we want to feel fulfilled by the lessons we painstakingly create, and we want to impact lives. The routes teachers take to achieve lightbulb moments vary, but for me the path I choose to follow was educator-scientist-student partnerships. In my 18 years in the education profession I have had the privilege of participating in several of these partnerships and the resulting lightbulb moments have been plentiful.

The purpose of this chapter is to show the long lasting impacts of educator-scientist-student partnerships on both educators and students. It includes my own personal thoughts and experiences written as a chronological reflection of my path as an educator before, during and after my PolarTREC experience. Throughout the chapter I will address several programs I have been involved with in the last ten years. The formal descriptions of the programs are as follows:

PolarTREC: PolarTREC is created and administered by the Arctic Research Consortium of the United States and funded by the National Science Foundation. PolarTREC educators (formal and informal) come

DOI: 10.4018/978-1-7998-4966-7.ch001

from the United States and spend 3-6 weeks participating in hands-on field research experiences in the Arctic or Antarctica, working side by side with scientists. The program has been around for over 10 years and has provided over 170 U.S. teachers a unique opportunity to work with scientists in the polar regions.

NOAA Teacher at Sea: The mission of the National Oceanic and Atmospheric Administration's (NOAA) Teacher at Sea Program is to provide educators hands-on, real-world research experience working at sea with world-renowned NOAA scientists, thereby giving them unique insight into oceanic and atmospheric research crucial to the nation. The program provides a unique opportunity for informal and formal educators, pre-kindergarten through college-level, to sail aboard NOAA research ships to work under the tutelage of scientists and crew. Since its inception in 1990, NOAA's Teacher at Sea Program has enabled over 800 teachers to gain first-hand experience of science and life at sea.

Albert Einstein Distinguished Educator Fellowship: The Albert Einstein Distinguished Educator Fellowship (AEF) Program provides a unique opportunity for accomplished K-12 educators in the fields of science, technology, engineering, and mathematics (STEM) to serve in the national education arena. Fellows spend eleven months working in Federal agencies or in U.S. Congressional offices, applying their extensive knowledge and classroom experiences to national education programs and/or education policy efforts.

Ocean Exploration Trust Science Communications Fellowship: The Science Communication Fellowship immerses formal and informal educators in the *Nautilus* Corps of Exploration as communicators for exploration and empowers them to bring the excitement of ocean exploration to students and public audiences around the world. Fellows share accounts of ocean research, expedition operations, STEM careers, and daily life with audiences of all ages through live audio commentary on the *Nautilus* Live website and through outreach to classrooms and communities via ship-to-shore live interactions.

My hope is for my reflection to be an inspiration to others to participate in educator-scientist-student partnerships. Formal and informal educators, their students, and researchers should pursue partnerships because there are huge benefits to be gained by all. Although I have never had an issue with school administrators being unsupportive, I would point out to leery administrators that research has shown that teachers' participation in research results in improvements in students' achievement in state science benchmark tests (Silverstein et al., 2009). There are also immense benefits for students and teachers as seen in the formal evaluations of both the PolarTREC and NOAA Teacher at Sea programs. Participating educators report that their students are more engaged in their science learning, know more about science as it applies to the world's oceans/atmosphere/climate, and have an appreciation for the relevance of scientific research (Simmons, 2010). Teachers' content knowledge has increased, teacher efficacy in science instruction has increased, and the programs have continued to impact teacher instruction well after their expeditions have ended (ARCUS, 2015). Wherever your motivations lie, you will never know what the benefits of an educator-scientist-student partnerships will bring, unless you try. As you will read in my following reflection, take a risk and your path might be changed forever.

MY START IN EDUCATION

When did you know you wanted to be a teacher? This is a question I have been asked countless number of times by my curious students. Many teachers would be able to respond that they knew from a very early age. My answer is not that straightforward. Growing up I was not a model student and being a teacher and spending more time in school was not on my list of what I wanted to be when I grew up.

Why then would I choose to spend the last 18 years of my life in a classroom? The answer is simple and not simple at the same time and lies in a series of experiences.

Growing up I loved science and nature. I had a million questions about the world around me and I was dead set on finding answers to those questions. It should be no surprise that throughout my childhood I had dreams of being a scientist and I went on to study biology in college. It was not until I had the opportunity to actually conduct field research as a part of the National Science Foundation's Research Experiences for Undergraduates (REU) program that I learned my favorite part about research science was communicating the science, not doing the science. I loved talking to people about my passion for science and getting them excited about it too. That fall, when I returned to Michigan State University, I started taking education classes and changed my career goal to science education. I had worked as a camp counselor for several summers leading up to that point and becoming a science teacher turned out to be the perfect blend of my loves for communicating science and influencing the lives of teenagers.

Through my 18 years in the education profession my teaching philosophy has changed. After graduating from Michigan State, I got my first job as a biology teacher in a school district just outside of Detroit, Michigan. My goals were lofty, and I had dreams of influencing all my students to be the next generation of scientists. Reality set in as I struggled through those first few years and learned important lessons about classroom management, the lack of funding for education, and the wants and needs of my students. I thought all my students would love science as much as me, but unfortunately many entered my classroom not liking science at all. I had to change my philosophy on teaching science and today it can be summarized by a phrase I use often: "science is a verb". Students are not inspired by learning about science, they are inspired by doing science.

Fast forward ten years and I currently teach ninth grade Environmental Earth Science in a school district outside of Charlotte, North Carolina. It is a new state, but the story is still the same, my students have lost the love of science they all had as elementary students and my goal is to reacquaint them with that love. I hook them by letting them get their hands dirty, taking them outside, and incorporating current science interesting to them into my lessons. As an environmental science teacher, I also have the goal of motivating my students to be change-makers. I tell my students on the first day of class that my ultimate goal for them is to be good humans. Through service learning each year my students design and implement projects to better our community while also learning about social and environmental issues in our town. As a result, many of my students have found their love of science again and have been motivated to effect change in our community.

Although I did not intend to be a science teacher, there is nothing else I would rather be. My proudest accomplishments have come as a result of being a teacher. There is no greater joy than to see that lightbulb moment when a child finally understands a concept or smiles when they enter your classroom. I am immensely proud when I see my students overcome disabilities and hardships, make a difference in the lives of others, or become first generation college students. I am also proud to work amongst amazing educators on a daily basis and I have been honored to represent my profession as the Teacher of the Year in both districts I have taught. How I got here has been both a challenging and rewarding journey, but the most impactful experiences on my career have been my partnerships with students and scientists.

POLARTREC: MY FIRST TEACHER RESEARCH EXPERIENCE

On a daily basis I receive countless emails informing me on practices, resources, and opportunities for educators and like most educators I skim and delete most of the emails as I try to stay on top of my growing inbox while simultaneously teaching thirty students. The day I received a notice about the PolarTREC program was different. I actually had time to open the email and when I did I saw a picture of a smiling teacher surrounded by snow and ice. It immediately caught my attention. I read on and learned that the program paired educators with scientists conducting research in the polar regions. It sounded amazing and the teacher even got to accompany the researcher into the field. I instantly wanted to apply, but I was intimidated. I had only been teaching for 5 years and I had never applied for something like this before. It took the encouragement of my friends and family to realize that the benefits of the program for both my students and myself far outweighed my fear of failure.

As I read the PolarTREC application, I thought of my 10th grade Biology teacher. To this day I can still remember her and her travels to Europe and South America. She was an amazing teacher and one of the few teachers I listened to as a high school student. This is dating me, but she would get out her slide projector and click through the slides of her adventures and I was mesmerized. The farthest I had ever traveled was to Florida to see my grandparents. She had been to so many places and had seen so many things. I wanted to be just like her, and she is a big reason why I graduated with a bachelor's degree in Biology and a master's degree in Environmental Science. The simple act of showing a slide show changed my future.

As I thought about my Biology teacher and the impact she had on me, I saw PolarTREC as my opportunity to do the same for my students. Many of my students had never been outside of the city limits of Detroit and had very few role models in the science community. Most of my students thought scientists were only on the Discovery Channel and there was no way they could ever be one. I wanted to show them I could take a risk, overcome my fears, and accomplish a goal I had never even dreamed of. I wanted to be their role model and introduce them to a diverse set of role models so they would see themselves as scientists. I also wanted to apply because of the science being conducted in the polar regions. *An Inconvenient Truth* had just been released and climate change was quickly becoming a topic my students and myself were interested in. I saw PolarTREC as a way to both inspire and educate my students. Additionally, I saw PolarTREC as an opportunity for my own personal growth. Though I only had five years in the profession, I was already feeling stale. I wanted to apply to PolarTREC to meet other like-minded educators, develop new lessons based on current scientific knowledge, and see what education was like beyond my classroom walls.

Although my motivations were genuine, to make them a reality I would need to be selected into the program and the application process was not easy. PolarTREC is a very competitive program with hundreds of teachers applying for only a few dozen positions. The application consisted of several essays, letters of recommendation, and even interviews. To say I was intimidated would be an understatement. I had never applied to a program like PolarTREC and I knew I needed help. The first person I went to was my school principal. I was worried that he might not approve of me applying, but as I told him about the program and the benefits to our students I knew he was onboard. He encouraged me to apply, helped me brainstorm ideas for expanding the reach of my experience throughout the entire district and into the community, and he wrote an amazing recommendation letter. With his help I was able to transform all the ideas dancing in my head on the benefits of PolarTREC for my students and myself onto paper. When I completed my essays, I was proud of myself for taking a risk, trying something new, and for being brave enough to approach my principal and seek help. As I hit submit on my application, nervous butterflies welled up in my belly. I had done my best and I had represented myself well. Now it was up to the review committee of scientists and educators to decide and for the waiting to begin.

I can remember the phone call telling me I was accepted like it was yesterday. I was in my classroom after school, and I screamed so loud the teacher next door ran in to see if I was okay. It had been months since I had hit submit on my application and I was completely taken by surprise. I was convinced I would never be selected. I immediately ran down the hallway and burst into my principal's office to tell him the good news. We called the superintendent, and she was equally excited. Our school district was going through some difficulties and my administration was elated to have something positive to talk about and to share with the community. When I called my husband and my parents, I could tell they were surprised but they were also immensely proud. I was blessed to have the unwavering support of both my school district and family. They all played a part in me being selected as a PolarTREC teacher and we were headed on this adventure together.

My excitement was still overflowing the next day at school. I was so eager to tell my students I was going to Antarctica, and I expected them to be equally excited. Their response was far less than what I expected. After getting out a map and showing them where Antarctica was, many thought I was crazy. They didn't understand why I would want to go there and why the research I would be doing was important. I had my work cut out for me if I was going to make this a meaningful experience. Thankfully, I had plenty of time to prepare myself and my students for my expedition. I was accepted into the program in January and I would not depart for Antarctica until October. My preparation started in February when I attended training with all the other PolarTREC teachers in Fairbanks, Alaska. It was one of my first professional development experiences where I was surrounded by other like minded educators. I was one of the youngest teachers and I had so much to learn from the veteran teachers. I soaked up everything like a sponge and I left Fairbanks feeling prepared for my expedition, having tons of ideas for my classroom, and having built my first professional learning community with educators I still remain in contact and continue to collaborate with today.

After Fairbanks, I had lofty goals for how I was going to make my PolarTREC experience meaningful for my students and others beyond my classroom. Step one was to better understand the research I would be conducting and to establish a relationship with the scientists I would be working with. To do this, I was able to visit my researcher at his lab at Montana State University. I met his team, received lots of reading material, visited his lab to see where our samples would be analyzed, and got to know my researcher by going on a hike. I left Bozeman feeling more comfortable with the people I would be in the field with and I felt more prepared to be a communicator of their research. Step two was getting my students excited and engaged. I collaborated with other PolarTREC teachers going to Antarctica to create lessons on Antarctica, the science of ice, climate change in the polar regions, and for my expedition microbiology and glaciology. The more my students learned about Antarctica and the science I would be doing, the more excited they got. My students also worked with the researchers to develop scientific questions and experiments I could do in the field and video for them to view while I was gone. Examples of these experiments included seeing which would melt first dirty or clean glacial ice, swabbing different surfaces to see if bacteria would grow and reproduce in Antarctica, collecting daily weather data, and measuring the pH and conductivity of glacial melt water. Lastly, I wanted to expand my reach beyond my classroom to impact others in my community. I visited elementary school classrooms and told them about my expedition, how to follow me online, and even let them try on the Antarctic extreme cold weather gear. I also promoted my expedition to local newspapers and news outlets. My favorite was a live newscast from my classroom where their tagline was "she would go to the end of the Earth for her students".

The night before my flight to Christchurch, New Zealand I could barely sleep. I had so much nervous energy about the adventure I would begin the next morning. Time seemed to inch by until it was finally time to go to the airport. It would take over 24 hours of traveling to get to Christchurch, but I was ready with my giant duffel bag, backpack full of communication equipment, and my favorite pillow. When I finally arrived in Christchurch there was barely time for sleep. We only had a few short days to prepare for our final leg of our journey to Antarctica and, unfortunately, all of our baggage was still in Sydney after missing the connecting flight. Luckily, the US Antarctic Program provided all the cold weather gear so we would not be completely left unprepared. The first thing we did was go to the Clothing Distribution Center and get geared up. I am not the most organized person and it was a stretch for me to make sure I had all the correct gear and it was sorted into five bags. One bag would meet me in Antarctica, one would hold all my extreme weather gear that I would wear on the plane to Antarctica, another would hold my carry-on items, one would be my boomerang bag and have gear for New Zealand if the plane had to turn around because of weather, and the last was any gear I didn't want to go with me and would stay in New Zealand. Thankfully, most of my team had been to Antarctica before and they walked me through the process and made sure I was ready for the adventure that would begin the next morning.

To call my time in Antarctica an adventure is putting it lightly. I would call it life-altering. From the moment I stepped off the C-17 and onto the ice I was struck by the beauty of Antarctica. It was incredible. There was ice in every direction, beautiful blue skies, and a line of jagged mountains along the horizon. However, I was not in Antarctica for a sightseeing trip, I was there to work. My job was to communicate the science being done by my research team to my students and community as well as students and teachers across the world. I also had to support my team and be an active participant in the research they were doing and be a productive member of our field camp. Before I could do any of those things I had to complete several safety trainings. My favorite was a two-day training called "happy camper school". During this training I learned how to light a camp stove, build a shelter, cook food, use radios, and recognize the signs of frostbite and hypothermia so I would be prepared to live in a remote field camp. In addition to training, we also had to prepare all our gear. We spent several days in McMurdo assembling and packing our field equipment, sleeping gear, tents, and food. We would be in a very remote location only accessible by helicopter during good weather for the next month, so it was very important to make sure we had everything and it was working. Any mistakes could have had severe consequences and I was thankful my team members were veterans at this process.

As a Biology teacher one of the topics I taught was the parameters necessary for life. Prior to my PolarTREC experience my students and I would discuss the "things" life needed for survival and in our textbook it would list parameters for temperature, elevation, salinity, oxygen, etc… That lesson would never be the same after I stepped foot in the Dry Valleys of Antarctica. The Dry Valleys are one of the driest, coldest, windiest places on Earth making it more similar to Mars than anywhere else on our planet. It was once thought that the Dry Valleys were devoid of life. How could life possibly exist there in such extreme conditions and if life does exist there, what is it? Those questions are what brought us to Antarctica; and I told my students the goal of our expedition was to prove our textbook wrong. By taking samples of ice from inside a glacier we were going to show that life could exist in extremely cold, low oxygen, low nutrient conditions without the presence of light. My students didn't think we could do it, but we proved them and our textbook wrong.

For four weeks we chainsawed and jack hammered our way inside the glacier until we had a 15 meter long tunnel perfect for collecting ice samples. It was hard work. I had never used a chainsaw or jackhammer before, but I was determined to be a functioning member of my team and do everything they did. I pushed myself mentally, physically, and emotionally over the course of that month and it was amazing. I was proud of myself, and I knew my students would be proud of me too. I had so many stories to tell them when I returned. I had experienced living in a cold desert with only my tent, sleeping bag, and hot water bottle to keep me warm in -40 degree temperatures. I had cooked my own meals and melted glacier ice for drinking water. I saw nights where the sun looked like it was going to set only to kiss the horizon and rise again. A favorite of my students was that I lived without a bathroom for a month and I had to pee in a bottle and poop in a bucket. It was all in the name of science and the experiences I had are ingrained in me for the rest of my life as well as my students'.

Though, I must note that not all the hours in my day were spent chainsawing and jackhammering a tunnel into a glacier. I had several other responsibilities. There were several chores to get done in camp like cooking, collecting ice for drinking water, sorting and collecting waste, and washing dishes. I also kept up with a daily blog describing our field camp and what we were doing, my experience at McMurdo Base, and the people I met, their careers, and their stories of what brought them to Antarctica. I wanted my students and other readers to feel like they were there with me and I wanted to connect them to the different career paths and diversity of the people working in Antarctica. Additionally, I conducted the experiments my students designed, collected data, and videoed the results. I even wrote lesson plans and made student worksheets so other teachers could have their students repeat our experiments or use our data. These lessons and several other amazing lessons created by PolarTREC teachers are all available on the PolarTREC website (https://www.polartrec.com/). With so much to do I was thankful that the sun was up for more than 20 hours a day or else I would have never experienced so much. After four weeks in the field, I was exhausted, but it was worth every minute.

When the time came to head home, I was sad to say goodbye to my home on the ice, but I was excited to see my family, friends, and students. I had been out of the classroom for two months so there was a lot of catching up to do, but it was worth it. My students were so excited to see me and we spent several class periods telling stories and looking at pictures and watching videos I took in Antarctica. Just as I had been engaged by my Biology teacher's stories, my students were drawn into my every word and had tons of questions. It was awesome. Over the course of the rest of the school year we looked at data we had collected, replicated experiments we had done in the field, Skyped with my research team members, and went to the elementary school classrooms to lead polar science activities. As teachers, we are not supposed to have favorites, but the groups of students I had that year were some of my favorites and my PolarTREC experience had a lot to do with it. I was rejuvenated as an educator and my students were inspired by science and by their teacher.

Two students and their stories of how my PolarTREC experience impacted them stand out in my mind. At the school I was working at during that time not many of my students went on to college. In a blue-collar community many of my students' parents had not gone to college and with so many manufacturing jobs it was possible to make a good wage without going to college. Some never graduated, and many went directly into the workforce. It wasn't a priority and many of my students hadn't even considered a career or furthering their education in the sciences. One such student was Katie. She had a love for the sciences and took every science class and participated in every science club we offered at the high school. She wanted to continue her education in the sciences, but she didn't know how. No one in her family had been to college and she wasn't sure what science looked like as a career. She lacked a role model to help guide her in the direction she wanted to go. After my PolarTREC experience she saw me as that person. Instead of seeing me as a science teacher, she saw me as a scientist. I helped her navigate applying to and selecting a college, obtaining financial aid, and moving away from home for the first time. She went on to earn her degree in Environmental Science and moved to California to work for the National Park System. It has been ten years and we are still in touch and I am so proud of her accomplishments.

Another student impacted by my PolarTREC experience was Trey. He was in my freshman Biology class when I went to Antarctica. He was a great kid and did well in my class, but he was more interested in business and history for a future career. Like many of my students he had never been outside the city of Detroit, had never been on a plane, and had only considered attending the local community college or university for college. After my experience he wanted to travel and see the world too, but it was not financially possible. In his eyes he was destined to stay in Detroit for the rest of his life. About a year after my PolarTREC experience I learned about the Joint Science Education Project (JSEP) and I knew Trey would be perfect. JSEP brings together students from Greenland, Denmark and the US to learn about polar science in Greenland. When I first told Trey about the opportunity he thought I was crazy, but I convinced him to apply. Several weeks later he learned he was accepted and he was headed to Greenland that summer. It was definitely a stretch for him. He would be traveling by plane, venturing to another country, and interacting with kids from all over the world. He did not have any of the gear for traveling to a polar climate and could not afford to buy any, so staff members stepped up and donated clothing so he would be prepared. It turned out to be a transformative experience for Trey. He ended up finding his love for science in Greenland and went on to study engineering at a college 12 hours away from Detroit. My PolarTREC experience inspired Trey to take a risk of his own which resulted in personal growth and a career as a mechanical engineer.

PolarTREC was not only transformational for my students, it was transformational for me as a teacher. I had been afraid to teach beyond the textbook and beyond my classroom walls; now that fear was gone. I thought that being a successful teacher meant that I taught my students everything in a book. Now I know that students are inspired by what is beyond their textbooks and success can be measured in so many different ways. They are inspired by actual stories, pictures and videos.They are inspired by meeting scientists, using their data, and collecting data of their own. They are inspired by getting out into their community and learning about their own environment and helping their community solve environmental problems. I also learned that my professional learning community of teachers was not just in my school or state, there were like-minded teachers across the globe with similar passions to myself. Through my PolarTREC experience I made lasting connections with other educators which have broadened my perspective as an educator and opened doors to countless resources for my students. It is amazing how one experience can completely change your life and the lives of others. It started with an email with a teacher smiling in the snow tempting me to apply to a program I knew nothing about and two years later I was that smiling teacher in the snow on the PolarTREC advertisements.

POST-POLARTREC: SO MANY DOORS TO CHOOSE FROM

After one teacher research experience and having had success partnering with scientists and my students, I was hooked. I loved the growth I saw in myself and the growth I saw in my students. I loved that when my students looked at me they viewed me as a scientist and as an information source versus a person regurgitating facts. It changed their viewpoint of what a scientist looked like, and they had role-models beyond the Discovery Channel or their pediatrician. I loved that my students started seeing science as a career option for the future. My classroom had changed, and they were doing science versus learning about science. They were using data, collecting data, collaborating with other students and researchers, and loving every minute. Lastly, I love how one experience started me down the road to becoming the teacher leader I am today. I went from being a somewhat shy beginning-teacher in Detroit to being a vocal teacher leader at the national level.

How I got from point A to point B was not an easy road. The biggest bump in that road was losing my job. While I was in Antarctica, the United States was at the lowest point of the Great Recession, and Detroit was hit hardest. For several years prior, I had seen the enrollment in our school drop as families moved to other areas when factory jobs became scarce. There had been layoffs every year, and now I was the teacher with the least seniority in a union state. I knew it was coming, but nothing could prepare you for the day when you actually get pink slipped. My school had become my safe place. I loved the staff, I loved the kids, and I had built classes and a science program I was proud of. I was afraid to move on, but I didn't have a choice. As much as it hurt, that pink slip (it actually wasn't even pink!) forced me to challenge myself and take another risk.

As a colleague was consoling me, they said "“where one door shuts, another one opens”. At the time I was not amused or comforted, but the cliche came true. While in Antarctica I met another educator who had been an Albert Einstein Distinguished Educator Fellow. She told me about her experience in the federal government and I was in awe and thought of her as a rockstar of science teaching. Her experience had opened so many doors for her and one had led her to Antarctica for several seasons. I loved my experience working with scientists in Antarctica so much, I applied with a position in the National Science Foundation Office of Polar Programs in mind. I had fallen in love with the polar regions and my goal was to use my new passion and PolarTREC experience to create polar science opportunities and resources for teachers and students. A week after I learned I would be unemployed at the end of the school year, I received an email letting me know I was a finalist for the Einstein Fellowship and that I would be headed to Washington, D.C. to interview with several government agencies.

Instead of the door opening to a position at the NSF Office of Polar Programs, it opened to what I call, oceans of opportunities with the National Oceanic and Atmospheric Administration (NOAA) Office of Education. I accepted a position with the NOAA Office of Education, moved to Washington, D.C. and spent two amazing years as a fellow. When others ask me about my fellowship, I tell them I was a voice for educators in the federal government. Whenever my office was doing anything pertaining to teachers or students whether it was grant opportunities, websites, curriculum, policy, or even speeches, they asked my opinion. I used my PolarTREC experience to help make NOAA science, data, and scientists obtainable to educators and students. I also spread my newfound NOAA love to other educators by making presentations at local, state, and national education conferences. I started to develop my voice as a teacher leader. Before the fellowship I thought my voice did not matter, but I learned it did and that it was my responsibility to use it. Other educators, policy makers, and government agencies cared what I had to say and I wanted to change the way science was taught in classrooms the same way PolarTREC had changed my classroom.

Upon completion of my fellowship, I was at a fork in the road. I had several opportunities to take leadership positions outside of the classroom or I could return to the classroom. I chose to follow my heart and it brought me to a classroom in North Carolina. After being a PolarTREC teacher and an Einstein Fellow, I was changed as an educator. I had a breadth of knowledge on teaching science and resources available for teachers and students. This helped me to become an immediate teacher-leader within my district and I took on the role of a veteran teacher mentoring beginning teachers. I had developed my voice, but I still yearned to get back in the field and connect my students with scientists. I had seen what my PolarTREC experience had done for my students in Detroit and I wanted the same for my new students in North Carolina.

It didn't take long for the road I was following to lead me to another teacher research experience. During my fellowship I learned of many opportunities for educators and I looked forward to the day when I was a classroom educator again to apply. One of those opportunities was the NOAA Teacher at Sea program. I was lucky enough to be accepted into the program and I had an incredible experience onboard a NOAA vessel in the Bering Sea as a part of a long term study looking at the impacts of climate change on pollock and salmon populations. I spent two weeks sorting and counting fish, looking at plankton under microscopes, and analyzing conductivity, temperature, and depth data. Similar to my PolarTREC experience, I did everything the research team did plus I wrote daily blogs and designed lesson plans to go along with the science we were doing on the ship. My students were comparable to the students I had in Detroit in terms of diversity, motivation and achievement, so I focused on the variety of jobs and people onboard the ship. I wanted my students to see there was a place for them on a NOAA ship and race, gender, and economic status didn't matter. My students loved it. I immediately established my credibility with them as a teacher who actually does science. Enrollment in my marine science class increased and students, especially girls and those of color, saw marine science as a future career. I have had several students go on to study marine science in college and I have my NOAA Teacher at Sea experience to thank for helping me encourage their career trajectory. I also broadened my professional network. The NOAA Teacher at Sea program has an amazing alumni network and through the network I have been able to connect with other Teacher at Sea alumni to share experiences and lessons, participate in professional development opportunities, receive mini-grants for classroom supplies, and travel to regional and national conferences to share my experience and encourage other teachers to apply.

After my NOAA Teacher at Sea experience, I wanted to get back on a vessel and my journey led me to the Ocean Exploration Trust Science Communications Fellowship. This fellowship would be unlike anything I had ever experienced. We would be sending Remotely Operated Vehicles (ROVs) to the places on the ocean floor that had never been mapped and human eyes had never seen. It was truly an adventure and my job would be to communicate the science going on onboard the ship. We live streamed video from the ROVs to the entire world and I facilitated the live audio and took questions from the public to ask the rest of the team. It was amazing. I loved that you never knew what you were going to see that day. The ROVs might reach the bottom and all you see is mud in every direction or you could see a hydrothermal vent community with crazy creatures you never thought could exist beyond a sci-fi film. What was even more amazing was that my students could follow along with me, hear my voice, and get wrapped up in the excitement of exploration. They thought I was a rockstar. As a fellow I also hosted live interactions with my students, other classrooms, museums, aquariums, and science centers around the world. We would show videos of where we were at and what we were doing and then the viewers could ask the scientists questions. My students had so many questions and knew all the scientists by name. There were so many diverse role-models of various ages, genders, and races that my students all connected to someone. I loved my experience so much that I returned the following summer for six weeks onboard the ship as a lead science communications fellow. I mentored the new fellows and I continued to inspire my students with live ocean exploration. The most rewarding part was returning for a third summer as an honors research program lead. Through the honors research program I had the opportunity to work with high school students to develop research projects and then take a group of girls onboard the vessel for a week to conduct their research and learn firsthand what it is like to be an ocean explorer. The experience was incredible, and all of my girls went on to study marine science in college.

Through all of my experiences I can truly say the road I have traveled has been blessed. I worked hard, but there were huge rewards for both myself and my students. Building relationships with scientists and connecting them to my classroom has been extremely beneficial for myself, my students, and the scientists we partnered with. I have stories, pictures, scientific data, and lessons that apply to actual field research to share with my students and other teachers. I have become a leader in my field and I have built networks of researchers and educators I can draw upon for help in my classroom or as role-models for my students. My students have been inspired by science and many have pursued careers or taken risks or traveled as a result of my experiences. The researchers we have worked with have gotten a better insight on working with teachers and students and have been inspired to reach out to more classrooms. I have seen firsthand how teacher, student scientists can be beneficial to all involved and I have been inspired to encourage others to follow in my footsteps.

Becoming a Voice for Student-Teacher-Scientist Partnerships

In my pursuit of encouraging others to follow in my footsteps I first had to encourage more scientists to get involved with programs like PolarTREC. I feared that if more scientists did not volunteer to host educators like my researcher had, teacher research experiences would no longer exist, and other educators would not have the opportunities that I had. Many funding opportunities require broader impacts, but I found many scientists were unsure of how to involve teachers and students in their plans beyond a blog or classroom visit and I wanted to help. Since my PolarTREC experience I have made several presentations and I hosted webinars for scientists on how to establish meaningful partnerships with teachers and students. I created a framework for how to find teachers to partner with, have a successful visit, extend a partnership beyond one visit, and how to successfully bring teachers or students into the field. I especially enjoyed working with early-career scientist groups. The young researchers were so excited about working with teachers and students and they wanted to give back to the schools and teachers who helped them to get to where they are. For some, however, it was not an easy sell. Writing a teacher into a grant, taking a student or teacher into the field, or making time to visit a classroom is not easy. It can be expensive, berths on ships and space in field camps can be limited, and it can be challenging to make the research applicable or understandable at the classroom level. Nonetheless, I can say all those challenges pale in comparison to the rewards of a meaningful partnership between scientists, educators and students.

The next challenge was to encourage more educators to apply for and seek out opportunities to partner with scientists. Just as it can be uncomfortable for scientists to partner with teachers, it also takes a leap of faith by the teacher. I often ask teachers, "What is stopping you?", and I hear a variety of responses. Fear of failure in writing a competitive application, lack of comfort in the subject area, unease of the unknown and going outside their comfort zone, lack of time, having small children and family concerns, and doubts about approval from school administration. To help teachers take that leap of faith I have made several presentations at the local, state and national level on expanding classroom walls and applying to teacher research experiences. I start with asking the question "What is stopping you?", but I proceed to tell teachers about my initial insecurities and how I overcame them to have the amazing experiences I have had. I also describe how my experiences have been influential for both me and my students and tips and tricks for how they can seek partnerships and successfully apply to programs. As a result of my presentations, many teachers have been inspired to follow in my footsteps and I have helped several to successfully apply and be selected to teacher research experience programs.

Regardless of the group I am speaking to, I stress the reason I have been successful is that all partnerships are a two-way street. In a partnership between scientists, teachers, and students all sides need to be held accountable to hold up their end of the bargain. Communication, flexibility, and understanding of the needs of others are incredibly important. Scientists using teachers and students for their broader impacts but not involving them in their research will struggle to have success. Teachers who want to travel to cool locations but aren't willing to help out their team, will find that it will be their first and last partnership. Students who choose to not take advantage of the opportunities their teachers present will be missing out on an amazing experience. Partnerships are not easy but as I have stated before, they can be transformative for all involved. Until I retire, I will continue to pursue opportunities to partner with scientists and be involved in teacher research experiences. I will always continue to encourage other teachers and scientists to form partnerships with the ultimate goals of showing kids that science is a verb and inspiring them to be the next generation of scientists.

ACKNOWLEDGMENT

I am just one piece of an educator-scientist-student partnership. The stories of growth I talked about in my chapter would not have been possible without the tireless efforts of the program managers and staff of the programs I mentioned. It would also not have been possible without the amazing scientists who invited me to their field camps, laboratories, and ships to be a member of their teams. They treated me like one of their own and gave me the opportunity to share their science. Lastly, I would like to thank my students. Their enthusiasm for learning and to understand the world around them is contagious and has been an inspiration to me to never stop learning.

Additionally, I would like to acknowledge the support of the South Lake School District and the Mooresville Graded School District, in particular my colleagues, schools, school administrators, students, parents, and communities. Without their support and guidance, I would not have been able to participate in these life changing professional development experiences. I would also like to thank my family, friends, and husband for their unwavering support. Leaving your classroom and home for even short periods of time is not an easy thing to do and I have been blessed to teach in supportive school districts and to have the support of my loved ones.

NOTE

Any opinions, findings, and conclusions or recommendations expressed in this chapter are my own and do not necessarily reflect the views of the programs I participated in or their funding organizations; including the National Science Foundation and National Oceanic and Atmospheric Administration. The views and opinions of the author expressed herein do not necessarily state or reflect those of the U.S. Government or any agency thereof. Opinions expressed here are personal and not representative of the positions or policies of the Albert Einstein Fellowship (AEF) Program.

REFERENCES

Arctic Research Consortium of the United States (ARCUS). (2015). *In House Report: PolarTREC 2010-2014 Final Evaluation Report*. Prepared by Goldstream Group Inc. (NSF PLR Award #0956825).

Silverstein, S. C., Dubner, J., Miller, J., Glied, S., & Loike, J. D. (2009). Teachers' participation in research programs improves their students' achievement in science. *Science*, *326*(5951), 440–442. doi:10.1126cience.1177344 PMID:19833969

Simmons, B. (2010). *Evaluation Report of the NOAA Teacher at Sea Program: 2005-2009*. National Oceanic and Atmospheric Administration - National Marine Fisheries Service - NOAA Teacher at Sea Program. (OMB 0648-0600). Available at https://teacheratsea.noaa.gov/#/Evaluation/

KEY TERMS AND DEFINITIONS

Albert Einstein Distinguished Educator Fellowship: The Albert Einstein Distinguished Educator Fellowship (AEF) Program provides a unique opportunity for accomplished K-12 educators in the fields of science, technology, engineering, and mathematics (STEM) to serve in the national education arena. Fellows spend eleven months working in Federal agencies or in U.S. Congressional offices, applying their extensive knowledge and classroom experiences to national education programs and/or education policy efforts.

NOAA Teacher at Sea: The mission of the National Oceanic and Atmospheric Administration's (NOAA) Teacher at Sea Program is to provide educators hands-on, real-world research experience working at sea with world-renowned NOAA scientists, thereby giving them unique insight into oceanic and atmospheric research crucial to the nation. The program provides a unique opportunity for informal and formal educators, pre-kindergarten through college-level, to sail aboard NOAA research ships to work under the tutelage of scientists and crew. Since its inception in 1990, NOAA's Teacher at Sea Program has enabled over 800 teachers to gain first-hand experience of science and life at sea.

Ocean Exploration Trust Science Communications Fellowship: The Science Communication Fellowship immerses formal and informal educators in the *Nautilus* Corps of Exploration as communicators for exploration and empowers them to bring the excitement of ocean exploration to students and public audiences around the world. Fellows share accounts of ocean research, expedition operations, STEM careers, and daily life with audiences of all ages through live audio commentary on the *Nautilus* Live website and through outreach to classrooms and communities via ship-to-shore live interactions.

PolarTREC: PolarTREC is created and administered by the Arctic Research Consortium of the United States and funded by the National Science Foundation. PolarTREC educators (formal and informal) come from the United States and spend 3-6 weeks participating in hands-on field research experiences in the Arctic or Antarctica, working side by side with scientists. The program has been around for over 10 years and has provided over 170 U.S. teachers a unique opportunity to work with scientists in the polar regions.

Chapter 2
Meet Them Where They Are (A Perspective That Works Both Ways): Reflections on Scientist–Educator Collaborations

Lawrence Krissek
The Ohio State University, USA

ABSTRACT

Partnerships between educators and science faculty can be professionally productive and personally fulfilling, but developing such partnerships can be challenging. This chapter describes the career trajectory of a geoscience faculty member who evolved from a focus on research within his science discipline to a portfolio of activities including disciplinary science research, geoscience education activities, and geoscience education research. The author provides suggestions for how to identify potential science faculty partners and how to develop and maintain successful partnerships.

INTRODUCTION

Some U.S. university science departments presently include faculty members who have been trained and hired specifically to: 1) participate in discipline-based pre-service teacher preparation programs, and/or 2) conduct discipline-based science education research. Where present, these science department faculty members are prime candidates for partnerships with teachers and teacher-educators. Such discipline-based education specialists form a small fraction of the overall science faculty population, however, so most science faculty presently involved in partnerships with educators have followed a different path to those partnerships. I am an example of the latter. In this essay I describe the path that led me to several decades of professionally productive and personally fulfilling collaborations with educators, while identifying internal and external factors that affected that path. Similar factors still are in

DOI: 10.4018/978-1-7998-4966-7.ch002

action today, and likely influence today's science faculty as they consider whether to become involved in partnerships with educators. By describing these factors and their effects along my career path, I hope to help educators: 1) effectively identify potential collaborators within science departments; 2) better understand the discipline-specific expectations facing science faculty, so that the commitments to a potential partnership can be negotiated effectively; and 3) work effectively with science faculty in partnerships characterized by mutual respect.

In this essay, I have used several terms in specific ways, as defined here:

1) "teacher" refers to K-12 classroom practioners, whether pre-service or in-service;
2) "teacher-educator" refers to someone who trains teachers and conducts science education research while based in a non-science department, such as a Department of Teaching and Learning;
3) "educator" is used as a more general term, incorporating both "teachers" and "teacher-educators" as defined above.

At later points in this discussion, the terms "geoscience education" and "geoscience education research" are used, also with specific meanings as derived from distinctions made in the structure of our professional societies. The term "geoscience education" generally refers to the practice of teaching the geosciences, whereas "geoscience education research" refers to the scholarly study of geoscience teaching and learning as well as geoscience discipline-based educational research.

EXPERIENCE WITH SCIENCE-EDUCATOR PARTNERSHIPS

During 35+ years as a faculty member in what is now the School of Earth Sciences at The Ohio State University, I have been fortunate to have the opportunity to collaborate with teachers and teacher-educators in a variety of settings. In addition to teaching educator-focused introductory-level Earth Science courses, these collaborations have included: 1) helping develop and co-lead two externally-funded teacher preparation programs for pre-service middle school and high school science teachers; 2) helping instruct a science methods course for pre-service adolescent and young adult teachers; 3) leading professional development programs for in-service K-12 teachers, as well as for faculty from minority-serving community colleges and undergraduate institutions; and 4) conducting teaching and research activities with teacher-educator colleagues. I also have worked with, and in some cases mentored, in-service teachers or teacher-educators who were embedded with scientists in large science research programs, such as aboard a research vessel or in Antarctica. None of these collaborations provided an opportunity for me to work with both an educator and her/his students, so my reflections here are focused on collaborations between a scientist and a teacher or teacher-educator.

REFLECTIONS on SCIENTIST-EDUCATOR PARTNERSHIPS, THEN and NOW

Thoughts about my partnerships with educators are arranged chronologically in the following discussion, reflecting stages in my academic training and career path that affected the trajectory of my collaborations with teachers and teacher-educators (Figure 1). In addition, I provide observations about the present impact of some of these influences, primarily based on personal perceptions. The points made here are not unique to me, however; many of my science-discipline colleagues who have developed and maintained long-running interactions with teachers and teacher-educators have followed a similar path toward that involvement. As a result, teachers and teacher-educators may find that thinking about potential science-discipline collaborators along this template of career evolution helps them identify productive science collaborators more efficiently.

The proverbial "bottom line" – it's important for the teacher/teacher-educator to work toward meeting the scientist "where she/he is", both in content expertise and in career trajectory, while also finding the scientist who's willing to work toward meeting the teacher/teacher-educator "where he/she is" for the needs of the project at hand. An expectation and practice of mutual respect is an important part of having each partner meet the other "where she/he is"; this can be expressed by being respectful of each other's time, respectful of the expertise that each partner brings to the partnership, and in any other ways that either partner feels is needed for open communication.

Figure 1. Stages in my career trajectory from a discipline-focused scientist to a scientist with strong interests in geoscience education and geoscience education research.

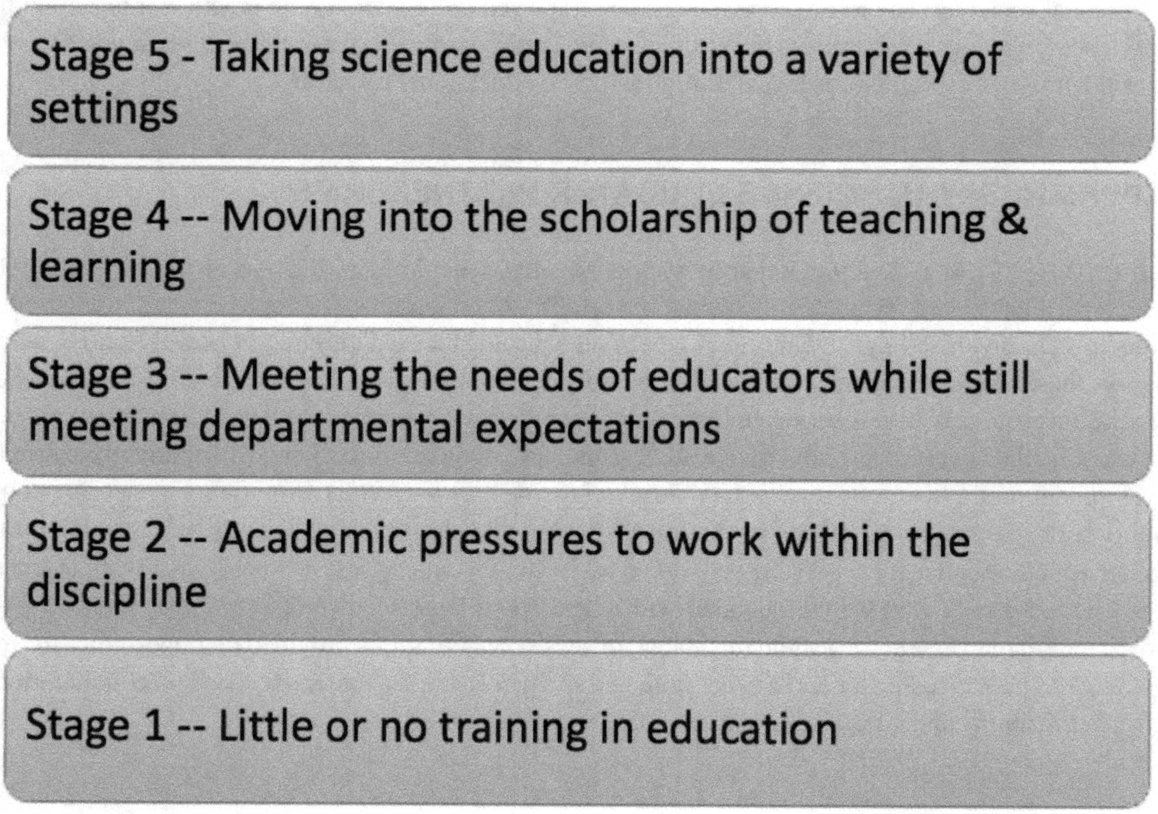

Stage 1) Little or no Training in the Field of Education – Meeting the Scientist "Where she/he is"

When I began my academic career at The Ohio State University in 1982, as an untenured assistant professor in the then-Department of Geology & Mineralogy, I had no formal training in education and very little practical experience in leading a classroom. My experience in education was almost entirely as a student. A lack of formal training in teaching and learning among new faculty members was very common at that time, and – despite some encouraging trends -- remains relatively common among faculty in the sciences today. Among geoscience faculty, at least, these encouraging trends have been led by concerted efforts by professional societies, which have provided training in teaching and learning through early career programs and dedicated workshops. As a result of such efforts, a recent survey (Manduca et al., 2017) finds that ~20% of all U.S geoscience faculty have participated in such programs to some extent. In addition, an increasing number of early career science faculty have been introduced to discipline-based education research and to the scholarship of teaching and learning, either as graduate students or once in their faculty positions. As a result, potential collaborators may be present within a science department, but teachers and teacher-educators will likely need to make the effort to find those potential collaborators. In the best-case scenario, those potential collaborators already will have knowledge of, and an interest in, education-focused partnerships. In many cases, however, the teachers and teacher-educators will have to educate those potential collaborators about opportunities to work together, as well as the accompanying expectations from the science education side.

Stage 2) Academic Pressures to Work Within the Science – Meeting the Scientist "Where he/she is"

When I began my academic career, I had no plans of becoming involved in any formal way with teachers or with the teacher-educators on-campus; my focus, outside the need to do my own teaching and some service work, was on establishing and growing my scientific research activities. In part, this was because I had little-to-no understanding of what opportunities might have existed in the education realm; in large part, however, this was because my senior departmental colleagues emphasized the need to meet expectations for research grants and publications in the science in order to be promoted and tenured.

From my observations and discussions with colleagues, similar discipline-focused pressures remain strong today, especially at research-focused institutions. The good news is that most science disciplines now recognize science education as an important area of effort, as indicated by the presence of science education and science education research divisions within their professional societies. The value given to science education efforts varies significantly, however -- between science disciplines, between institutions, and even between departments within an institution. As a result, teachers and teacher-educators who are looking for collaborators within science departments need to **be sensitive to the discipline-, institution-, and department-specific expectations for the science faculty**, and especially for untenured science faculty. In order to maximize the chances for a long, collegial, and productive partnership, the potential partners should have an open and honest discussion about the external expectations and the time and intellectual commitments each faces before either partner fully commits to the partnership.

Stage 3) Meeting the Needs of Teachers/Teacher-Educators While Still Meeting Departmental Expectations – When "Meeting them where they are" Works Both ways

Within my first few years of teaching introductory courses, especially during the summer, I met both pre-service and in-service teachers who told me about their need for, and interest in, additional geoscience content courses to meet licensure/certification requirements. Several tenured Associate Professors in my department, who had chosen career paths linked primarily to the existing teacher preparation and teacher professional development programs on campus, also helped me understand more about the curricular needs of those populations, and introduced me to potential colleagues in the teacher-educator network on our campus. As a result, I experimented with offering educator-focused versions of several introductory courses – specifically modified to accommodate the schedules of the educators, and evolving, over several years, to emphasize content that was especially relevant in Ohio. In this effort, I was fortunate that I was supported by my academic department, as long as I also continued to meet the standard expectations for research, service, and teaching (including teaching a mix of large-enrollment introductory courses, undergraduate major program courses, and graduate courses).

This stage marked the beginning of a positive feedback cycle that I found very fulfilling; I learned from the educators what their needs were, such as the importance of place-based learning and of information they could transfer directly into their classrooms, and I was rewarded by their interest and what they reported as a positive impact on their teaching as I adjusted course content and pedagogy to meet their needs. This portion of my personal trajectory of involvement with teachers and teacher-educators expanded in the early 1990s, when I developed two full-time, 1 week-long field-based summer courses for educators, due directly to input from teachers and teacher-educators. Those courses continued for ~20 years.

An important lesson from this part of my experience was that my **contributions to educator-focused courses had to be made <u>in addition to</u> meeting the standard expectations for research, teaching, and service in my department, rather than as a substitute for my standard teaching load**. I chose to make these contributions for personal reasons, rather than for explicit professional advancement; I felt that I was having a positive impact, and I was willing to spend the time and energy to have that impact. An unfortunate reality is that a similar situation still exists today for science faculty in many departments and at many institutions, so that the decision about whether to collaborate with teachers/teacher-educators often is determined by the individual scientist's personal evaluation of these costs and benefits. Not surprisingly, the individual scientist's evaluation of these costs and benefits also can be affected significantly by that individual's career situation; for example, an untenured assistant professor may, on a personal level, see the value of collaborating with a teacher/teacher-educator, but may choose – or be strongly advised to choose – to avoid such commitments outside the scientific discipline until at least having been promoted and tenured. In contrast, a science faculty member who already is tenured may feel much freer to choose to collaborate with educators.

Stage 4) Moving into Formal Professional Development and Science Education Research – Meeting the Teachers and Teacher-Educators "Where They Are", While Learning Much More About Science Education from "where I was"

For ~20 years, my involvement with teachers/teacher-educators consisted almost entirely of offering educator-focused science content courses. In the early 2000s, however, my involvement deepened when a new faculty member, with research interests in earth/space science education, joined OSU's School of Teaching and Learning. At her invitation, we began a very fulfilling and professionally productive collaboration at a variety of levels – on organized training for preservice educators, on professional development for in-service educators, on science education research projects, and on graduate student committees. In all of these efforts, I contributed science expertise, calibrated to the appropriate level, while I learned a tremendous amount about pedagogy, assessment methods, qualitative research, and educational theory. As with my earlier efforts with individual educator-focused courses, I found a very fulfilling positive feedback cycle in this work, although taken to the proverbial "next level" intellectually; I learned from the educators/teacher-educators what their needs were, and I was rewarded by their interest and the reported positive impacts as I adjusted content and presentation mode to meet those needs. I also benefited professionally from what I learned about pedagogy, assessment methods, educational theory, and qualitative research, as I was able to apply that knowledge both to my own teaching and to situations where science colleagues were considering initial efforts into the world of earth science education.

Colleagues have suggested that at this stage in my career I was moving into the field of "scholarship of teaching and learning" – but I'll re-emphasize that reaching this point took ~20 years from the start of my time as a university faculty member. As a result, educators and teacher-educators may find it useful to focus on two populations when looking for potential partners among science department faculty: 1) early career faculty who already have been introduced to discipline-based education research and to the value of scholarship in teaching and learning (especially if they were hired to work specifically in those areas), and 2) senior faculty whose personal journeys have led them to value efforts in science education and science education research.

Another important lesson from this part of my experience was a continuation of an issue identified previously – that many of my science colleagues had difficulty understanding and recognizing the value of teaching, service, and research contributions that overlapped into science education, so those **contributions generally were expected to be <u>in addition to</u> meeting the standard departmental expectations for research, teaching, and service in the discipline.** As a result, grant funding and publications focused on science education and science education research sometimes were judged as inferior to science-focused funding and publications, with accompanying impacts on annual performance evaluations and salary adjustments. Based on discussions with colleagues at other institutions, as well as a recently published editorial (Dolan et al., 2018), this situation appears to still be relatively common within science departments.

Stage 5) Taking Science Education into a Variety of Settings – Meeting the Various Audiences "Where they are" (While Working to Improve Those Positions)

I find it difficult to overstate how much my view of science education and science education research – and how it should be done in order to be done well – has changed as a result of deeper engagement with the topic in the last ~20 years. For that, I am very grateful to my collaborators in the area of science education, and to my collaborators and colleagues who practice geoscience education as a field of rigorous scholarship and research, rather than as an easy "add on" to meet the "broader impacts" requirements for grant funding. In addition, the lessons I've learned about qualitative research and assessment methods from science education researchers have been important tools when I've had to think about curricular issues on a larger scale.

A primary lesson I've learned is that science education isn't a "one size fits all" exercise; the nature of the activity should be aligned to the goals of the project, which means that the goals have to be defined before the activity is designed. (My apologies if I'm stating something that's very obvious to you. For many scientists who think only rarely and/or superficially about "science education", this statement isn't at all obvious.) For example, the time involved, the pedagogies employed, and the scientific depth considered are – or should be – quite different if the goal of the activity is exposure to scientific facts as compared to a goal of identifying and addressing misconceptions. My personal experience has been that many of my science colleagues – in my home department, in other geology/earth science departments in the U.S. and in other countries, and associated with large multinational research programs – primarily think of science education as the process of exposing people to a suite of scientific facts and the technologies that allow the collection of such data; this is the proverbial "dog and pony show". Few of these scientists are aware of the need to think more carefully about defining learner-centered goals, and even fewer are equipped to design the assessment necessary to adequately evaluate whether those goals have been met. And as a result, the science education components of many science-discipline research projects rarely take full advantage of the scientist-educator collaborations that could have occurred if learning goals had been defined with input from educators and education researchers from the first stages of program design.

A primary lesson from these experiences is **the importance of including science education and/or science education research expertise from the earliest stages of designing a science-discipline research project** – and writing a research proposal – in order to maximize both the science/education collaboration within the project and the science education impacts coming from the project. Unfortunately, many science research proposals are not developed this way. As a result, potential science education collaborators often are left in the challenging situation of not knowing that a science research proposal is being written until it is too late to help design a well-integrated science education and science education research component. This is a difficult problem to solve; the best cases I've seen have involved teams of scientists and science educators/science education researchers who work together on multiple projects, with the science education component gradually becoming more integrated with the science over that time.

LESSONS LEARNED

Based on the points discussed in this essay, how can an educator improve her/his chances of developing a productive partnership with a science faculty member? Here are some steps that my experience suggest are important, listed ~chronologically through the lifespan of a partnership:

1. **Find an interested science-discipline colleague**, by
    a. Networking on a personal level, both locally and regionally, with university science faculty. Explore options with early-career faculty who already are trained in discipline-based education research and/or the scholarship of teaching and learning; with more senior faculty who already express an interest in science education; and with faculty who are recognized within their department for a commitment to good and effective teaching;
    b. Joining and being active in the education and education research divisions of science-discipline societies (in the geosciences, for example, the science education and/or science education research divisions of the National Association of Geoscience Teachers (NAGT), the Geological Society of America (GSA), and the American Geophysical Union (AGU));
    c. Reading the published literature on science education research in your discipline, so that you can identify scientists who are more likely to be interested in collaborating; and

   d. Using national and international networks that are designed to foster collaborations between scientists, large science programs, and science educators.
2. **Negotiate the terms of the potential partnership**, with an expectation and practice of mutual respect. Become familiar with the discipline-specific expectations facing your potential science-faculty partner, and be prepared to explain your expectations for the contributions from the science-faculty partner. Be prepared to explain the types of data to be collected, the types of analysis to be used, and any training that the potential science-faculty partner will have to complete (e.g., training for working with human subjects).
3. Once a partnership is established, **maintain good lines of communication** with your science-faculty partner. Be prepared to continue educating him/her about the intellectual underpinnings of science education research, such as educational theory, and the collection and analysis of qualitative data.
4. **Enjoy the collaboration**!

ACKNOWLEDGMENT

I thank Donna Farland-Smith for inviting me to contribute this essay. I also thank the various groups of students, teachers, and teacher-educators I have interacted with through the years; their suggestions, advice, and willingness to tolerate instructional experiments have helped me develop as an educator. Kathy Cabe Trundle played a major role in guiding me into the world of science education research, for which I am grateful. Kristen St. John and Mark Leckie have been steadfast and supportive colleagues in the world of geoscience education and geoscience education research, as well as within our scientific discipline. Kristen St. John reviewed an earlier version of this essay and provided many helpful comments, especially about the present status of earth science education and earth science education research.

REFERENCES

Dolan, E. L., Elliott, S. L., Henderson, C., & Curran-Everett, D. S. (2018). Evaluating discipline-based education research for promotion and tenure. *Innovative Higher Education*, *43*, 31–39. doi:10.100710755-017-9406-y

Manduca, C. A., Iverson, E. R., Luxenberg, M., Macdonald, R. H., McConnell, D. A., Mogk, D. W., & Tewksbury, B. J. (2017). Improving undergraduate STEM education: The efficacy of discipline-based professional development. *Science Advances*, *3*(2), e1600193. doi:10.1126ciadv.1600193 PMID:28246629

Section 2

Chapter 3
The Role of the Knowledgeable Other in Student-Scientist-Teacher Partnerships:
Examining Triads in Second Grade Classrooms

Sarah L. McClusky
Ohio Northern University, USA

Donna Farland-Smith
The Ohio State University, USA

ABSTRACT

This chapter focuses on the language and interactions of a visiting scientist and a group of second-grade students (N=18). Guiding this study was the understanding that a student-scientist-teacher partnership supports a social cultural perspective that provides a zone of proximal development in a multifaceted and effective exchange of knowledge for all members of the partnership. The study examined how these interactions can be understood in terms of the zone of proximal development at different levels. It was determined that all three members of the triad can be the more knowledgeable other and move the knowledge between them.

INTRODUCTION

"Okay, friends," teacher Ms. Smith said, clapping her hands in a rhythmic pattern to get the student's attention. "Dr. Z just informed me he is coming in the next thirty minutes!" So that means we have a change in our plans for today." A one-eighty was never a problem for Ms. Smith; she was grateful when Dr. Z would visit her classroom. She never minded the interruption. Ms. Smith knew the benefit to her young students learning science from Dr. Z far outweighed the imposition and remained flexible. Upon

DOI: 10.4018/978-1-7998-4966-7.ch003

her announcement, they instantly filled with excitement! As if rehearsing for a fire drill, they cleared their desks and cleaned up their workspaces with no further direction from their teacher. The students knew what a visit from Dr. Z meant; they wondered what he might bring today. Even though they had no idea what Dr. Z might bring today, and they were certain it was going to be exciting!

Today's durable student-scientist-teacher partnerships have evolved from yesterday's one-shot visiting-scientist programs. Lessons learned about the strengths and weaknesses in elementary classrooms have improved and refined the way educators think about these partnerships. Little research exists to document short-term scientists' effectiveness in classroom programs (Larson, Liston, Thiry, & Graf, 2007). The assumption is that a visit from a scientist will improve children's perceptions of scientists. Bozdin and Gerhinger (2001) and Flick (1990) reported that visiting scientists decreased stereotypical beliefs about scientists. However, Buck, Leslie-Pelecky & Kirby (2002) reported no change in students' stereotypical images of scientists when three female scientists visited an upper elementary classroom. Farland-Smith (2009) found that interactions with various scientists over a week-long period at a science camp demonstrated a positive impact on middle school girls' perceptions of scientists. Grossman & Farland-Smith (2020) evaluated the impact of virtual, discussion-based interactions with five scientists on middle school students' perceptions of scientists. They discovered through technology students' perceptions of what scientists do, where scientists work, and what scientists look like could be broadened. Therefore, classroom technologies may have opened up new opportunities for disrupting scientists' problematic stereotypes and have supported students in developing more expansive perceptions of science and scientists. Therefore, the kinds of visits and relationships between students, scientists, and teachers can be multifaceted and effective in providing benefits for all involved and thus, worthy of investigation.

Student-Scientist-Teacher Partnerships a Multifaceted Exchange of Knowledge

Student-scientist-teacher partnerships can serve as a vehicle for exchanging knowledge via a more "knowledgeable other" in an effective and fluid process. Lewis et. al (1998) claimed the expertise provided by an outside "other" in the classroom was noticeable in making lesson study effective (Lewis, Perry, Hurd, & O'Connell, 2006; Lewis & Tsuchida, 1998; Takahashi & Yoshida, 2004). Vygotsky (1978) also discussed the more knowledgeable other in terms of a child's learning predicated by the social influences and interactions of a more knowledgeable "other".

Additionally, Vygotsky's zone of proximal development (ZPD) describes this exchange more deeply when it comes to problem-solving. Quite simply, ZPD is "the distance between the actual development level as determined by independent problem solving and the level of potential development as determined through problem-solving under adult guidance or in collaboration with more capable peers" (Vygotsky, 1978). More recently, Goos, Galbraith, & Renshaw (2002) described how the exchange of knowledge in ZPD is dynamic. It can be a bi-directional and collaborative process as both learners possess knowledge that requires the others to exchange it. An interactional achievement allows participants to become both teachers and learners; roles change at any given time and become apparent as each engages with each other and their environment (Roth & Radford, 2010). Vygotsky (1978) described the relationship between tools and symbols and individuals' lives as culturally significant. "These tools and symbols have meaning and serve as mediational markers setting frames of reference for individual thinking in context. The most omnipresent and important tool/symbol in the life of the individual is of course, language p.7 (Glassman, 2001).

Assuming the student-scientist-teacher partnership can serve as an effective exchange of knowledge regarding academic disciplines, we had questions about the role of language, specifically science vocabulary. Cosset-Lent (2016) claimed that learning involves engaging in discipline-specific academic language as students become aware of the content language to support their understandings within the discipline. Nagy & Townsend (2012) defined academic language as "the specialized language, both oral and written of academic settings that facilitates the communication and thinking about disciplinary content." As language is exchanged the manner in which that knowledge is exchanged can be identified within a framework.

Triads are used to examine three members of a collective working together, and for this purpose, the authors sought to use the exchange of knowledge in a student-scientist-teacher partnership the framework described by Simmel in 1950 as cited in Shiltaloppi & Vargo (2017):

Triads examine the associations among three actors, involving, at a minimum, the analysis of two dyadic ties among three interrelated actors. By making apparent the indirect ties bearing on actors, the triad is the smallest unit of analysis for a network. Literature characterizes triads as systems of three actors in which each actor interacts directly and reciprocally with another actor and also operates as an intermediary between the other two (p. 395).

We were interested in finding out if the student-scientist-teacher-partnership could serve as an interactional achievement that allows all participants to become teachers and learners. If so, the triad would enable each member to espouse the role of both a teacher and a learner at any given time and become apparent as each engages with each other and the environment.

Student-Scientist-Teacher Partnership an Effective Exchange of Knowledge

While the student-scientist-teacher partnership can reveal a multifaceted exchange of knowledge it also highlights an effective process to deliver new knowledge included in the *Next Generation Science Standards* (*NGSS*) to each individual in the partnership. The *Next Generation Science Standards* (*NGSS*) provide an opportunity to enhance learning and demonstrate how science is practiced in the real world. Through integrated, three-dimensional performance expectations of the standards (NRC, 2012), disciplinary core ideas (DCIs), science and engineering practices (SEPs or often called 'Practices'), and crosscutting concepts (CC), knowledge is acquired. The main content categories for the DCIs are physical science, life science, earth and space science, and engineering design. The *NGSS* practices are: Asking Questions & Defining Problems, Developing & Using Models, Planning & Conducting Investigations, Analyzing & Interpreting Data, Using Mathematics & Computational Thinking, Constructing Explanations and Designing Solutions, Engaging in Argument from Evidence, Obtaining, Evaluating, & Communicating Information. The science and engineering practice dimension list describes the "practices in science" as "behaviors that scientists engage in as they investigate and build models and theories about the natural world" (*NGSS*, 2013, p. 41). The crosscutting concepts are the ideas and practices that cut across the science disciplines such as structure and function, cause & effect, stability & change, patterns, energy & matter, systems & system models. These three dimensions (DCIs, SEPs or practices, and CCs) and their connections with the nature of science as outlined by the *Framework* (NRC, 2012) are the components of a high-quality science education program. However, one challenge for elementary classroom teachers can be understanding how these three dimensions work together during instruction. It becomes more complicated if a visitor is involved, like a visiting scientist, simply because they may or may not be aware of the standards. These practices demonstrate how science looks in the real world.

Guiding this study was the understanding that a student-scientist-teacher partnership supports a socio-cultural perspective that provides a zone of proximal development in a multifaceted and effective exchange of knowledge for all members of the partnership. Specifically, this study examined student-scientist-teacher interactions in a second-grade classroom and how these interactions can be understood in terms of the zone of proximal development at different levels. For example, if the teacher is the students' more knowledgeable other, and the scientist is the teacher's more knowledgeable other, *how* do these three interact when together, to move the knowledge between them. Additionally, this study examined how effective the more knowledgeable other was in delivering new science learning. The research questions selected for this study were:

1. What does the exchange of knowledge look like in a student-scientist-teacher partnership within an elementary classroom?
2. What interactions serve as evidence to demonstrate the effectiveness of "the more knowledgeable other" in a student-scientist-teacher partnership within an elementary classroom?

Science in 2nd Grade Classrooms

Those not familiar with second-grade classrooms might be surprised to learn that today's science classrooms' typical practices do not reflect the most recent findings regarding effective science teaching and learning. According to *Ready Set Science* (NRC, 2008) "current curricula tend to cover too many disparate topics in a superficial manner, and many are based on an outdated understanding of how children learn" (Banilower et al., 2013) concluded that science receives far fewer instructional minutes than other core subjects. A growing body of evidence shows that what a teacher knows about science influences instruction quality and has a powerful effect on the success and type of discussions that teachers can engage in and sustain with students. It is commonly accepted that both elementary and middle schools do not provide teachers either the time for practice-embedded professional development or the resources, materials or pedagogical/content knowledge needed to learn science themselves and teach it well (NRC, 2008).

The National Science Teaching Association (NSTA) (2014) identified the following key principles to guide effective science learning in the elementary grades: 1) The elementary educational environment plays a key role in student learning; 2) Elementary students have the capacity to engage in scientific and engineering practices as they develop conceptual understandings over time; 3) Elementary students can and should engage in science within the broader community of science; 4) There must be adequate time in every school day to engage elementary students in high-quality science instruction that actively involves them in the processes of science.

It is no surprise that teachers and scientists view science education differently based on their backgrounds. However, in order to teach science well, teachers need to understand science differently from the way that scientists do (NRC, 2008). A scientist understands scientific theory and scientific content, along with the ways in which questions are investigated in his or her field and may or may not have pedagogical knowledge. An elementary teacher has pedagogical knowledge, but not necessarily the content and ability to convey scientific knowledge to children. To convey these ideas to children and set up scientific experiences as recommended by NSTA, teachers need to know science and understand science. It is then, when they can select appropriate instructional materials and investigations to foster scientific language and reasoning. This process requires teachers to engage in an internal dialogue between disciplinary science goals and the pedagogical means of determining what children know and how to move their understanding forward (NRC, 2008). It requires administrative support, and sometimes a knowledgeable other to serve in the role to teach elementary teachers how to teach science effectively. This is why the idea of teachers and scientists collaborating together are so promising for children; they each have strengths the other may not.

This Unique Student-Scientist-Teacher Partnership

The research site was a second-grade classroom within a public K-5 school in the United States. The teacher was a self-contained teacher, who we will call Ms. Smith and teaches English Language Arts (ELA), Math, Science, and Social Studies. A total of 18 students were in her classroom, with 8 females and 10 males. A few important things to mention for this particular year were the students' first year of looping with Ms. Smith. Which meant the students would have Ms. Smith as their second-grade teacher and then their 3rd grade teacher next year. Ms. Smith had surgery, and Dr. Z did not attend for 2 months; in fact, his first visit was not until December (see Appendix A for the list of his visits topics and durations over the last one and half years). When Dr. Z visited, he brought artifacts of animals, such as bones and or live organisms, to support the engagement and experience for Ms. Smith's young students. Ms. Smith had been teaching for over twenty years and had no formal training in science. The partnership formed between Ms. Smith and Dr. Z. (a local veterinarian) when Dr. Z's daughter was in her class five years ago. Dr. Z was not just a scientist but a father of two of the students in Ms. Smith's classroom in this particular year. This initially drew Dr. Z to serve as a visiting scientist, because his daughter was in the class. After the first year Ms. Smith was driven to continue the partnership because she quickly realized he brought a more in-depth knowledge base of science than she could provide to her students. Dr. Z, was an avid collector of animal bones and used these as props to provide inspiration and focus for his visits to Ms. Smith's classroom. What began as a way to enrich his daughter's class remained as a student-scientist-teacher partnership year after year. Visiting Ms. Smith's class once a month became his "hobby," and talking with students about his diverse collection of animal bones became their science class. The focus of his visits helped students understand the different types of animals, how they live, the care and maintenance of animals, and their understandings of the other species of animals.

The partnership is now in its sixth year of existence. The scientist and the teacher each know their roles and have found multiple opportunities to practice with the students. Most often, Dr. Z took the leadership role in teaching the class; however, the partnership was marked by a constant back and forth of switching responsibilities between the classroom teacher. The relationship between Ms. Smith and Dr. Z was one in which she released her role of leading the experiences with a continual exchange of expertise between the two. At times, Ms. Smith took a back seat, allowing Dr. Z to fully lead the experience. During such times, their partnership was strengthened by Ms. Smith's constant support of Dr. Z and his interactions with the students. Ms. Smith jumped in when necessary to offer support in managing classroom behaviors. The partnership between Ms. Smith and Dr. Z also included a collaborative approach by engaging in discussions before his visits (i.e., this is the existence of co-planning as defined by Dr. Z and Ms. Smith) and co-teaching lessons.

Dr. Z's visits throughout the year were consistent: 1) bringing bones or live organisms which always appeared to promote language rich discussions; and 2) allowing for ample time for observation of the props so that students could question and reason their way through the identification of observable characteristics. Based on his pattern, the students knew that every time Dr. Z visited, he would be bringing in interesting items, talking about them, and fielding their questions. This goes back to the vignette about Ms. Smith's class that provides a visualization for the elementary classroom when it is announced that their visiting scientist will be making an appearance. This group of young students had discovered his pattern, and it kept them engaged in science. So, with the announcement that Dr. Z was coming, their minds raced with the thoughts of what he might bring today! Dr. Z. attended Ms. Smith's class ten times per year for approximately one hour or more each visit, and because he was a veterinarian, he always chose to teach about life science.

Among the observations from video files and transcripts, observing and questioning were the most frequently occurring behaviors throughout the year. Typically, he would begin with a description of the animals, and either show the class a skull or two from that particular animal as he asked students questions about it. Then, he would allow time for their questions and because the partnership had been established for the ten-month period. Dr. Z did not always have to prompt students with guiding questions to begin their observations as they were familiar with this kind of back and forth 'science talk.' Such student behaviors are recognized as practices and identified as the work of scientists (*NGSS*, 2013; Jennings & Mills, 2009). The students engage in the lessons over and over again and see them as creative endeavors. In this case, science engagement is 'learning by doing' scientific practices, such as to make and record close, detailed observations of objects and phenomena. These behaviors are also regarded significant from the socio-cultural perspectives in that "[r]egular patterns of activity in a [classroom], in which individuals participate" (Greeno, 1998, p. 6), is analogous to the practices of a community of practitioners – e.g., a community of scientists (Crawford, 2005; Jennings & Mills, 2009; Cavagnetto et al., 2010; Cobb et al., 1991).

METHODS

All participants have been assigned a pseudonym, including the scientist and the teacher. In this study, data came from three primary sources. The first data set was video recordings of lessons that included student, scientist, and teacher interactions. To explore how the students and Dr. Z exchanged ideas and promoted each other's science thinking, learning and language we took video recordings of classroom interactions. These recordings were audio-recorded and transcribed. The second set of data was the researcher's field notes from the science class sessions and video recordings. The third set of data came from interviews with the teacher and scientist. Once each data set was analyzed; triangulation occurred between the transcripts, field notes and interview data to answer the research questions. Careful consideration was given to conduct this research with deep sensitivity in a bias-free manner.

Analysis

The field notes, videos, and transcripts were triangulated to identify three significant exchanges of knowledge within the Student-Scientist-Teacher Partnership:

- *Knowledgeable Other Exchange #1:* The scientist and students worked together to benefit the teacher in an understanding of how *NGSS* look in a classroom

- *Knowledgeable Other Exchange #2:* The scientist and teacher worked together to benefit the students in the area of vocabulary
- *Knowledgeable Other Exchange #3*: The students and teacher worked together to benefit the scientist in the area of classroom culture

Figure 1. An example of one of Dr. Z's specimen.

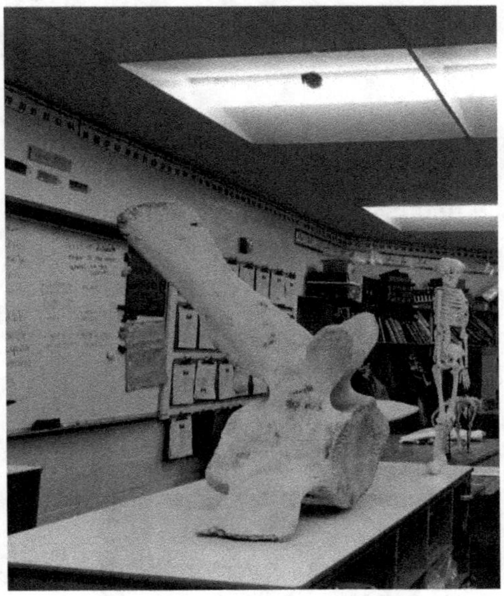

RESULTS

Results indicated three notable "knowledgeable other exchanges" with specific examples of how the role of the "more knowledgeable other" is both multifaceted and effective. *Figure 1* is an example of a specimen Dr. Z might bring to class during one of his visits.

Knowledgeable Other Exchange #1 The scientist and students worked together to benefit the teacher in an understanding of how NGSS practices look in an elementary classroom.

Dr. Z presented the students with opportunities to construct explanations when he showed students two nostrums from sawfish, one smaller and the other larger, and asked them to identify whether or not the nostrums were from different species or were which were a parent/child. He first asked students to raise their hands if they think the nostrums were from the same species and then asked students to raise their hands if they thought the nostrums were from a different species. Once the students answered in mind, he assisted them through a debate. In the lesson, the students and Dr. Z made observations of the sawfish nostrums and then use them as evidence to reason the justification for or against their support for why they thought the sawfish were either from different species were parent/child. Below is a partial excerpt:

Dr. Z: *"Now we're going to ask some of the people on the different species side. It's the first time we hear from the different species side to convince them to come over."*

Student: *"They're different shapes."*

Figure 2. Knowledgeable Other Exchange #1

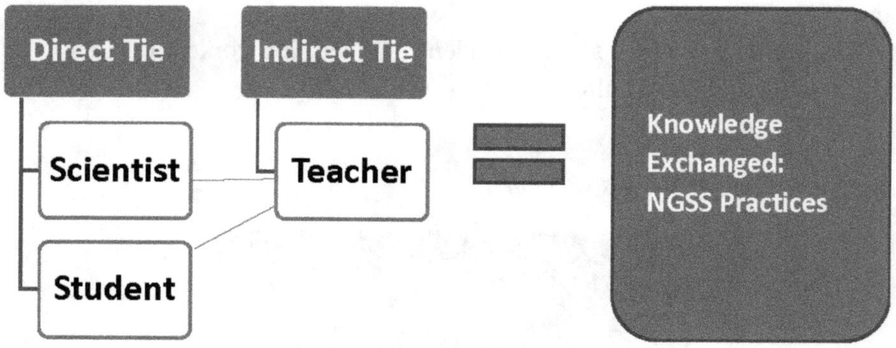

Dr. Z: *"They're different shapes? What's a--?"*
Student: *"No."*
Dr. Z: *"Hold on. What's a different shape?"*
Student: *"It looks kind of rounder up at the top."*
Dr. Z: *"It looks rounder at the top?"*
Student: *"A little."*
Students: *"These [inaudible]."*
Dr. Z: *"Hold on. Meghan is going to explain."*
Student: *"This one looks a little bit flatter than this one."*
Dr. Z: *"Okay, you convinced what she's saying? You hear what she's saying? Does that make you still think it's the same species [crosstalk]? Anybody here wants to change sides? You're what? You're convinced based on that, huh? Okay, go ahead. Why are these different species?*
Student: *"Because they both have different amounts of teeth so they're probably different species because the amount of teeth maybe means that they're different."*
Dr. Z: *"Okay. Who likes that answer? You guys not convinced? Think different numbers of teeth it's still okay to be the same species?"*
Student: *"That might be a baby, and that might be a mommy. It might be—"*
Dr. Z: *"Okay. So. Stella's still convinced. She's still convinced same species. Anyone else? A reason. Convince a couple people to come over. You know the answer, right?"*
Student: *"If the teeth are the same size—"*
Ms. Smith: *"Maddie's convinced [inaudible]."*
Student: *"Since the teeth are the same size, if it grew over time, the teeth, I mean, would have to grow with it so that [crosstalk]."*

Additionally, when Ms. Smith was interviewed, she made room in her curriculum for Dr. Z even though he could not commit to a consistent time because of his schedule at the veterinary clinic and home responsibilities. Ms. Smith stated that the reason she is so flexible is that his visits not only benefit the students, the visit benefit's her:

Every single time Dr. Z comes to my classroom, I learn something new. I learn about all kinds of animals, but more importantly, the way he thinks [deductively] and gets the students to reason. I am just hoping that if I watch him enough one day, I will do it like him.

Ms. Smith recounted that on one occasion, Dr. Z presented students with a large live tortoise and asked students to observe the front of the tortoise's legs and shared that the legs of the tortoise have 'scoops.' He then asked students to think about what the 'scoops' may be used for in the wild. He helped lead their thinking by asking what the tortoise may use to dig and asked them to further observe the 'scoops.' Dr. Z explained that the 'scoops' served as shovels scooping dirt to make a home or to protect the tortoise against hot climates and predators. She shared this example for several reasons, first, because she did not know about tortoises or turtles that Dr. Z had, and second, she did not have the live specimens. And if she did, would she have pushed her students to "infer" how the scoops may be used and then offer reasons why this may help the tortoise? Not likely, because this is not the kind of lesson a typical second-grade teacher would perform.

Ms. Smith reminisced about the first time she saw Dr. Z teach about the difference between carnivores and herbivores. Dr. Z asked the students to observe a skull providing the student with an opportunity to *obtain* the information necessary to determine an *evaluation* of whether the skull is an herbivore or carnivore animal skull. He then asked the student to *communicate* this information (the student's response/ reason) to the class on why he/she thinks the skull is that of an herbivore. The student responds by explaining that they have "grinders" while pointing to the skull's teeth. Dr. Z, then pushes the student's thinking further, asking him to continue explaining what he means by "grinders." The student responds by saying the "animal uses the teeth to 'grind' grass, and then they swallow it as food." Dr. Z continues to engage the student further by asking the student to elaborate on what he means by "grinders."

"Grinders!" She went on, "I have never, ever used that vocabulary word in relation to teaching about carnivores and herbivores. I was having the students just color pictures and do a matching worksheet about herbivores and carnivores. His lesson was so much more powerful. I knew it would stay with the kids much longer than my science lessons. He was simply more knowledgeable, and I am darn lucky to have him in my classroom.

In this more knowledgeable other exchange, we see the scientist use the sawfish nostrum (the tool) to engage the students with the *NGSS* practices. In this exchange, the scientist is the more knowledgeable other for both the teacher and the students. For the students, "knowing" as a result of this activity refers to the possibilities that become available to the participants as a way of thinking, reflecting and arguing about the sawfish (Roth & Radford, 2010). Additionally, as the scientist is serving as the more knowledgeable other, the teacher is also gaining new information. Thus, revealing the exchange of knowledge (*NGSS* Practices) from scientist to students and to the teacher. These observations also highlight the relationship of triads and are not defined simply as systems of three actors. Instead, existing research converges on a view that triads are defined, at minimum, by the coexistence of two ties between three associated actors (Vedel et al., 2016). Here, the two ties can be seen as the "direct" exchange between two parties and the "indirect" exchange between the others, both influencing the dynamic of the three-actor system as a whole (Chandler & Vargo, 2011). *Figure 2*. Knowledgeable Other Exchange #1 displays this triad relationship between the Scientist and Students as the "direct tie" and the Teacher "indirect tie" and the *NGSS* Practices as the exchange of knowledge gained.

Knowledgeable Other Exchange #2 The Scientist and Teacher Worked Together to Benefit the Students in the Area of Vocabulary

In this exchange, Dr. Z and Ms. Smith worked together to introduce the term "endotherm" when referencing warm-blooded animals. They suggested the students "break-apart" the word to determine the meaning. The teacher uses (the tool) prefixes and suffixes to engage the students with language. Below is an excerpt from the exchange.

Figure 3. Knowledgeable Other Exchange #2

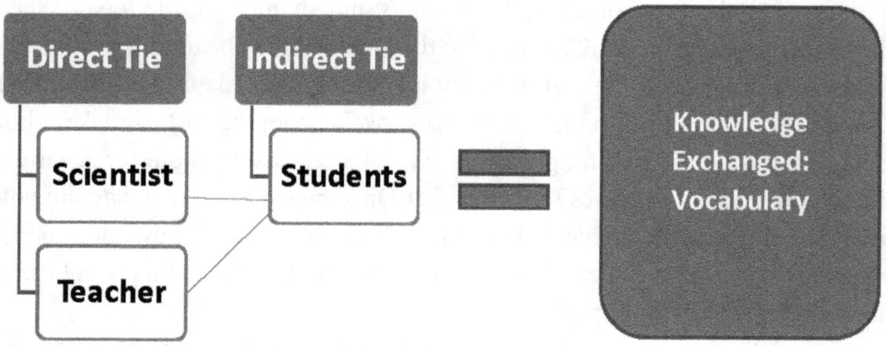

Dr. Z: *"We can learn what this word means by breaking apart the word. Let us look at the last part of the word 'therm.' Who can tell me the name of the instrument that tells us the temperature?"*

A student responds by saying a thermometer.

Dr. Z *then asked, "What does thermometer start with?"*

A student responds by saying 'therm.'

Dr. Z *continues by saying 'therm' means temperature and endo mean inside. Ms. Smith can you explain to them a little bit about prefixes and suffixes?"*

Ms. Smith *begins by saying, "scientists know and use a lot of vocabulary and one way to learn new words is to break them apart like Dr. Z said. Remember the word "unicycle its prefix is "uni" which means one. Bicycle where the prefix "bi" means two, and tricycle the prefix "tri" means three? Same thing here with the word "endotherm." Do you remember boys and girls, when we were talking about those vehicles that have four big wheels that go over dirt called a 'quad runner' where the prefix "quad" means four and there are four tires on that vehicle?"*

Dr. Z *then adds "thank you Ms. Smith. How about the word Quadricep? Thinking about what*

Ms. Smith *just said who can guess what the word quadricep means? Quadricep is in your thigh. How about this one...... How many muscles make up a bicep? Triceps?"*

Students respond by saying 2 then 3.

Dr. Z *continued and asked the class how many horns does a triceratops have? The class responded by saying three.*

Dr. Z *then added, "See you might not know what a word means but breaking it apart allows you to understand what the word means by looking at the prefixes and suffixes of words".*

When interviewed Dr. Z spoke about the fact that he routinely assessed the progress of the students by asking a series of questions in order for them to recall what they may have learned from the last session. However, it was his summative assessment, that he was most excited to talk with us about. Dr. Z described the formal assessment at the end of the year, a paper-pencil "test" consisting of a series of questions relating to any of the life science experiences in which the students had engaged during the year. He was most proud of the vocabulary that students were exposed to and learned. It became the talk of the class to receive a beaver skull for getting 100% on the end of the year assessment. Just another way Dr. Z motivated the students to further apply their knowledge by offering a reward.

On the day we spoke with Dr. Z he had just finished a lesson about Owl Pellets. He referenced this lesson in our interview when asked about building students vocabulary.

"Did you hear me say it during the lesson? Yes, I told them that they are going to know more words than fourth or fifth graders if they pay attention. Words like humerus, keratin, metatarsals to name a few."

He believes vocabulary is one of the keys to doing well in science and so it started early on rather than later in school, like in middle school.

"This is where Ms. Smith and I agree, that vocab[ulary] is so important, if they understand one or two science words it can help them understand like 6-8 other words. Like the lesson we did about prefixes. I did it the science way and then I passed it off to Ms. Smith because I know she teaches it the second-grade way."

This exchange is unique because it included another layer of transferring knowledge in the beginning where the scientist (Dr. Z) is the more knowledge other and is leading the students to connect between parts of words and their meanings (prefixes and suffixes). It is interesting and worth mentioning that the role of "more knowledgeable" other shifted twice during this exchange. Once Dr. Z showed the students how to break apart the word, the teacher becomes the more knowledgeable other by affirming (and making the application developmentally appropriate by simplifying) and identifying his word placement's correctness. Ms. Smith then asks a series of questions to the students and extends the conversation with prefixes and suffixes. In this exchange, the teacher finishes as the more knowledgeable other for both the scientist and the students. Overall, the "direct" tie is between the scientist and the teacher because they worked together to move the students' knowledge as the "indirect" tie. As a result, the knowledge exchanged to the students is vocabulary. Both Dr. Z and Ms. Smith serving as knowledgeable others, allowed an exchange of vocabulary knowledge and ways to determine their meaning. This relationship revealed a team approach to the exchange of knowledge for the students.

Knowledgeable Other Exchange #3 The Students and Teacher Worked Together to Benefit the Scientist in Understanding How to Engage Students in an Inclusive Way

In this exchange, the teacher began by using the sticks (the tool) to engage the students using an inclusive strategy. Here we see the teacher as the more knowledge other at the beginning of the exchange

Figure 4. Knowledgeable Other Exchange #3

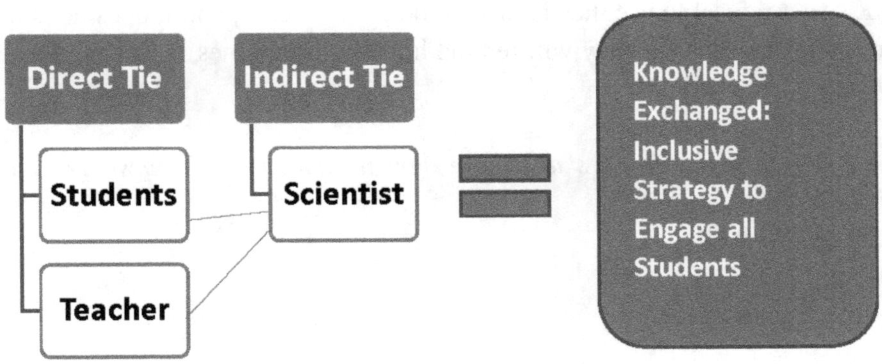

as she interrupts and stands in front of the class with the container of sticks. Once Dr. Z violated the rules set forth, the students became the more knowledgeable other by affirming (and reminding him, repeatedly) and identifying the sticks as means to engage all of the students. The scientist then asked a series of questions to the students and extended the conversation regarding each particular stick in each case. Below is an excerpt from the exchange.

Ms. Smith interjects, *"boys and girls, I have one more thing to add because I want every person listening and being ready to answer. This comes from a book called Teach Like a Champion that's used in New York City. We use our sticks here* [holding up a bucket containing popsicle sticks to the class]. *Mary and Kali find yourself seated, correctly, and ready to learn.*

Dr. Z: *Paces back and forth as she speaks, checks the overhead, adjusts the anchor chart, and folds his arms, waiting for Ms. Smith to finish speaking to the class, not realizing she is talking to him too.*

Ms. Smith*: Everybody is ready to answer. If you don't know the answer, someone will come along side you and help you, and Dr. Z will help you, so you do not have to feel bad if you do not know the answer. Everybody is here, and rather than raising hands, we just take a stick." [teacher models]*

DR. Z: *"I like that."*

Ms. Smith: *"That way, everyone has a chance to answer."*

DR. Z: *"Remember, there are no goofy answers; everyone gets a chance, and we don't laugh at anyone."*

He continued with the science lesson, and at 18:36 (precisely one minute later) this exchange occurs between Dr. Z and the students:

DR. Z: *"Who knows what this is [holding up a skull], raise your hand, what this is?*

Students*: Aren't you going to use the sticks?*

DR. Z: *Oh, [grabs a stick and reads a name] Caden*

Another minute later, the students have identified that Dr. Z is holding an alligator skull. He points to the back of the skull and asks students to look where he is pointing, and this exchange happens:

DR. Z: *Who thinks they know, using their power of observation, who can tell me why this animal died?*

A student calls out "sticks"

DR. Z: *"Yeah, [He turn and faces the stick container.] Oh no, this is a tough one! I want to call on someone who is using their powers of observation."*

This is Dr. Z's first refusal to use the classroom tool, "the sticks", and it does not stop the students reminding him to use the sticks as they will remind him ten more times in just this lesson. Here is an example of how they reminded him:

DR. Z*: "This is a herbivore [holding a skull] so we know it's a mammal. Who wants to guess what type of an animal it is?"*

Students: [pointing] "Sticks!"

DR. Z: [repeating] "Sticks!"

Dr. Z refused several times, claiming he either did not have time or a tough one or just wanted to call on someone. So, he kept pointing at students. The students kept reminding him of the agreed-upon tool of communication to engage all students through a fair and equitable process. Next, the students started *educating* Dr. Z about how the sticks serve as tools in their classroom. See the following excerpts as examples:

DR. Z: *"Do you see the difference? Look at the differences in the teeth?"*
DR. Z: *"Oh ya! Sticks! Thanks for reminding me." [He picks a stick] and then says Thomas!*
Students: *"No! it's Morgan."*
DR. Z: *[He looks down at the stick] "Thomas!"*
Students: *[giggling] "They have two sides!"*
DR. Z: *"Do they all have two sides?"*
Students: *[giggling] "Yes!"*
DR. Z: *"Now let me ask you, based on the size of this [skull] do you think this is a full-sized horse or a baby horse?"*
Student: *"A baby horse!"*
Student: *"A full-sized!"*
DR. Z: *"I did not see hands."*
Students: *"Sticks!" (pointing)*
Ms. Smith: *[laughing]*
DR. Z: *Okay, [grabbing a stick] Nancy!"*
Students: *S in unison "NANCY!"*
DR. Z: *"Okay, well if you are going to pick on me for using sticks, I am not going to use them."*
Student: *"I think it is full-sized."*
DR. Z: *"Hold on, What?... Raise your hand."*
Students: *[lots of noise] inaudible.*
DR. Z: *"I will use the sticks."*
DR. Z: *Nathan*
Students: *[in unison] He is not here he went home!*
DR. Z: *"Oh Nathan! [drops stick and proceeds to take another stick]*
DR. Z: *"What? Do you think this is a full-sized horse? Okay, who would think that if you found these two skulls they are the same species? [Dr. Z placed the two skulls down next to each other.] Who thinks they are the same species? Who thinks they are different species? Raise your hand if you think they are different species.*
Students: *[inaudible] lots of noise*
Student: *"Sticks!"*
Ms. Smith: *[laughs] "Sticks!"*
DR. Z: *"I am just gonna hold them [sticks]so I don't forget to use them."*
DR. Z: *"Hayley! Where is Hayley?"*
Students: *"Which one? E-Y or I-G-H?*
DR. Z: *"You mean there are two? Alright then, E-Y"*
Students: *[laughing] that's Hayley D.*
DR. Z: *Hayley, do you think this is the same species or different species? Do they like pretty much identical except in size?*

Dr. Z is a scientist that wants to ask questions. In this particular lesson which was 97 minutes long, he asked 239 questions (Farland-Smith & McClusky, manuscript in progress). His frustration was evident on camera and we followed up in the interview after with questions about why he thought the students were so focused on the sticks. Dr. Z commented on his frustrations, saying:

"I am asking all these terrific questions while holding the skulls, or beaks, like guess what type of animal this might be, who thinks they know what kind of bird this is? And they are worried about who I call on. I just did not get it. Since I am not here regularly, it's just not enough to know their names. It must have been necessary to them, but it was not as important to me as the science and the observation and questioning I was trying to generate. "

When the teacher was interviewed, she was not surprised by how her students took ownership of their classroom and guided Dr. Z to use the tools. From her perspective, they all wanted to answer, and when he would say, "don't call out", it was like a reminder to them to use an element such as the sticks (an inclusive practice strategy) as an artifact of the culture of their classroom and their space. They were educating him, and they were not going to get to the science learning until he used the tools of the classroom. In one particular lesson, we even had visitors, the researcher's children and one student got out of her seat and came to me about ten minutes in and said they will need their own sticks to participate. The sticks represent equity, and inclusiveness and the students know this by being active members of the classroom community.

Ms. Smith added saying: *"You see the reason I got up and got them [the sticks] was that I noticed that all the students were not being called on and the noise level was rising."*

Ms. Smith was referring to a technique named "Cold Call", where the sticks represent the tool to carry out a "cold call" in the book, *Teach like a Champion* (2013). Strategies such as this can be used to increase rigor and engagement in a classroom inclusively and equitably. They can be used by almost any teacher and in concert with almost any other approach. It sets an expectation where students may be called on at any time and expected to participate positively. The author intended for use when all the children eagerly want to participate, such as a day when a scientist is visiting, and thus levels the playing field.

However, what Ms. Smith did not anticipate was Dr. Z's aversion to the technique and her students' persistence that he follows the classroom's order. This type of tool makes it explicit that he was *not* making intentional decisions about who to call on and appeared to leave him uncomfortable. Perhaps he just would have rather chosen intentionally. The technique of "Cold Call" is to build accountability and inclusiveness, but Dr. Z wanted to reward passion and enthusiasm. He appeared frustrated as the act of stick pulling interrupted the flow of observation and inquiry.

DISCUSSION

This study revealed how the role of a knowledgeable other in a student-scientist-teacher partnership can be multifaceted and effective in its exchange of knowledge. We will discuss each research question separately. The first research question: *What does the exchange of knowledge look like in a student-scientist-teacher partnership within an elementary classroom?*

This question was supported by a few key findings. First, *the role of the knowledgeable other shifted and was shared by more than one individual within the partnership* in each of the exchanges. Goos, Galbraith & Renshaw (2002) claim that interactions are both bi-directional and are a collaborative process of moving knowledge within a zone of proximal development as both individuals have knowledge but at times require the other to move it. For example, in the first exchange, both the scientist and the students assumed the role of the more knowledgeable other to exchange the knowledge of the *NGSS* practices to the teacher. To visualize this shift and reflect upon it in practical terms, the role of knowledgeable others was like a ball passed (and in some cases held on to for mere minutes) from one individual to the other within the triad (framework).

Second, when examined in a triad framework, *the student-scientist-teacher partnership is not an equally distributed relationship as the name suggests.* Rather, two parts of dyad worked to bring understanding to the third part making a triad relationship. Chandler and Vargo (2011), Vedel et al., (2016) and Simmel (1950) describe the relationship between members of the exchange as either "direct" or "indirect" tied together to influence all three members of the system. We see the constant exchange of direct and indirect ties within the student-scientist-teacher partnership in each of the interactions.

Third, *the more knowledgeable others used tools to exchange the knowledge between individuals in the student-scientist-teacher partnership.* Vygotsky asserted that social interactions exist within a zone of proximal development (ZPD) where tools and signs could result as artifacts of the interactions (Vygotsky, 1978). Throughout each of the interactions we clearly saw the more knowledgeable others using tools to promote good learning within the ZPD to exchange the knowledge. The scientists used *NGSS* as a tool and the students and teacher both used craft sticks as a tool to exchange knowledge. Next, *the relationship between the teacher and the scientist evolved naturally where there was an observed reciprocity.* Gonzales et al., (2005) defined this as "an attempt to establish a social relationship on an enduring basis that is formed by the assumption of *confianza* (mutual trust)". Ms. Smith and Dr. Z's relationship seemed to unfold effortlessly over a course of six years, hallmarked by a common understanding and trust that proved to be beneficial in enhancing the learning experience for students. During interviews, Dr. Z stated *"This is where Ms. Smith and I agree, that vocab is so important if they understand one or two science words it can help them understand like 6-8 other words. Like the lesson we did about prefixes. I did it the science way, and then I passed it off to Ms. Smith because I know she teaches it the second-grade way."* Dr. Z's statements revealed his trust in Ms. Smith's teaching knowledge.

Fourth, *growth in knowledge was exhibited by all members of the partnership.* Dr. Z's knowledge was situated in his professional career as a veterinarian yet extended way beyond 'hobbyism'. Dr. Z also used the visits to implement explicit and structured instructional strategies like questioning and the explicit instruction of vocabulary consistent with most well-trained and seasoned teachers. Additionally, Dr. Z exhibited growth in knowledge in understanding how strategies such as "Cold Call" can serve as a valuable and useful tool to engage students inclusively. For Ms. Smith, the exchanges proved beneficial and aided in her conservative content knowledge of science. Ms. Smith knew her limits when it comes to science knowledge and knew that partnering with Dr. Z would enrich her students' lives by enhancing and supporting their meaning of science. Ms. Smith brought the field trip experience into the classroom when traditional protocols or teaching methods would not have allowed for such an opportunity. She was driven to provide the students with science content and experiences that may or may not have been available to most students. The students appeared to be more competent and capable science learners. Their capabilities and knowledge are a resource that can and should be accessed and built upon during science instruction. Children offer tremendous promise for science learning when we compare their knowledge and skills to what scientists do in their work. Students were also able to grow in their knowledge of real scientists and see them as real people just like them, which could serve as means to increase their awareness and interests in science content. Students were also given an opportunity to engage in science content not generally provided in the mainstream curricula.

The second question we looked at in this study was: *What interactions serve as evidence to demonstrate the effectiveness of "the more knowledgeable other" in a student-scientist-teacher partnership within an elementary classroom?*

This question was also supported by a few key findings. First, *student-scientist-teacher partnerships can offer elementary students the capacity to engage in scientific and engineering practices as they develop conceptual understandings over time.* The evidence to support this finding is that over the ten months of instruction in which we observed this student-scientist-teacher partnership deepened their content knowledge and their scientific vocabulary and conceptual understanding of the life sciences. For example, in a lesson about Owl Pellets Dr. Z's very first comment was, *"When you dissect the owl pellet, you should be able to pick out all the femurs. They all look this tiny with the little femoral head, right?* This example is evidence of students' ability to retain a certain amount of knowledge from past lessons (i.e., femur and femoral head) before the actual instruction began. This is consistent with the recommendation from NSTA (2014) *that elementary students can engage in scientific and engineering practices as they develop conceptual understandings over time*.

Second, *student-scientist-teacher partnerships beginning at the youngest levels can offer elementary students' opportunities to observe phenomena, engage in problem-solving, and explain their thinking outside the typical activities done in the classroom.* The evidence to support this finding was observed during an interview with the teacher which she stated, "*Dr. Z's visits afforded my students the kind of questioning and deductive reasoning I don't even think to try, like why a tortoise might have legs shaped like scoops. On one visit, I told him, we are doing state and district testing tomorrow on opinion and reasoning.*" Katz (2010) spoke of the need for opportunities early in science education to observe phenomena, engage in problem-solving and explain their thinking. As the earlier students learn these skills, the more likely they will be to use them as they move through upper elementary and build on skills to enable them to 'think like a scientist'. Moulding, Bybee and Paulson (2015) described high quality as moving from curiosity to interest to reasoning. The frequency and intentionality of the visits gave students opportunities to practice observe phenomena, engaging in problem solving and explaining their thinking in such a way that these skills were expected each visit and therefore became somewhat 'normalized' by age 8.

Third, *student-scientist-teacher partnerships can offer elementary students the opportunity to engage in science within the broader community of science*. Tapping into volunteers in the classroom community (a parent who was a scientist) allowed the students to become active participants within the culture of science. Recall the assumption mentioned at the beginning of this chapter was that a scientist's visit would improve students' perceptions of scientists. While this study did not measure the students' perceptions of scientists, it was clear from the transcriptions that the long-term relationship the class developed with the scientist allowed for differing values and priorities throughout the extent of the partnership. As a result, the relationship with an individual in the "broader community of science" certainly helped shape children's ideas about those individuals who choose to study science as a career. Only a longitudinal study could provide data to support how those experiences impacted the 7 or 8-year-old's science identity.

The Role of the More Knowledgeable Other and Implications for the Elementary Classroom

The results of this study revealed useful lessons learned in the elementary science classroom.

The provision of a student-teacher scientist partnership and the more knowledgeable other exchanging knowledge proves to be an effective model for students to experience the *NGSS* Practices. Crawford (2005), Mercer et al., (2004) Roth (1993) and Vygotsky (1978) all agree that science learning is the process of "doing" scientific practices and procedures grounded in dialogue and socio-cultural interactions. Consistent with this view is the understanding that students establish standard interaction practices with each other through action. These behaviors define what counts as learning in the classroom. In this study, behaviors such as "observing," "testing," and "explaining" are worth mentioning as they are beneficial for this group of students who were introduced to and gained a deeper understanding of the practices identified as necessary by the *NGSS*.

Just as scientists do, the students in this classroom gained new knowledge and science vocabulary while being involved in a student-scientist-teacher-partnership that helped them make evidence-based conclusions. The teacher worked with scientist and the children to gain new knowledge and vocabulary while supporting the students in making their classroom equitable. The scientist made it all possible by providing engaging discussion, opportunities for decision making, and argumentation, with evidence with both the students and the teacher as members of this partnership.

REFERENCES

Banilower, E. R., Smith, S. P., Weiss, I. R., Malzahn, K. A., Campbell, K. M., & Weis, A. M. (2013). *Report of the 2012 national survey of science and mathematics*. Horizon Research, Inc.

Bodzin, A., & Gehringer, M. (2001). Breaking science stereotypes: Can meeting actual scientists change students' perceptions of scientists? *Science and Children*, *38*(4), 36–41.

Buck, G. A., Leslie-Pelecky, D., & Kirby, S. K. (2002). Bringing female scientists into the elementary classroom: Confronting the strength of elementary students' stereotypical images of scientists. *Journal of Elementary Science Education*, *14*(2), 1–9. doi:10.1007/BF03173844

Cavagnetto, A., Hand, B., & Meier, L. N. (2010). The nature of elementary student science discourse in the context of the science writing heuristic approach. *International Journal of Science Education*, *32*(4), 427–449. doi:10.1080/09500690802627277

Chandler, J. D., & Vargo, S. L. (2011). Contextualization and Value-in-context: How Context Frames Exchange. *Marketing Theory*, *11*(1), 35–49. doi:10.1177/1470593110393713

Cobb, P., Wood, T., & Yackel, E. (1991). Analogies from the philosophy and sociology of science for understanding classroom life. *Science Education*, *75*(1), 23–44. doi:10.1002ce.3730750104

Cossett-Lent, R. (2016). *This is Disciplinary Literacy: Reading, Writing, Thinking and Doing- Content Area by Content Area*. Corwin Literacy.

Crawford, T. (2005). What counts as knowing: Constructing a communicative repertoire for student demonstration of knowledge in science. *Journal of Research in Science Teaching*, *42*(2), 139–165. doi:10.1002/tea.20047

Dr. Z (2017). *Powers of Observation-year long visiting scientists experiences*. Academic Press.

Farland-Smith, D. (2009). Exploring middle school girls' science identities: Examining attitudes and perceptions of scientists when working "side-by-side" with scientists. *School Science and Mathematics*, *109*(7), 415–427. doi:10.1111/j.1949-8594.2009.tb17872.x

Farland-Smith, D. & McClusky, S. (manuscript in progress). Why are rabbits not rodents? Examining the role questions across multiple student-scientist-teacher visits in a second grade classroom.

Flick, L. (1990). Scientist in Residence program: Improving children's images of science and scientists. *School Science and Mathematics*, *90*(3), 204–214. doi:10.1111/j.1949-8594.1990.tb15536.x

Glassman, M. (2001). Dewey and Vygotsky: Society, experience, and inquiry in educational practice. *Educational Researcher*, *30*(4), 4–14. doi:10.3102/0013189X030004003

Gonzales, N., Moll, L., & Amanti, C. (2005). *Funds of Knowledge: Theorizing Practices in Households, Communities and Classrooms*. Lawrence Erlbaum Associates, Publishers.

Goos, M., Galbraith, P., & Renshaw, P. (2002). Establishing a community of practice in a secondary mathematics classroom. In L. Burton (Ed.), *Learning Mathematics: From Hierarchies to Networks* (pp. 36–61). Falmer Press.

Greeno, J. G. (1998). The situativity of knowing, learning, and research. *The American Psychologist*, *53*(1), 5–26. doi:10.1037/0003-066X.53.1.5

Grossman, B. T., & Farland-Smith, D. (2020). Collaborating with virtual visiting scientists to address students' perceptions of scientists and their work. *Innovations in Science Teacher Education, 5*(3). Retrieved from https://innovations.theaste.org/collaborating- with-virtual-visiting-scientists-to-address-students-perceptions-of-scientists-and-their- work/

Jennings, L. B., & Mills, H. (2009). Constructing a discourse of inquiry: Findings from a five- year ethnography at one elementary school. *Teachers College Record, 111*(7), 1583–1618.

Katz, L. G. (2010). *STEM in the early years*. Collected Papers from the SEED Conference Fall 2010.

Kirch, S. A. (2007). Skepticism and open-mindedness for learning, teaching and criticality in science. *Cultural Studies of Science Education*, 2.

Larsen, S., Liston, C., Thiry, H., & Graf, J. (2007). What good is a scientist in the classroom? Participant outcomes and program design features for a short- duration science outreach intervention in K–12 classrooms. *Life Sci Educ*, *6*(1), 49–64. doi:10.1187/cbe.06-05-0165 PMID:17339394

Lemov, D. (2013). *Teach like a champion: 49 Techniques that put students on the path to college*. Wiley & Sons.

Lewis, C., Perry, R., Hurd, J., & O'Connell, M. P. (2006). Lesson study comes of age in North America. *Phi Delta Kappan*, *88*(4), 273–281. doi:10.1177/003172170608800406

Lewis, C., & Tsuchida, I. (1998). A lesson is like a swiftly flowing river:How research lessons improve Japanese education. *American Educator*, *22*(4), 12–17, 50–52. lessonresearch.net/lesson.pdf

Mercer, N., Dawes, L., Wegerif, R., & Sams, C. (2004). Reasoning as a scientist: Ways of helping children to use language to learn science. *British Educational Research Journal*, *30*(3), 359–377. doi:10.1080/01411920410001689689

Moulding, B., Bybee, R., & Paulson, N. (2015). *A vision and plan for science teaching and learning*. Essential Teaching and Learning Publications.

Nagy, W., & Townsend, D. (2012). Words as tools: Learning academic vocabulary as language acquisition. *Reading Research Quarterly*, *47*(1), 91–108. doi:10.1002/RRQ.011

National Research Council. (2007). Taking science to school: Learning and teaching in science grades k-8. Committee on science learning, kindergarten through eighth grade. Board of Science Education, Center for Education. Division of Behavioral and Social Sciences and Education. Washington, DC: The National Academies Press.

National Research Council (NRC). (2008). *Ready, set, Science!: Putting research to work in K–8 science classrooms*. National Academies Press.

National Science Teachers Association (NSTA). (2014). *Early Childhood Science Education*. NSTA.

National Science Teachers Association (NSTA). (2018). *NSTA Position Statement: Elementary Science Education*. NSTA.

NGSS Lead States. (2013). *Next Generation Science Standards: For states, by states*. National Academies Press.

Roth, W. (1993). Construction sites: Science labs and classrooms. In K. Tobin (Ed.), *The practice of constructivism in science education* (pp. 143–170). Lawrence Erlbaum Associates, Publishers.

Roth, W. M., & Radford, L. (2010). Rethinking the zone of proximal development: (Symmetrically). *Mind, Culture, and Activity*, *17*(4), 299–307. doi:10.1080/10749031003775038

Simmel, G. (1950). *The Sociology of Georg Simmel*. Free Press.

Stilaloppi, J., & Vargo, S. (2017). Triads: A review and analytical framework. *Marketing Theory*, *17*(4), 395–414. doi:10.1177/1470593117705694

Takahashi, A., & Yoshida, M. (2004). How can we start lesson study? Ideas establishing lesson study communities. *Teaching Children Mathematics*, *10*(9), 436–443. doi:10.5951/TCM.10.9.0436

Vedel, M., Holma, A.-M., & Havila, V. (2016). Conceptualizing Inter-organizational Triads. *Industrial Marketing Management*, *57*, 139–147. doi:10.1016/j.indmarman.2016.01.005

Vygotsky, L. S. (1978). *Mind in society: The development of higher psychological processes*. Harvard University Press.

ADDITIONAL READING

Abell, S. K., Anderson, G., & Chezem, J. (2000). Science as argument and explanation: Exploring concepts of sound in third grade. In J. Minstrell & E. H. van Zee (Eds.), *Inquiring into inquiry learning and teaching in science* (pp. 65–79). American Association for the Advancement of Science.

Carr, D. (1998). The art of asking questions in the teaching of science. *The School Science Review*, *79*, 47–50.

Farland-Smith, D. (2011). Personal & social interactions between young girls and scientists: Examining critical aspects for identity construction. *Journal of Science Teacher Education*, *23*(1), 1–18. doi:10.100710972-011-9259-7

Farland-Smith, D. (2016). My daughter the scientist? Mothers' perceptions of the shift in Their daughter's personal science identities. *Journal of Educational Issues*, *2*(1), 1–1. doi:10.5296/jei.v2i1.8670

Gnyawali, D. R., & Jinyu, H. (2004). Two's Company, Three's a Crowd? Triads in Cooperative- competitive Networks'. *Academy of Management Journal*, *47*(6), 918–927.

Kowalkowski, C., Kindstro¨m, D., & Carlborg, P. (2016). Triadic Value Propositions: When It Takes More Than Two to Tango. *Service Science*, *8*(3), 282–299. doi:10.1287erv.2016.0145

Lowrey, T. M., Otnes, C. C., & Ruth, J. A. (2004). Social Influences on Dyadic Giving over Time: A Taxonomy from the Giver's Perspective. *The Journal of Consumer Research*, *30*(4), 547–558. doi:10.1086/380288

KEY TERMS AND DEFINITIONS

Elementary School: Formalized education between the ages of 5-10.

Knowledgeable Other: An individual who serves as "the teacher" and dispenses knowledge to others.

Science Vocabulary: Words and or language specific to the discipline of science.

Student-Scientist-Teacher Partnership: Integrative partnerships in which students, teachers, and scientists are involved and benefit.

Triad: Triads are systems of three parties in which each party interacts directly and reciprocally with another party and can also operate as an intermediary between the other two parties.

Visiting Scientists: Individuals who may be professional scientists or non-professional scientists and visit classrooms to share their knowledge.

Zone of Proximal Development: The distance between the actual developmental level (determined by independent problem solving) and the level of potential development (determined through problem solving) under guidance from an adult/teacher or in collaboration with more capable peers.

Chapter 4
"The Way I Think About Scientists Has Changed": Measuring the Impact of Virtual Scientists on Middle School Students' Perceptions of Scientists

Brandon Grossman
University of Colorado, USA

Donna Farland-Smith
The Ohio State University, USA

ABSTRACT

The purpose of this research was to assess the impact of virtual, discussion-based interactions with five scientists on middle school students' (N=48) perceptions of scientists. Based on students' self-reflections and results from Enhanced-Drawings of Scientists Test (E-DAST), the week of virtual discussion-based interactions broadened students' perceptions of what scientists do (activity), where scientists work (location), and what scientists look like (appearance). The virtual discussion-based experiences had the most significant impact on middle school students' perceptions of the work scientists do (activity). As this important finding demonstrates—through middle school students' own words—students' perceptions of the appearance and location of scientists may be 'gatekeepers' or secondary to the their ideas of what scientists do for activities. In other words, until students can 'get past' or debunk the stereotype of what scientists look like (appearance) and where they work (location), students may not truly understand the work scientists do (activity).

DOI: 10.4018/978-1-7998-4966-7.ch004

INTRODUCTION

Bridging the disconnect between students and scientists has been a focus of science education for years. Despite decades of work by researchers and educators, stereotypic perceptions of scientists still persist and are contrary to what students should understand about science and scientists as outlined by the *Next Generation Science Standards (NGSS)* (2013). For this reason, science teachers often struggle as to how to humanize the content they teach to their students as recommended in the following documents: *National Science Education Standards* (NRC,1996), *A Framework for K-12 Science Education: Practices, Crosscutting Concepts, and Core Ideas* (NRC, 2012) and *NGSS* (NGSS, 2013). In each of these documents, science as a human endeavor emphasizes the ways in which men and women of different races and ethnicities engage with science and technology.

Engaging classrooms filled with children with science professionals is not a new idea for teachers. However, practical ways helping students understand the nature of science careers is limited (Wyss, Heulskamp, & Siebert, 2012). Advances in classroom technologies have created new possibilities for scientists to engage with students in creative ways, thus creating more opportunities for broadening students' perceptions of science and scientists. For example, the idea of the virtual "field trip" has emerged as one possible way to connect scientists and students (Adedokun et al, 2010). Niemitz et al (2008) stated that such virtual connections with scientists can provide students with opportunities to experience and understand scientific processes and enhance students' aspirations for becoming scientists. The motivation behind meeting, engaging, and working with career scientists is to help disrupt students' stereotypic perceptions and beliefs about science and scientists. These stereotypical perceptions of scientists are well documented in the literature (Barman, 1997; Beardsley & O'Dowd, 1961; Bozdin & Gehringer, 2001; Chambers, 1983; Etzioni, & Numm, 1974; Finson, 2002; Fort & Varney,1989; Mason, Kahle & Gardener, 1991; Mead & Metraux, 1957) and typically include representations of scientists as 'nerdy' looking older white men with 'crazy hair' wearing white lab coats and working by themselves in a laboratory.

If only teachers knew best how to dispel these stereotypic perceptions, they would be willing to set aside precious classroom time to once and for all rid them from students. If this information is offered in a school setting it can be presented to hundreds of students at once, and ultimately students will be better informed to make decisions about course selections and a variety of STEM careers. But it's no simple task, otherwise researchers would have answers as to how help increase the STEM pipeline. In order to try new innovative ways to teach about the nature of scientists, teachers need administrative support and permission to break down cultural barriers within the confines of their classroom walls. Partnerships between students, scientists and teachers have the potential to fill this gap and provide more benefits than previously thought and are worthy of investigation. This chapter examines how virtual classroom visits by scientists were beneficial to broadening middle school students' perceptions of scientists in terms of their understanding of what scientists look like, where scientists work, and what scientists do.

LITERATURE BACKGROUND

Why Middle School?

Students perceptions scientists is a process in which students form positive or negative associations from within their own culture (Farland-Smith, 2017). It is natural for students to look to their culture to help reinforce or redefine their perceptions of scientists. Bruner (1996) labeled this effect culturalism, and it applies to the people middle school students see in their culture that they connect to a science career. By middle school, students have constructed perceptions of scientists, which are unlikely to change until they have personal contact with a scientist or experience a situation that causes a change in perceptions. The problem occurs when students form perceptions about scientists that are limiting and as a result, they are turned off from the idea of becoming scientists. When in reality they don't know much about scientists or science as a discipline and students are unable to identify with science professions.

Middle school students are especially vulnerable as they are constructing their own identities during adolescence. Since the early 1980s it has been established that middle school students begin to limit their career choices as early as fifth grade in the areas of science and technology (Erb, 1981). Joyce & Farenga (1999) argue by age 9, some children will decide whether or not they will select to study science as a career. Maltese & Tai (2010) found that scientists developed their interest in science before middle school. Additionally, Lindahl (2003) found that students' subject interest is also a key factor in their educational and career choices and Silver & Rushton (2008) contend that the stereotypical perceptions children hold of scientists can prevent children from selecting science as a career interest. The literature points to students' perceptions about science and scientists as a key factor for students' subject (Chen et al, 2020) and career interests (Ozgelen, 2017).

Why Five Scientists in One Week?

Today's student-teacher scientist partnerships are still evolving. They are built on the assumption that a scientist will improve children's perceptions of who scientists are, where they work and what scientists do for work. Bozdin and Gerhinger (2001) and Flick (1990) reported that visits from scientists in formal classroom settings resulted in a decrease in many stereotypical beliefs about scientists, indicating that students' perceptions of scientists can be influenced by visits from professional scientists. However, the practical applications for the interventions used in this study, i.e. STEM professionals visiting classrooms is not likely because schools do not have enough access to professional who do not reinforce the stereotypical scientist. Buck, Leslie-Pelecky & Kirby (2002) reported no change in the stereotypical image of a scientist when three female scientists visited an upper elementary classroom. A science camp with week-long interactions with a variety of scientists demonstrated a positive impact on middle school girls' perceptions of scientists (Farland-Smith, 2009).

The educational researcher the classroom teacher had contacted had experience because of this experience with middle school students and five university scientists, which demonstrated that middle school girls' perception of science and scientists were successfully improved (Farland-Smith, 2009). Unlike the previous studies mentioned, this study connected middle school students to five professional scientists over a period of a few days in laboratory and field exercises. An analyses of student journal entries and E-DAST, and attitude surveys, revealed a keen appreciation of the sciences among the majority of participants and both a heightened and broadened awareness of Science as a Human Endeavor.

OVERVIEW OF THE PROJECT

The classroom teacher initiated the project, and it began when he reached out to the educational researcher in an email explaining that a few scientists and engineers would be visiting his classroom during the upcoming schoolyear. The classroom teacher had experience bringing in scientists to meet his students and was familiar with the array of past challenges associated with bringing such professionals into the classroom. He was planning on holding a "Skype a Scientist Week" or a "Skype an Engineer Week" for his students, where students would video-chat over the internet with a different scientist or engineer every day (please note this was pre COVID-19). The teacher envisioned that during each session, scientists or engineers would describe their work (and their life outside work) and participate in a question and answer session of about 45-minutes. The educational researcher was contacted because the teacher was interested in determining if this inexpensive, straightforward, and scalable method of bringing scientists into the classroom would have a measurable impact on student perceptions of scientists or engineers and the work that they do.

Five professional scientists (two females and three males) interacted with middle-school students during this one-week experience. One of the advantages to conducting a solely online video-based conference platform was that the teacher was able to reach and invite scientists who were not bound by geographic location. The participating scientists were intentionally selected so that they represented a range of diversity in terms of scientific, educational, racial, ethnic, and linguistic backgrounds. They came from a variety of colleges and universities in the United States and held various degrees from doctoral students to tenured faculty. It is important to mention that all participating scientists were willing and eager to participate. More than one scientist had prior experience working with middle school students and used the opportunity to fulfill their own service requirement.

The virtual sessions were held in a middle school classroom and required minimal technology upgrades, such as speakers, microphone, and a camera. The students (N=48) came from one of the largest public-school districts in the United States. There were twenty-five females and twenty-three males who participated. At the time of study, there were 1308 total students enrolled in the middle school (60% Hispanic, 19% Asian, 11% White, and 10% Black). 81% of students were eligible for free or reduced lunch. 27% of students were listed as students with disabilities. 14% of students were English language learners. The most common languages spoken at this particular school, other than English, included Spanish, Mandarin, Arabic, Tagalog, and Urdu.

All scientists worked with the classroom teacher to develop lessons/discussions before each virtual experience. In which the scientist suggested a topic, and or lesson or discussion to the classroom teacher. Then, working with the teacher to ensure science content and activities were relevant to the *NGSS* and age-appropriate. The teacher spent time with the students before the scientists visited virtually, helping them to develop questions to ask the scientists. More details about the planning and collaboration involved in this study were published in Grossman & Farland-Smith (2020).

OVERVIEW OF THE WEEK

Day #1

On the first day, students visited with a Caucasian white female professor who was joined by four undergraduate research students that included one black female, one white male, and two white females. The session began with each describing how they became scientists, what they love about science, and what scientific questions were interesting and meaningful to them. The research team then discussed their most current research project which involved studying mosquitos and their preference to prey on one species of bird over another. Then, for the remaining portion the class engaged in an informal student-led discussion.

Day #2

On the second day, students visited with an Asian female research scientist. The session with the scientist began by describing the role her family and other personal relationships played in developing her interests in science. She showed the students how she analyzes water resource risk and large-scale climate variability. Then, for the remaining portion the class engaged in an informal student-led discussion.

Day #3

One the third day, students visited with a Latino male doctoral student in physics. He described his current interest in using mathematics to study weather patterns. Before engaging in informal student-led discussions, he described how his secondary school science experiences shaped the development of his interests in science. Then, for the remaining portion the class engaged in an informal student-led discussion.

Day #4

On the fourth day, students visited with a Latino male research scientist. His work involved applying computer science and mathematics to environmental issues. He described the way in which he used technology to identify parts of large mines around the world that could be potentially harmful to the Earth. He also described his interests outside of science, including music and dancing. Then, for the remaining portion the class engaged in an informal student-led discussion.

Day #5

On the fifth day, students visited with a Latino male post-doctoral researcher. This scientist was a neurobiologist that studies the neural basis of animal behavior. He shared videos of experiments examining the factors that help mosquitoes track their prey. He described how his interest in science came from his love of animals as a child and early interest in wondering what was happening inside of their heads. He said he continues to ask the same question in his science lab, where he studies how the brain enables mosquitos to make decisions. Then, for the remaining portion the class engaged in an informal student-led discussion.

RESEARCH QUESTIONS

1. In what ways do weeklong virtual, discussion-based interactions with scientists influence students' perceptions of what scientists look like (*appearance*), where scientists work (*location*) and the work scientists do (*activity*)?
2. How did students self-report their experience of the sessions?

METHODS

The classroom teacher made the pedagogical decision to arrange the visits in the span of one week so that a pre/post assessment of how students' perceptions Enhanced Draw-A-Scientist Test (E-DAST) developed by Farland-Smith & McComas (2009) could be utilized. However, logistically, the virtual visiting scientist project needed to be conducted over two weeks simply because of the scientists' and middle school students' schedules. During the virtual visiting scientist week, the student generated questions developed prior to the visit were asked as part of the methods for exploring and understanding the kind of work scientists do. For the 'Do Now' portion, at the end of the virtual visiting scientist week, students wrote answers to the four questions in their science journals from the electronic prompt the week following the experience.

DATA COLLECTION

Data collection was mixed methods. The quantitative portion came from pictures students drawn in the E-DAST and qualitative portion was the "Do Now" responses in which the teacher posed a series of questions, included in the next section. In the quantitative portion students' pictures of scientists were given a numerical score and then a mean score was assigned before comparing pre, post and delayed posttests. Therefore, no pictures are included in this article because the analyses were focused on the quantitative measure and not the qualitative nature of the pictures themselves. The discussion of the pictures will be limited to their numerical score on the DAST Rubric (Farland-Smith, 2012).

Instruments

E-DAST

The Enhanced Draw-a-Scientist Test (E-DAST) asks students to draw three pictures of scientist at work consecutively. Originally developed by Chambers (1983), the Draw-a-Scientist Test (DAST) provides teachers and researchers with important information regarding students' perceptions of science and scientists. The DAST Rubric was developed specifically to score these pictures in the categories of *appearance*, *location*, and *activity*.

'Do Now' Responses

Students wrote answers to these four questions in their science journals: 1) What did you think about last week's [virtual visiting scientist] program? 2) Did you think it was a valuable experience? Explain why or why not? 3) Did the program change how you think about scientists and what they do? Explain how. 4) What was the most interesting thing you learned during last week's session?

Analysis

The pre, post, and, delayed posttest drawings were analyzed as described in this section and includes only the results of the students who completed all three administrations of the E-DAST. The Shapiro-

Wilk Test of Normality showed that the distribution of the sample was not normal, so Wilcoxon Signed Rank Test (non-parametric test) was used to analyze the data. The data from each group is listed in subsequent Tables 1, 2, and 3.

The E-DAST Rubric was used to evaluate the three pictures provided by each student. The E-DAST Rubric assigns a rating of 0, 1, 2, or 3 in the categories of "*appearance*", "*location*" and "*activity*".

'Do Now' Responses

The student responses to each of the four questions were obtained electronically by the researcher. Each of the four questions were analyzed separately. The responses were coded by the researcher. When appropriate, the researcher counted and logged reoccurring themes. The emerging themes from the 'Do Now' Responses were *appearance*, *location*, and *activity*. When students made comments about how a scientist looked or what kind of clothes they wore or how they acted the letters APP were logged to establish the category of *appearance* of the scientist. When students made comments about where a scientist worked the letters LOC were logged to establish the category of *location* of the scientist. Anytime students made comments about what kinds of work a scientist does the letters ACT were logged to establish the category of *activity* of the scientist. The results from these reflections were then compared to the results from the E-DAST for consistency once they were scored.

RESULTS

E-DAST

The E-DAST (pre, post, and delayed post) were scored using the DAST Rubric. The DAST Rubric was scored by the researcher who developed the DAST Rubric and scoring protocol. The mean scores for each of the categories: *appearance*, *location* and *activity* are shown below in Table.1. They demonstrate change from pretest to posttest at the .05 level. The results from Table 1 indicate that there was a statistically significant difference in the *appearance* and activity of *scientists*. However, there was not a statistically significant difference in the pretests and posttests of the mean scores of the *location*.

Table 1. Pretest-Posttest

	Pretest Mean	Posttest Mean	Related-Samples Wilcoxon Signed Rank Test
APP	7.67	8.42	.004*
LOC	6.25	6.46	.302
ACT	5.29	5.96	.045*

The mean scores for each category, *appearance*, *location*, and *activity* are shown below in Table 2 and demonstrate the change from posttest to delayed posttest at the .05 level. The results from Table 2 indicate that there was no statistically significant difference in the *appearance* and *activity* of scientists. However, there was a significant difference in the pretests and posttests of the mean scores of the *location*.

The mean scores for each category, *appearance*, *location* and *activity*, are shown below in Table 3 and demonstrate the change from pretest to delayed posttest at the .05 level. The results from Table 3 indicate that there was statistically significant difference in the *appearance* and *location* of scientists. However, there was no statistically significant difference in the pretest and delayed posttest of the mean scores of *activity*.

Table 2. Posttest-Delayed Posttest

	Posttest Mean	Delayed Posttest Mean	Related-Samples Wilcoxon Signed Rank Test
APP	8.42	7.92	.405
LOC	6.46	6.54	.029*
ACT	5.96	5.5	.579

E-DAST DISCUSSION

Students' perceptions of the *appearance* and *activity* of scientists changed in a way that was statistically significant following the virtual visiting scientist program. There was not initial a statistically significant difference in the category of *location* from pretest to posttest. However, from posttest to delayed posttest students' perceptions of where scientists work, i.e., *location* did change. So, while the sample size in this study is small, the only thing demonstrated for certain through the E-DAST is that this intervention, (virtual discussion-based interactions with five different scientists over a period of five days), with this group of students did have an impact on these students' perceptions. This was evidence by a statically significant change that showed up immediately following the experience in terms of what scientists look like and the work they do (*appearance & activity*) and where scientists work weeks later (*location*). However, what might have caused the change in what students think scientists look like (*appearance*), versus where they work (*location*) versus what they do (*activity)* is still unknown by using this instrument. This is a limitation of the instrument which balances out the strength of the instrument, that there was enough of a shift from the intervention that could be measured when quantifying students' pictures of scientists.

Table 3. Pretest-Delayed Posttest

	Pretest Mean	Del Posttest Mean	Related-Samples Wilcoxon Signed Rank Test
APP	7.67	7.92	.007*
LOC	6.25	6.54	.010*
ACT	5.29	5.5	.206

FINDINGS FROM 'DO NOW' RESPONSES

Question #1 What did you Think About Last Week's [Virtual Visiting Scientist] Program?

Forty-eight out of the forty-eight responses were positive. Students used words like cool, useful, interesting, and fun to describe the program. *"I didn't know the scientist were like how I expected. I don't really know how to explain it. Like in the beginning I thought they could wear white coats. I thought they were cool when we finished Skyping with them"*. Even after one student initially admitted they found scientists boring in there comment. *"I thought that they were these boring people who search up the same stuff over and over again. Now I know that they are these cool people who search up anything to find out what's behind the curtain."* Several students alluded to meeting scientists who were 'famous' and how exciting that was for them. While others focused on the content, or what they learned about science content, tools, and practices from the week. It is worth mentioning that many students expressed a desire to do the program again. One student related to what it felt to be a scientist as demonstrated by this quote, *"I think that the program was a good because kids could learn what it felt like to be a scientist."*

Question #2 Do you Think the Virtual Visiting Scientist Program was a Valuable Experience? Explain why or why not?

Forty-eight out of forty-eight responses reported that the virtual visiting scientist program was a valuable experience. Several students mentioned meeting scientists and learning about the work the work they did was valuable for them because this was information they did not know before the program. *"I now [know] they can work inside and out if they need to and they are completely normal no crazy white hair or thirst for knowledge enough to bring the dead back to life. The way I think about scientists has changed because I always thought the scientists always worked in labs no matter what..."* For other students, the most valuable part was realizing they felt unique to be participating in such a program as they mentioned this did not happen in all schools. Some students even commented about how much work their teacher put into planning the week. Here is one example from a student about the value of the program, *"I think this experience was valuable because I learned a lot about science. I also learned what scientist do"*. While another student documented how their perception changed from before to after the experience, *"I think that this program we did was a great experience for us because it actually improved our knowledge of what we really think of scientists. For example, before we thought of a scientist as someone who works in a lab and investigates different things like rocks and animals, but it [is] actually more than just that"*.

Question #3 Did the Program Change how you Think About Scientists and What They do?

Forty-five out of forty-eight responses reported that the virtual visiting scientist program did change them in a positive way. Of the three students remaining, one student indicated they did not know if the program had changed them. Another student reported *"I don't think the program changed the way I thought it would because I really think a lot of different scientists do different stuff than other people"* and the third student said, *"I don't think the program change[d] me but it's cool what they study."* A number of students discussed in their responses the idea that the program broadened their perception from a single idea about a scientist to many ideas of scientists. For example, one student wrote, *"My thoughts change[d] about a scientist because I thought they only do experiments, but they do more than experiments they study the world which that's cool."* Recall the comments were coded into three categories: *appearance* or behavior of scientists, *location*, or places where scientists work, and things scientists do or *activities*. Nine responses were specific to the *appearances* of scientists. Only one comment was specific about the gender of scientists, "*It did because I mainly thought scientist[s] were guys, but I learned there [are] a lot of girls that are also scientists.*"

In response to this question, six comments were specific to *location* of where scientists worked. "*I thought scientists were people who stood in labs and just focused on one thing, now I know they do so many things to figure out one thing.*"

Twenty-two responses were specific to the *activity* of scientists. Most of them discussed how students' ideas expanded for one idea to multiple ideas about the kinds of work scientists do. For examples "*The program did change [me] about how a scientist works. I thought they only did experiments the whole time.*"

DISCUSSION

The purpose of the virtual visiting scientist week was to allow middle school students to explore a variety of areas of science, with a different scientist on each of the five days of the week and study the effect on students' perceptions of scientists. The intention was that multiple interactions with five scientists would encourage middle-school students to understand the work of scientists do and sustain an interest in science. This was demonstrated based on students' reflections of the experiences and E-DAST results. Some students addressed more than one category in a comment. For example, in these comments the student addresses both the *activity* and the *location* of where scientists work "*I will say it has changed a bit because I thought other things that a scientist did like experiments and there they be in a lab or something but there not all the time.*" Likewise, another comment address both the *location* and *activity*, but the student makes reference to the fact that the program has changed 'what scientists look like' without the supporting details. This is an example where a comment would have been coded for all three themes it reads, "*This program changed what scientist looked like because not scientist work at a lab and not all scientist experiment on the same task.*"

The fact that students mentioned simultaneous changes in their perceptions after the intervention may seem obvious and irrelevant, however, a closer look is warranted. Before this study, we knew that students could believe the stereotype that typically includes representations of scientists as 'nerdy' looking older white men with 'crazy hair' wearing white lab coats and working by themselves in a laboratory (Barman, 1997; Beardsley & O'Dowd, 1961; Bozdin & Gehringer, 2001; Chambers, 1983; Etzioni, & Numm, 1974; Finson, 2002; Fort & Varney,1989; Mason, Kahle & Gardener, 1991; Mead & Metraux, 1957) and it was also documented that the streotype could change with the intervention of visiting scientists (Bozdin & Gerhinger, 2001; Flick,1990), however, what has not been documented is the simultaneous changes in their perceptions in more than one category, where they work, what they and what they look like. It is important to mention that student comments about the laboratory setting were most mentioned, it is worth considering that this part of the stereotype might be most predominantly associated with the label or word scientist for students.

Multiple Interactions with Scientists

The exposure to different individuals with different personalities, areas or work, places they work were held together only through the label of the word 'scientist'. As the students reflected on their experiences with the scientists they began see to the label as the only common feature that had the ability to encompass more differences than likenesses. This is where the results from the 'Do Now' coupled with the evidence of change as documented in the E-DAST make this study unique. The results included students' own awareness of the change. This was insightful, as the relatively short amount of interaction time (45 minutes times 5 days) led to middle school students to form take-away impressions of the scientists that were more positive and they were able to articulate. The words the student used in describing the scientists could not have been used unless they had 'met' them and interacted with them. For example, one student reported they had never recalled meeting a scientist before this week, "*It was pretty cool last week because I meet a scientist in person and I never met a scientist in real life*". This is worth mentioning as it was noted by the student as a monumental shift in their life.

Several students used the word "REAL" to describe the scientists they were interacting with in all caps. When given the opportunity, described how a real scientist competed with the way scientists were portrayed in movies. "*I know that they work, and work, and work to discover new things or uncover things that are confusing or has not enough information on the topic. But one thing that did change about how I think about scientists is their work and how they do their work. They also helped me figure out how REAL scientists looked like and where they live and what they do in the lab.*" When given the opportunity, two students described how a real scientist competed with the way scientists were portrayed in movies. "*I thought scientists did like more crazy things like you see in the movies. I used to think they worked in very cool and big labs but, their labs are really just a laptop in their house. I thought all of them worked by cutting animals everyday but no in reality.*"

Human Endeavor

Three students also reported value in taking a break from the content in order to connect to the human aspect of science by meeting several people who have chosen careers in science. One student reported that, "*we talked about personal stuff and science too.*" For another student, hearing about the scientists' experiences in seventh grade was powerful, "*I think it is very valuable because this scientist took days off just to see us. It was also valuable because we were all able to learn from the scientist, what they were doing, what was going on there personal life, how 7th grade was for them, were their struggles that they had to face, what did they like about science so much....it was very valuable because we stopped them from working to show us what they did in their job and we got to learn about things that people did aside of teaching we got to learn cool and fun things about water and the most deadly animal to humans... .*" Even discussing scientists display of emotions, the human aspect that they are in fact real people, "*I thought that they were some real non-emotional people who only care about their work and will do anything to make their life's work complete*". Reinforcing the fact that students will not enter the field if they do not fit that stereotype (citation).

Releasing the Stereotype

However, what was particularly interesting in this study was students' ability to document their stereotypic perception of scientists before and after the virtual visiting scientist week. Participants in this study were able to identify their own shifts in stereotypic perceptions of scientists. As one student reported, age and gender stereotype "*I now know for a fact all scientists were not all guys who are white and old.*" Further, the middle school students' perception of scientists appeared to change in real-time as a result of this experience, as one student wrote,"*the program changed the mind set of scientists that they only wear glasses, white jackets and crazy hair but now I know what they look like they are just normal.*" The descriptions offered by several students through their reflections were vivid. Recall The stereotypical perceptions of scientists are well documented in the literature (Barman, 1997; Beardsley & O'Dowd, 1961; Bozdin & Gehringer, 2001; Chambers, 1983; Etzioni, & Numm, 1974; Finson, 2002; Fort & Varney,1989; Mason, Kahle & Gardener, 1991; Mead & Metraux, 1957) and typically include representations of scientists as 'nerdy' looking older white men with 'crazy hair' wearing white lab coats and working by themselves in a laboratory. We certainly found evidence of students' perceptions in their own words in this study. Many students view scientists as quite different from those in other jobs and, at the same time, often hold highly stereotypical and inaccurate views of scientists and the work they do (Farland-Smith, 2017).

A reference to working alone in the lab and wearing a lab coat "*I thought scientists were people who stood in labs and just focused on one thing, now I know they do so many things to figure one thing out.*" An example of a lab coat, "*This changed the way I think about scientists. I thought that they just put on their jacket and that is that. But they actually study stuff work with partners and read about the animal or something they are doing....*" A reference to the age "*I know for a fact all scientists are not guys who are white and old but it's just how I knew scientists [they] were before.*" Even a reference to scientists being a genius, "*I had always thought of a scientist as someone who has a lot of knowledge in something but it's actually more than what I think it is. Scientists are actually learning new things as they work just like us but more fun.*"

These comments are important as they are evidence that students are moving beyond stereotypic ideas to more expansive understandings about the work of scientists. In the comment about scientists just standing in a lab discusses the students' preconceived idea about where scientists work and what they do there (just standing) and moves toward less stereotypic ideas of what scientists do for work. In terms of the preconceived idea of the lab coat, or jacket and also moves to a more expansive notion of what scientists do for work. What emerges from the comments above is the possibility that change in one category (*appearance, location,* or activity) may lead to change in another category. Analysis of data from this study suggests that *appearance* and *location* may serve as gatekeepers for shifts in perceptions regarding the *activity* of scientists—until students can get past the stereotype of what scientists look like and where they work, they may struggle to develop more expansive notions of the work scientists engage in. For example, the student who was discussing the lab *(location*) and then they moved on to realize scientists do not just focus on one thing, rather, they do [what is needed] or 'many things' to figure out one thing. They focus on every little thing to find out what they need to find out…". In the second comment the student was discussing their lab coat (*appearance*) and somehow that translated for them to the work scientist do (*activity*).

Unplanned Outcomes

Virtual discussion-based interactions with scientists had two benefits we did not anticipate. One scientist participated in the session with their dog, which prompted a student to write, *"they work at home all day next to their dogs they have a pretty chill and fun job"*. This is an experience unique to virtual interactions and certainly dispels the myth of the mad scientist in ways we have not planned. We could have never guessed the dog showing up on the video session would have caused such a reaction with the students. It was like, "of course scientists would own a pet?"

Recall the entire motivation for the project was the classroom teacher was interested in determining if this inexpensive, straightforward, and scalable method of bringing scientists into the classroom would have a measurable impact on student perceptions of scientists or engineers and the work that they do. What the teacher did not anticipate was the immediate and noticeable "shift" in the students' attitudes the following the scientists' visits. The teacher reported that the students' attitudes toward science had shifted in way that science had now become accessible to them, meaning many students now carried the belief that they too could be scientists. While this was not measured in any particular way by the teacher the way the students leveraged their new understanding of science and scientists frequently returned to class discussions without the teacher's prompting. For example, the students expressed an interested taking more time away from standard science content in class in order to explore personal and historical accounts of diverse scientists (Grossman & Farland-Smith, 2020). The teacher also noticed but was unable to measure a commitment to continuing to take class time to focus on the "human aspect" of the subject they were learning.

Comparing Students' Perceptions in Drawings Versus Words

To date the researcher has performed over thirty studies using the E-DAST and DAST, therefore, the researcher used their own experience to organize these categories: time, numbers, scoring. Students typically need about 30-45 minutes to complete their illustrations. Whereas the reflections in this study required less than 30 minutes. In order to analyze the E-DAST, all portions of the pretest, posttest, and delayed posttest must be completed. In this particular study, only fourteen students completed all three sections out of forty-eight in the class. The numbers were lower because in this particular study the researcher and the teacher were not able to secure students to complete either all three sections of the E-DAST, and answer all the questions on the corresponding page and turn in a permission slip. Whereas, a writing prompt for self-reflection can be easily done by classroom teacher in this case, yielded 100% participation.

Most importantly, writing reflections allowed the reader to 'hear' the middle school student in their own words what they thought before and after based on the intervention, which may have gotten lost in in E-DAST. *"Yes, the program did change about what I think about science a lot because before the program I thought that all scientist do is use the computer in their office and build stuff"*.

Therefore, the writing tool was in a lot of ways easier to administer, obtain, analyze and provided a window into students thinking in a way not previously analyzed. The purpose of analyzing them together is the evidence of the shift (E-DAST) along with insights into student's on thinking in their own words ('Do Now' Responses).

CONCLUSION

Recall that the initial motivation for the study was the teachers desire to what the value of short, discussion-based interactions over the internet with scientists for the development of students' perceptions of scientists and their work? Knowing how much time classroom teacher dedicated to this project with administrative support, the question was worth asking.

The opportunity to meet more than one scientist in a short period of time (one week) was fruitful for this group of students. Would the same outcome have happened if we stopped at one scientist, or perhaps on day three after three scientists? Is there a saturation point with meeting a diverse group of scientists from scientific, educational, racial, ethnic, and linguistic backgrounds? These questions are still unanswered.

Rather than leaving students with a singular impression of a scientist, the virtual visiting scientist week left students with a range of work done by a variety of people in the broad field of science. Bringing visiting scientists into the classroom has the potential to positively influence students' perceptions of scientists and their work. The Draw-a-Scientist Checklist (DAST-C) was first used in Finson, Beaver, & Crammond (1995). The checklist was created from the common aspects or features found in illustrations from previous studies and were based initially on the scientists, and broken down into "stereotypical" and "alternative" images shown in the drawings. The modiðed DAST and the DAST Rubric brings more reðnement to analyzing students' drawings as it enables clarities to emerge and subsequently increased detail to what one could ascertain from students about their mental images of scientists (Farland-Smith, 2012).

Could what scientists look like (*appearance*) and (*location*) hold the key to students' perceptions of what scientists do (*activity*)? As evidenced by two related factors in this study, first, the fact that students' own words supported the E-DAST data. The sample size of the reflections was much greater and easier to administer and analyze. Second, the significant finding from the E-DAST demonstrates the gain from pretest to delayed posttest in *appearance* and *location,* yet the majority of the written responses discuss *activity*. The DAST Rubric was developed in 2003 under the assumption that each of the categories, appearance, location, and activity were equal. In light of this new evidence, perhaps they are not? What if students could not develop a more expansive notion of the work scientists do until the they breakdown, debunk, address, confront, or dismantle their preconceived ideas about what scientists look like (*appearance*) and where they work (*location*)? This information would certainly be valuable to those in the field of science education, particularly in the context of student-teacher-scientist partnerships. Further questions abound: do students' perceptions about what scientists look like (*appearance*) and where they work (*location*) compete with each other or do they work to reinforce one another? Do students need to address one before the other or can they address both at the same time?

REFERENCES

Adedokun, O. A., Hetzel, K., Parker, L. C., Loizzo, J., Burgess, W. D., & Robinson, J. P. (2010). Using virtual field trips to connect students with university scientists: Core elements and evaluation of zipTrips™. *Journal of Science Education and Technology, 21*(5), 607–618. doi:10.100710956-011-9350-z

Barman, C. (1997). Students' views of scientists and science: Results from a national study. *Science and Children*, (September), 18–24.

Beardsley, D. C., & O'Dowd, D. D. (1961). The college-student image of the scientist. *Science, 122*(3457), 997–1001. doi:10.1126cience.133.3457.997 PMID:17743790

Bodzin, A., & Gehringer, M. (2001). Breaking science stereotypes: Can meeting actual scientists change students' perceptions of scientists? *Science and Children, 38*(4), 36–41.

Bruner, J. (1996). *The culture of education*. Harvard University Press.

Buck, G. A., Leslie-Pelecky, D., & Kirby, S. K. (2002). Bringing female scientists into the elementary classroom: Confronting the strength of elementary students' stereotypical images of scientists. *Journal of Elementary Science Education, 14*(2), 1–9. doi:10.1007/BF03173844

Chambers, D. W. (1983). Stereotypic images of the scientist: The draw-a-scientist test. *Science Education, 67*(2), 255–265. doi:10.1002ce.3730670213

Chen, S., Binning, K. R., Manke, K. J., Brady, S. T., McGreevy, E. M., Betancur, L., Limeri, L. B., & Kaufmann, N. (2020). Am I a science person? A strong science identity bolsters minority students' sense of belonging and performance in college. *Personality and Social Psychology Bulletin*. Advance online publication. doi:10.1177/0146167220936480 PMID:32659167

Erb, T. O. (1981). Attitudes of early adolescents toward science, women in science, and science careers. *Middle School Research Selected Studies, 6*(1), 108–118. doi:10.1080/08851700.1981.11670215

Etzioni, A., & Nunn, C. (1974). The public appreciation of science in contemporary America. *Daedalus, 103*(3), 191–205.

Farland-Smith, D. (2009). Exploring middle school girls' science identities: Examining attitudes and perceptions of scientists when working "side-by-side" with scientists. *School Science and Mathematics, 109*(7), 415–427. doi:10.1111/j.1949-8594.2009.tb17872.x

Farland-Smith, D. (2012). Development and field test of a rubric for assessing the draw-a-scientist test. *School Science and Mathematics, 112*(2), 109–116. doi:10.1111/j.1949-8594.2011.00124.x

Farland-Smith, D. (2017). The evolution of the analysis of the draw-a-scientist test. In P. Katz (Ed.), *Drawing for Science Education* (pp. 169–178). Brill Sense. doi:10.1007/978-94-6300-875-4_15

Farland-Smith, D., & McComas, W. F. (2009). Teaching the human dimension of science. *Science and Children, 46*(9), 32–35.

Finson, K. D. (2002). *A multicultural comparison of draw-a-scientist test drawings of eighth graders*. Paper presented at the Annual International Conference of the Association of Educators of Teachers of Science, Charlotte, NC.

Finson, K. D., Beaver, J. B., & Cramond, B. L. (1995). Development and field test of a checklist for the Draw-A-Scientist Test. *School Science and Mathematics, 95*(4), 195–205. doi:10.1111/j.1949-8594.1995.tb15762.x

Flick, L. (1990). Scientist in Residence program: Improving children's images of science and scientists. *School Science and Mathematics*, *90*(3), 204–214. doi:10.1111/j.1949-8594.1990.tb15536.x

Fort, D. C., & Varney, H. L. (1989). How students see scientists: Mostly male, mostly white, mostly benevolent. *Science and Children*, *26*(8), 8–13.

Grossman, B. T., & Farland-Smith, D. (2020). Collaborating with virtual visiting scientists to address students' perceptions of scientists and their work. *Innovations in Science Teacher Education, 5*(3). Retrieved from https://innovations.theaste.org/collaborating-with-virtual-visiting-scientists-to-address-students-perceptions-of-scientists-and-their-work/

Joyce, B. A., & Farenga, S. J. (1999). Intentions of young students to enroll in science courses in the future: An examination of gender differences. *Sci Educ*, *83*(1), 55–75. doi:10.1002/(SICI)1098-237X(199901)83:1<55::AID-SCE3>3.0.CO;2-O

Lindahl, B. (2003). Pupils' responses to school science and technology? A longitudinal study of pathways to upper secondary school. *Göteborg Studies in Educational Sciences, 196*.

Maltese, A. V., & Tai, R. H. (2010). Eyeballs in the fridge: Sources of early interest in science. *International Journal of Science Education*, *32*(5), 669–685. doi:10.1080/09500690902792385

Mason, K., Kahle, J. B., & Gardner, A. L. (1991). Draw-a-scientist-test: Future implications. *School Science and Mathematics*, *91*(5), 193–198. doi:10.1111/j.1949-8594.1991.tb12078.x

Mead, M., & Metraux, R. (1957). The image of the scientist amongst high school students. In B. Barbar & W. Hirsch (Eds.), *The Sociology of Science* (pp. 38–61). Falmer Press.

National Research Council (NRC). (1996). *National science education standards*. National Academy Press.

National Research Council (NRC). (2012). *A Framework for K-12 science education. Practices, Crosscutting Concepts, and core ideas*. National Academy Press.

NGSS Lead States. (2013). *Next generation science standards: For states, by states*. The National Academies Press.

Niemitz, M., Slough, S., Peart, L., Klaus, A., Leckie, R. M., & John, K. S. (2008). Interactive virtual expeditions as a learning tool: The school of rock expedition case study. *Journal of Educational Multimedia and Hypermedia*, *17*(4), 561–580.

Ozgelen, S. (2017). Primary school students' views on science and scientists. In P. Katz (Ed.), *Drawing for Science Education* (pp. 191–203). Brill Sense. doi:10.1007/978-94-6300-875-4_17

Silver, A., & Rushton, B. S. (2008). Primary-school children's attitudes towards science, engineering and technology and their images of scientists and engineers. *Education 3-13, 36*, 51-67. doi:10.1080/03004270701576786

Wyss, V. L., Heulskamp, D., & Siebert, C. J. (2012). Increasing middle school student interest in STEM careers with videos of scientists. *International Journal of Environmental and Science Education*, *7*(4), 501–522.

Chapter 5
Enriching the Lives of High School Teenagers With Science Cafés

Michelle K. Hall
https://orcid.org/0000-0002-4445-7788
Science Education Solutions, USA

Michael A. Mayhew
Science Education Solutions, USA

Jessica Sickler
J. Sickler Consulting, USA

ABSTRACT

This chapter provides a roadmap for any who are inspired to start a teen science café program as a member of the teen science café network. These out-of-school programs are a free, fun way for teens to explore the advances in science and technology affecting their lives. Teens and scientists engage in lively conversations and activities to explore a topic deeply. Teens get increased understanding of the nature of science and develop a realistic perception of scientists, science careers, and the lives they lead. Scientists gain skills in communicating science to the public and inspire youth to STEM careers. Adult leaders serve as guides for the teen leaders and are the linchpin of the program. Evaluation shows programs positively influenced teens understanding of science issues in the news, the ability to use facts to support scientific points of view and considering multiple sides of an issue before making a decision.

DOI: 10.4018/978-1-7998-4966-7.ch005

INTRODUCTION

In centuries past, people of all walks of life would gather in public places to discuss and debate the changes in their world to come to deeper understandings of what the future might **bring**. Books were not prevalent and most people received their news by word of mouth. Sitting in a pub or café, people would discuss changes in law, politics, philosophy, science, art, the future of medicine, and much more. Over time, these public gatherings diminished.

In 1992, the French philosopher Marc Sautet revived the idea of ordinary people engaging in rich philosophical discussions and founded the Café Philosophique in Paris, France. A mostly forgotten process of engaging in public discussions was reborn across Europe. Six years later, Duncan Dallas, an English scientist with a passion for communicating scientific ideas with the public, envisioned a "Café Scientifque" (Dallas, 2006; Café Scientifique, 2012) in which people come together in a social setting to explore the latest ideas and discoveries in science.

The model is popular with teens for the same reason that adult Cafés are popular; the blend of a collegial social setting where participants can interact with each other and satisfy their curiosity about a science-based topic through lively interaction with a scientist (Farmelo, 2004). Teen cafés differ only in the leadership and the addition of hands-on activities that correspond with the topic. The Café movement coincided with the growing importance in academia of the study of and practice of communicating scientific ideas. Thus, the science café model spread rapidly across every continent.

BACKGROUND

The teen science café model was constructed on programmatic answers to a series of key questions, then validated via rigorous evaluation.

The following questions guided the development of the teen science café program:

- How to make science engaging and accessible to more youth?
- How to help them see science and technology research and discoveries as relevant to their daily lives?
- How to develop a positive culture around thinking scientifically and maybe even pursuing STEM careers?
- How to prepare science and technology professionals to effectively engage youth in their research and discoveries through discussion and other activities?
- How to build a program that will be professionally rewarding to the scientist participants?
- If it is built, will the scientists and youth come?

Answers to the questions became key objectives in the program development and led to the following goals for the Cafe´ implementation. Teens will develop:

- an informal community that engages in scientific discourse, thought, and exploration;
- a deep and nuanced understanding of current issues in science and technology and be able to communicate that understanding confidently and expertly;

- skills and attitudes for lifelong learning and an appreciation of science as a career and as a way of thinking.

The program summative evaluation results (Hall et al., 2010; Foutz and Luke, 2010) indicates that the program was largely successful in achieving these goals.

IMPACT ON TEENS

The following two questions guided the teen impact evaluation:

(1) Does engagement in the program enhance teen participants'
  (a) understanding of current STEM content and issues?
  (b) sense of confidence and competence in their ability to develop and communicate factually supported positions on STEM issues?
  (c) appreciation for and interest in STEM careers?
  (d) attitudes toward science, the process of science, and scientists?
  (e) personal development, such as growth in character, perceived contribution to their community, and leadership skills.
(2) To what degree do teens feel a sense of belonging to and ownership of the Café community?

The evaluation drew from a framework described in the field of Positive Youth Development. The study used an outcomes-based framework in which six outcomes called the six "C's" (competence, confidence, connection, character, caring and compassion, and contribution) are the basis for a successful program. In addition, related participant impacts, such as attitudes toward science, were also important to the program model, and therefore were also investigated.

METHODS

Written quantitative surveys were used to investigate the program outcomes for teen participants. A quasi-experimental design was used, allowing for a matched (non-randomly selected) control sample. The design allowed for two types of comparisons: (1) between teens who participated in the Cafe´ program (treatment group) and teens who did not (control group), and (2) retrospective pre-program participation and post-program participation for teens who attended the program.

CONCLUSIONS

The most essential elements of the Cafe´ model are the people involved (program leaders, scientist-presenters, youth leaders and attendees). The strength of the model is that it has been implemented and evaluated at more than 130, diverse sites. The program is successful because the benefits of participation outweigh the challenges for both the teens and the scientist-presenters. Evaluation data show very clearly that cafe´ programs positively influenced participants' attitudes toward science, including interest in science and science careers, knowledge of scientists' work, awareness that science affects their daily lives, and interest in science careers. It also positively influenced their self-efficacy and cognitive competence toward science including understanding of the nature of scientific research, their understanding of science issues in the news, the ability to use facts to support scientific points of view, and considering multiple sides of an issue before making a decision. This was true both for analyses that compared the ratings of participants to non-participants and for the retrospective-pre-to-post ratings of participants.

Learning from experience, as well as the formative and summative evaluations, the authors have made continual improvements in the program—both on the teen side and the scientist-presenter side. The maturity of the program and its clear impacts make a strong case for other communities to start their own programs.

SUPPORTING TEENS' DEVELOPMENT IN SCIENCE, TECHNOLOGY, ENGINEERING AND MATH (STEM)

Teens are going through a unique and important time in their lives—transitioning from childhood to adulthood. They are in the process of defining who they are and how they fit into the world. Teens need safe, secure, and trusting relationships with peers and adults to help them explore new challenges and opportunities, as they become independent adults (McLaughlin, 2000). Science cafés can be an important element of this exploration period in teens' lives, and have been shown equally effective for girls and people of color (Foutz & Luke, 2010; Gambone et al., 2002; Gambone & Connell, 2004; Kekelis, L., 2005).

Teen science cafés open the eyes and minds of teens to the importance of understanding how science, technology, engineering, and mathematics (STEM) affect their lives now and into the future, and inspire lifelong learning (Mayhew & Hall, 2011). These programs have been shown to positively increase teens' interest in STEM and STEM careers. Teens often come away from a Café experience with new understanding about how the basic science they learn in school applies to solving real life challenges.

One of the most effective ways for this to happen in a science café is for teens to learn more about the day-to-day work of STEM experts, and to see first-hand the passion they have for their work. Thus, presenters share their personal story describing the journey they have taken to arrive at their current career, and revealing a pathway for teens. Teens are interested in how adults advanced from their first job in high school to the career they have today. They enjoy stories about the twists and turns in the scientists' career and the people who helped them find a path to success. They also are interested in learning about careers requiring varying levels of education. These stories help teens envision the opportunities and steps they will need to take on their own journey.

Launching Café Scientifique New Mexico

In 2006, the authors were inspired to start a novel science café program for high school teens in northern New Mexico after learning about the proliferation of adult science cafés world-wide. The popularity of science cafés stems from its stimulating informal conversation, intriguing topics, and cutting-edge experts who make the topics come alive and accessible (Nisbet & Mooney, 2007a; Nisbet & Mooney, 2007b). The programs are typically held in cafés or other social venues serving light refreshments; it includes a brief presentation by an expert on a topic, followed by discussion during which civil discourse and debate is encouraged. This blend of ingredients accounts for the café model's global popularity. The authors wondered if this science café model could appeal to that hard-to-reach segment of the public, the high school teen and serve an unmet need for high school age teens interested in peer-to-peer exploration and discussion of science issues affecting their everyday lives (Bell, et al, 2009).

The authors reached out to teens in four culturally, demographically and economically diverse towns and cities in northern New Mexico: Santa Fé, Los Alamos, Albuquerque, and Española. The authors sought the teen audience due to the importance of encouraging youth to appreciate the relevance of STEM to their lives and to spark their interest in STEM career paths.

In each community, the authors visited schools and shared the idea of starting a teen science café. Teens were overwhelmingly positive about the idea, especially meeting a scientist, and learning about STEM careers. The goal was to have teen leaders organize their programs with the help of a caring adult, so they would become personally invested in the program.

The authors proposed that the programs would be organized by teens and for teens, and that a group of volunteer teen leaders would have a voice in all decisions. These included suggesting topics and presenters, choosing and buying the refreshments, and providing critique on the scientist's presentation before the main event. The teens also wanted to have some form of active learning in the program. So, the teens and adult leaders work with the presenters to incorporate a hands-on component. It may be analyzing malaria outbreak datasets, unraveling a mock crime scene, identifying species of crickets by their sounds, or hosting a white-hat hackathon to explore careers in cyber-security. When some of the early presentations and discussions did not inspire, teens began coaching the presenters in what makes the program the exciting and engaging event they desired. Thus, the program is now led by teens, for teens.

Teachers were supportive of the concept and committed to help the teen leaders promote the program, and invited the adult leaders of the programs to make a pitch to their students to develop and lead their own program. By hosting teen cafés out of school, it allows any teen the free choice to expand their mind and make new connections with STEM experts.

However, transportation was a challenge for some, as not all teens can drive and public transportation was not feasible in the rural towns. So, the scientists volunteered to travel to the four different communities, rather than teens coming to the scientists.

The authors applied and received a grant from the U.S. National Science Foundation to try the experiment and develop a model for teen science cafés. The authors knew that teen science cafés were being held in schools in the United Kingdom, and connected with the leaders in the U.K. to learn from them. This was to be an out of school program, as teachers are pulled in too many directions already. This appears to have been a good choice, as the U.K. programs led by teachers have ended due to lack of time and resources. Thus, adult leaders only ask teachers to help the teens promote their program to their peers.

The primary goal was for the teens to get a significantly increased understanding of the nature of science and to develop a realistic perception of scientists and the lives they lead (NRC, 2010). The authors wanted teens to understand that a scientist is a real, complex, multidimensional human, like them, with his or her own unique set of motivations, delights, abilities, and challenges. The authors wanted presenters to convey that—because of a career in science—she or he has had, and is having, a particularly interesting life. The authors wanted to instill a passion for lifelong learning and an appreciation of the relevance of all kinds of science to their own lives.

The top reason the teens cited for attending the teen café was to meet a scientist. This surprised the authors at first, as the authors are scientists. However, there are 42 million teens in the U.S. and 7 million scientists and engineers (Sargent, 2017). For most, the only scientists that teens meet are their high school teachers. The secondary goal was to help scientist-presenters to effectively communicate their science to the public. And finally, the authors wanted to create an informal community of teens that engages in scientific discourse, thought and exploration. Youth in the focus groups were enthusiastic enough about the project that the majority volunteered to be on the site-based teen leadership teams.

Figure 1. Teens visit a scientist's lab and learn how to make 3-D Braille books. (Photo credit Teen Science Café Network by permission.)

The twelfth season of Café Scientifique New Mexico (cafénm.org) is now completed having served over 48,000 teens. The Teen Café model has evolved through trial and error and formal evaluation over those years. Contrary to the initial trepidation—what if there is a Teen Café party and no one comes?—the program has proven highly popular with the teens, and the evaluations indicate the program has been quite successful in achieving the goals. In fact, more than 10 sites are led exclusively by teens for teens, with no adult guide!

But there are other secrets of success associated with this particular audience. One is that the teens achieve a sense of ownership of the program thru participating in the Teen Leadership Teams in each program.

Teens are encouraged to lead all aspects of the program to the greatest extent possible, with adults in the background for support as needed. Another is that, with few unusual exceptions Cafés are deliberately not held Cafés in schools. However, the authors have made a point of building relationships with high school teachers, who have proven supportive and helpful in innumerable ways, such as lending scientific equipment for experiments at the Cafés and making connections between what teen learn in school with what they learn in the science cafés. And a third is that for every Café event, the authors work with the presenter to develop an activity or deep discussion that the teens can engage in as an adjunct to the interaction between the presenter and the teens—teens like to do stuff. Hands-on activities that actively engage the teens help to cement the science message.

The degree of engagement needed to spark interest in science topics—especially for the teen audience—is unfamiliar and daunting to many Café scientist-presenters, even those with experience in public speaking. Most have been trained to approach science communication in the "information deficit mode" (Nesbit and Mooney, 2007a.b). But with coaching from the program staff, after some initial trepidation the scientists typically rise to the occasion.

Finding and Preparing Presenters

The authors identify potential Café presenters within local organizations with a science mission and look particularly for scientists working in areas that Teen Leaders have expressed interest in learning about. Asking multiple people about the presentation skills of the potential presenter helps to ensure a great presentation. After informal conversations with a presenter in which the essential ingredients of a successful Café are conveyed, the authors provide him or her with "Guidance" documents intended to formally frame their preparation for and conduct of the Café sessions. It stresses the importance of knowing the audience. Teens will readily engage with a presenter on some hot science topic if it is accessible to them. The presentation organized around one essential provocative idea or concept are encouraged, and the presentation needs to be free of jargon, while full questions to engage the audience in discussion. Presenters are encouraged to *tell a story, but leave some mystery in the story*, and stress that *interactivity* is one of the most important ingredients of a Teen Café (Mayhew & Hall, 2011). In a Café presentation, communication—meaning two-way verbal interaction, supported by a few key graphics—is of the essence.

Finding great presenters who can make their research topic relevant and exciting to teens can be challenging. A good place to start is to ask colleagues for help. Explain what a teen science café is and is not...it is not a lecture! Describe the characteristics of the type of presenter you wish to recruit—someone who is funny, relaxed, flexible, and can explain the science behind their work in plain language.

Attend local adult cafés, and lectures at colleges, universities, and museums to observe potential presenters first hand. Ask the Public Relations Office of local STEM organizations about experts who have given particularly good interviews. Call the science editor of the local or regional newspaper or TV station and ask them for recommendations of STEM professionals that have given great interviews. Outreach centers at the local college may have recommendations. Ask staff of the U.S. Representative's office about names of scientists they use to help them keep abreast of current STEM research.

Science is everywhere! For teen science cafés in larger cities, finding presenters is often fairly easy. There is often a zoo or aquarium and there may be museums, colleges or universities, and science and technology companies. But one of the wonderful aspects of science and teen science cafés is that they have no limits on topics or places to convene. There are scientists everywhere and all you need is some dedicated teens and an adult leader to pull things together, even in small-town and rural areas. It might not be so obvious at first where to find scientists, but you may in fact have many hidden opportunities for both great presenters and really unique hands on opportunities.

Every community has utility systems – electric, water, sewage, waste and recycling, gas, cable, phone. Some homes and farms may have septic tanks and wells. How do they work? Where does the recycling go? There are scientists and engineers in each of these utilities that can answer that question and even show teens how they work with an onsite presentation. Then, there are civil engineers that build the roads and transportation engineers that manage traffic lights. Even small towns typically have their own radio stations and maybe a public television station. Small engineering firms and renewable energy companies may have individuals with interest in sharing their knowledge with youth.

Just asking around and talking with everyone you can about what you are trying to accomplish will probably turn up at least a name or two of someone you can follow up with later. Learning about the technology behind the communications systems that are taken for granted can be eye opening. The local weather forecaster is often trained as a meteorologist and would make for an interesting café.

Agriculture, in general, is ripe for making science connections for youth as well as connecting them to where their food comes from. Animal husbandry and plant breeding require knowledge of genetics and may get kids thinking about the relevance of what they are learning in the classroom. Cooperative Extension Agents and master gardeners that can share their knowledge about animals, seeds, crops, plant diseases and many other topics. Dairies, local cheese makers or breweries can be great sources of presentations on how to prepare these unique foods. Cooking in general—the science of baking or marinating or other aspect—could be explored. There may be a Civil Air Patrol group who can teach about flight and the nuances of flying in different types of weather conditions.

Funeral homes and crematoria have experts in preparing bodies after death and what teen isn't a bit intrigued by dead bodies? A county most likely has crime scene investigators, and Emergency Medical Technicians can talk about the opportunities and challenges of emergency medicine. Local dentists and doctors, physical therapists and yoga instructors can teach about the body. Psychiatrists can talk about brain and emotional disorders. Teens are keenly interested in the brain, especially their own.

Continue to vet them. Always get multiple independent recommendations for a potential presenter, unless you can observe the presenter in action first hand. When you feel confident you have a strong candidate, contact them to see if they can meet with you to explore the possibility of presenting at the teen café. The in-person invitation is critical to ensuring the presenter understands what you are seeking to ensure a great experience for everyone at the teen science café.

Make the pitch. A key element of recruiting a presenter is helping them understand what is in it for them. Why should they do this when their calendar is already overflowing with demands? Maybe they teach all day at the university and see this as just another lecture. Maybe they are afraid of committing because they know nothing about teens. Maybe they are not sure they can do a good job presenting to teens.

The pitch needs to include how you will help them better understand and prepare for this audience. The TSCN website has many resources and trainings to enhance adult leader' skills in coaching the scientists. The research shows (Hall et al, 2010) that presenters gain valuable skills in presenting their work to this audience that carries over to how they present their work to other non-technical audiences such as funding managers, the press, peers outside their narrow area of research and even family members. So, they have much to gain from this experience, if they take a chance and give it a try, they will be glad they did!

The Essential Rehearsal

It is highly valuable—and indeed essential—for presenters to do a rehearsal with a small group of Teen Leaders before presenting to a full house. This has proven exceedingly valuable in getting the presentations pitched at the right level and any graphics comprehensible. It also serves to overcome a certain intimidation factor for many presenters. The attitude of this presenter about the value of the dry run is typical: *"The dry run was immensely valuable. It helped me select appropriate verbiage and content for the presentation. It also helped me gauge the level of delivery....After the dry run, I made significant changes to the presentation, including the elimination of confusing content, identification of real-world connections, and simpler examples."*

Café Scientifique New Mexico has hosted over 100 presenters. Without exception, they considered their experience engaging with teens via the Café program to have been enjoyable and of value. They have improved and broadened their communication skills. Some have also benefited in an unexpected way: they have come to a new perspective on their own research. As one presenter put it, *"The Café*

Figure 2. Vetting and training presenters is often done through role play in groups, which gives them confidence in their presentation. (Photo credit Teen Science Café Network by permission.)

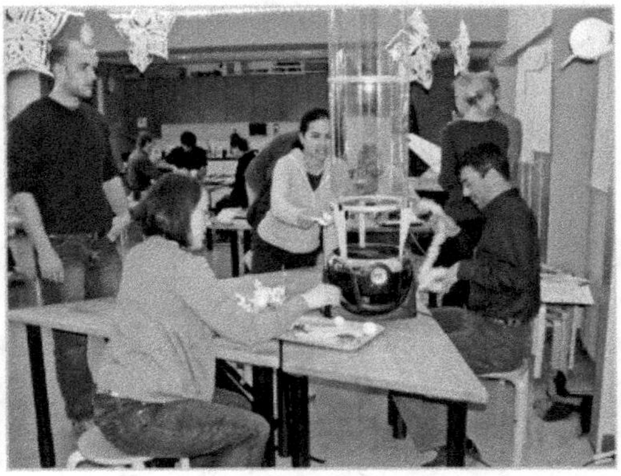

experience was beneficial to me as a scientist in that preparing an interactive talk for an audience of young people helped me identify the critical issues in my work: why I was doing it, why it is challenging, what we are trying to accomplish (vs. being bogged down in technical details). This really focused my thoughts about my work."

What is the Teen Science Café Network (TSCN)?

Based on the success of Café Scientifique New Mexico, the NSF provided funding to take the model nationally as the Teen Science Café Network (teensciencecafé.org). The TSCN is a dynamic, rapidly-growing community of practice (Wenger, 2009) with the mission of connecting high school teenagers with science and scientists. From its first phase, starting with 5 sites in 2012, the Network has grown at an almost geometric rate to ~133 sites and 46 states and Canada. The long-term goal of the Network is for it to become a widely recognized and self-sustaining movement among a diverse American culture, such that it provides many thousands of teens across the country each year with an understanding of the nature of science, an appreciation of scientists as real people leading interesting lives, and a motivation for lifelong learning in science as adult citizens, a significant number of whom will follow myriad pathways into the U.S. STEM workforce.

How do I Join the Teen Science Café Network and What Will I get out of it?

Any organization or person that wishes to do so can become a member of the Teen Science Café Network and start its own teen science café program; membership is free. Working closely with new members to ensure a successful start-up and continued operation is essential. Using video chats with prospective members, then follow up with more chats until they have hosted their first café event. As part of a simple application, prospective new members commit to seven "Core Design Principles" of the TSCN. Within that framework, a member organization has flexibility in designing the particulars of its program. Some Cafés are theme based, such as astronomy, oceanography, or emergency response, but most programs explore STEM topics more broadly.

Members become part of a *community of practice*, in which they share a common sense of purpose and are able to contribute to and receive benefit from participation; this sense of common purpose and community is something that the members have readily embraced. One of the best ways to promote this is at the annual workshops, at which members get to know each other on a personal basis, share knowledge, and form collaborations. Evaluations of the workshops document that they are perhaps the best way of all of fostering the community of practice.

What are the Core Design Principles?

Using the initial experiment in New Mexico and the formative and summative evaluation results, a set of Core Design Principles that ground the Teen Science Café model was developed. Rather than a strict recipe for a program, the Core Design Principles represent a framework that gives a prospective café leader flexibility in program design to be responsive to their context, teens, and community. A common attribute is that a teen café is a STEM program with a low barrier to entry that is primarily lead by teens, for teens. The principles are:

1. Café programs are highly engaging and interactive. *Cafés are structured to promote lively conversation and interaction among teens and the presenter; a Café program is not a lecture.*
2. Scientists are carefully vetted and provided guidance in communicating with teens. *Focusing on storytelling, the big ideas, why it matters, and how it will change the world—supported by a few key, jargon-free graphics—is of the essence.*
3. Teens gain a sense of ownership of *their* program through opportunities for leadership. *Teen Leadership Teams are responsible for all aspects of their program, with adult leaders providing support in the background.*
4. Each Café site has one or more adult leaders who are committed to the program. *The adult leaders have the energy and dedication to organize and support the work of teen leaders.*
5. Café programs seek to attract teens with diverse motivations and life experiences. *Diversity includes ethnicity, culture, and gender, but also motivations to learning about science. Teen Cafés are not just for the science geeks; they are for all curious teens.*
6. Café programs actively cultivate strong community relationships. *Building relationships with scientific and teen-serving organizations can increase participation of both teens and scientists, and support community awareness and sustainability of the Café program.*
7. Café organizers promote awareness of career trajectories and STEM opportunities beyond the Café experience itself.

What is the Format of a Typical Teen Science Café?

A teen science café begins with an icebreaker, and period of socializing. Teens have simple food and drink to create a welcoming and comfortable atmosphere. Next, the presenter engages in an interactive conversation with the teens aimed at framing the topic of the program and shares personal stories about experiences in doing the science that has shaped his or her career. A wide-ranging discussion can go back and forth between the presenter and the teens, and among the teens as well. When the conversation has reached its logical end, the teens and presenter engage in an active learning experience aligned with the topic. This helps the teens digest the new ideas and formulate additional questions for the presenter

Figure 3. Teens explore strategies for reducing climate change through games.

that can be answered in the wrap up conversation. In a typical 90-minute café, this usually works out to a presentation of 15 - 20 minutes, an activity for 10 to 40 minutes, and the remaining time for discussion.

Who can Host a Teen Science Café?

Any teen or adult who is motivated to organize the program. It may be easier to connect with partner organizations in the community to spread the word and work, and to get access to a venue, but teens have started programs on their own.

Where are Teen Science Cafés Located?

Network sites are located across the United States and in British Columbia. A directory of teen science cafés is located on teensciencecafe.org. They are held in libraries, museums, zoos, universities and community colleges, aquaria, Boys and Girls Clubs, 4-H, 21st Century Community Learning Centers, and many other organizations.

How to get Started?

1. Develop and make a pitch to diverse groups of teens and community partners about the value of cafés. Include why you want to make this happen in the community.

2. Develop contacts with both teens and scientists to market the program. Capture the email and phone numbers of all teens and their parents in an Emergency Response form to be able to respond in any emergency, and to be able contact the teens beyond the years they attend.
3. Motivate and inspire teen leaders to bring friends to a meeting to learn what it takes to create the program and what they will gain from attending a teen café. Pizza helps a lot!
4. Organize a teen leader training in which the teens roleplay every element of the event. Use resources on the TSCN site for ideas on leadership training,
5. Find a suitable venue for the event and determine schedules that best meet teens' needs. Seek out venues that are free to use and that have plenty of space for the teens to move around.
6. Explore the topics teens want to learn about and find high quality presenters by asking friends and colleagues.
7. Coach the scientists in how to work with teens in the café setting using the resources on the TSCN site.
8. Schedule the events well in advance and market, market, market the program. Marketing with social media, flyers on the STEM teachers bill boards, announcements at school sports events, notices on the school's marquee and more ideas on the TSCN site. This will help you in fundraising for refreshments.
9. Capture systematically the attendance and participation (as well as emergency contacts). Having parents contact info is crucial for safety and for following up with the teens years after the program.
10. Debrief with the teen leaders to ensure continuous improvement and buy in of the program.

Many more resources for launching the program are freely available on teensciencecafe.org.

What are the Crucial Roles of an Adult Leader?

Mentoring and inspiring teen leaders. The "secret of success" of a teen café is that it is primarily lead by teens, for teens. Teens embrace a program when they feel it is *theirs*, as opposed to one created and run by adults for their intended benefit. Sites in the Network achieve teen ownership of their program via teen leadership teams, led by a dedicated adult leader who fosters a sense of ownership and proactivity in the teen leaders, such that they take on responsibility for the operation of their program and in the process develop lasting leadership skills. Promoting positive mentoring relationships between teens and adults increases retention and commitment to the program (Gambone et al., 2002; Sullivan & Reed, 2009).

Teen leader responsibilities include engaging and introducing the presenter, conducting a closing ceremony, preparing and serving at the food table, welcoming attendees at the check-in table, making teen attendees feel at home, orchestrating an ice breaker, and promoting the program in their schools and communities. These responsibilities all require skill development through practice; most importantly, teen leaders need to feel empowered to step up to their responsibilities.

The role of the adult leader in mentoring teen leaders is critical. It is first and foremost to instill in the teens a proactive mind set and a sense of freedom to take charge, to take responsibility for doing what needs to be done to make their program a success, in short, to grow as leaders. With this mindset teen leaders will seek to improve and perfect their program. The teens will also have the freedom to make mistakes, but then learn from them and improve. Adult leaders can be "gateways" or "gatekeepers" and the outcomes from these two approaches are stark, based on the program evaluations.

"Adult leaders play a pivotal role in the teen leader experience in terms of how they treat teens, how much support they provide, and the kinds of guidance they give. They serve [either] as gateways [or as] gatekeepers of information about the program and the partner institutions.... As gateways, adult leaders provided teen leaders with information and resources needed to run cafés as well as to grow personally. One teen leader reflected, 'I think it's a really strong mutual support system, in that we support them [the adult leaders] in whatever they need us to do to get everything together and they support us into like forming it into what we want it to be'.... Teens who experienced the gatekeeper–type of adult leaders expressed frustration about the support they received from the adult leaders, particularly when the teens sought to take initiative beyond the more traditional roles of marketing, organizing, and running the cafés."

Thus, an adult leader will be there to provide resources and to regularly frame the work that needs to be done, but will allow the teens to take on full responsibility for carrying it out, staying in the background for selective guidance as needed. The qualities that an adult leader seeks to instill in teen leaders are much the same as those articulated by the 4-H: confident, responsible, team players, resilient, self-motivated, compassionate, problem solvers, decision makers.

As experience within teen café programs in the Network has demonstrated, an adult leader who is heavy handed in directing the teen leaders will discover that they have a group that is unmotivated, will tend to stand around waiting to be told what to do, and will likely carry out their tasks in a careless and sloppy manner. The program will not function well and its reputation will suffer. The teens will have little opportunity for personal growth.

The adult leader is thus the lynchpin of the café program. A whole section of the Resources library on the TSCN website is devoted to teen and adult leadership. The Network provides mentor training of adult leaders across the Network at the annual summer workshops.

Coaching Scientists in Effective Communication

Few scientists are well prepared to do an effective teen café; most need coaching. The "information deficit" (Nisbet, & Mooney, 2007a,b) approach to science communication too often used in the context of professional society meetings, lecture series, or the classroom—is ineffective. This one-way transmission of facts from an expert to an information-deficient lay audience is especially ineffective for teens.

An effective café presentation requires full engagement between the presenter and the teens, who need to be met where they are. The presenter needs to be able to calibrate his or her presentation—often on the fly—to existing knowledge. One way to achieve that is to ask questions of the audience. It is important to make a connection to their daily lives. Hands-on activities that actively engage the teens help to cement the science message. Ideally, these interactions take account of current knowledge, misconceptions, biases, and cultural and other affective responses.

The degree of engagement needed to spark teen interest in science topics is unfamiliar and daunting to many café scientist-presenters, even those with much experience in public speaking; most have been trained to approach science communication in the information deficit mode. However, with coaching, after some initial trepidation the scientists rise to the occasion.

Using small groups and think-pair-share techniques (Davis, 1993) to build teens confidence in speaking in a group can open up the conversations quickly. Having scientists implement wait time protocols (Swisher, 1994; Moore, 1987) when asking and answering questions can create a relaxed, respectful and supportive environment. These techniques are effective in encouraging discussion with all learners (Kekelis et al., 2005), especially Native American youth and with females.

Once a prospective presenter has been identified and vetted, he or she receives an email summarizing the program and indicating that they have been recommended as someone who could do a good café presentation. The email also emphasizes the importance of interactivity, meeting the teens where they are, and the desire to have an accompanying "hands-on" activity. Often there will be a follow-up phone call to answer initial questions. Rarely will a scientist simply say they are not interested—most are likely to say, "sounds like fun"—*provided they are contacted months in advance.*

Next, the presenter is sent the "Presenters Guide to the Essence of a Teen Science Café" document, which stresses the importance of knowing the audience. Teens will readily engage with a presenter on some hot science topic if it is accessible to them. The presentation needs to be free of jargon and delivered in an engaging manner at an entry level so that teens will be pulled in and be able to develop new mental images. It is important that presenters not try to cover the whole breadth of a science topic, thus creating too many new mental pictures for the teens to try to process at once. A better approach is to organize the presentation around one essential provocative idea or concept, and let everything flow to it. This is referred to as the *Most Important Thing*, an idea deliberately designed to be accessible to the teens. It is most effective if a presenter gets across the *Most Important Thing* by *telling a story.*

Presenters are provided with a simple checklist:

- ***Make it highly interactive.*** A Café is not a lecture! Make the presentation a ***conversation***.
- ***Watch your pacing.*** Pause frequently to allow them to mentally catch up with you.
- ***Be lively and animated...*** like you just <u>love</u> the subject!
- ***Meet them where they are.*** You will need to verbally paint very simple mental pictures for them.
- ***Tell a story.*** Story-telling will engage them emotionally and mentally. Weave in your own story!
- ***Keep it light.*** Include jokes, cartoons, funny stories, that is always engaging.
- ***Keep it simple***. Avoid technical words and jargon. If you can say it more simply, do.
- ***Keep it focused.*** Organize the presentation around ***a very few*** simple take-home messages.
- ***Take care with graphics***. Keep slides simple, colorful, jargon-free, and few.
- ***Make it personal.*** Convey the challenges, delights, and passions of your own life in science.
- ***Relax!*** You will find that interacting with teenagers is a lot of fun.

Next the adult leader arranges to meet with the presenter informally to talk over what he or she will present. This is an opportunity to help the presenter identify the *Most Important Thing*—plus no more than three take-away ideas or images—and thus focus his or her thinking. It is an opportunity to emphasize that it is best to assume the audience knows nothing at all about the topic. It is also an opportunity to review best practices in PowerPoint presentations. This is the time to brainstorm with the presenter about a hands-on activity and work with them. These meetings are also of great value in establishing a personal connection with the presenter, which will carry through the café series and beyond.

Very early on the authors came to believe that *a dry run is an essential part of scientists' preparation.* Dry runs are most effective if they are held with a small group of teen leaders; they are the ones that can give the most relevant feedback. The dry run has proven exceedingly valuable in getting the presentations pitched at the right level and the graphics comprehensible. It also serves to overcome a certain intimidation factor for many presenters concerning the prospect of presenting before an unfamiliar audience. While many presenters have initially told the authors they are experienced at presenting to the public and never do a rehearsal, <u>every</u> presenter has told us afterward that the experience was well worth their time. This presenter comment is typical: *"The dry run was immensely valuable. It helped me select appropriate verbiage and content for the presentation. It also helped me gauge the level of delivery. Furthermore, I found the student input extremely important in identifying what their peers would find interesting. After the dry run, I made significant changes to the presentation, including the elimination of confusing content, identification of real-world connections, and simpler examples."*

Figure 4. Teens engage in an icebreaker at a café event focused on Managing Stress Through Yoga
Photo credit: Teen Science Café Network by permission

What Professional Benefits can you Expect to Derive From Being an Adult Leader?

Our adult leaders participate in a yearlong professional development program with their peers that is focused on all the elements of leading a teen science café: science communication, leadership, organization, fundraising developing partnerships to support your program, and various other topics relevant to informal STEM education.

What can you Expect Teens to get out of it in the Short Term?

Teen café programs have proven popular with teens throughout the Network, from small towns to big cities, and the reason for their popularity is much the same as for adult cafés: the combination of the socializing with their peers—with food! —and the lively conversation with a scientist on some interesting topic. Indeed, the thing teens appreciate most is being able to interact face-to-face with a scientist. For most teens, exposure to practicing scientists and thus scientific thinking are rare.

When presenters convey their personal stories, the teens gain an appreciation of scientists as real people leading interesting lives in science. They get a better understanding of the true nature of science than they get in school, though they often find that what they learn in a café compliments what they learn in science class. Some teens may find the possibility of a life in science for themselves that had not occurred to them before. They find a relevance of science in their lives. The earliest evaluation in the New Mexico program documented the impact on teen; these impacts were later replicated in the TSCN:

- Teens gained a much more realistic perception of the nature of science, the interesting lives that scientists lead, and for some the possibility of a life in science.
- It positively influenced teens' interest in science, science careers, and the nature of scientific research.
- 71% of teens agree that the Café changed their view of the importance of science to their lives.
- The Café positively influenced teens' understanding of science issues in the news, their use of facts to support scientific points of view, and considering multiple sides of an issue.
- 90% of respondents to a survey said that they learn something new almost all the time in Cafés.
- Teens improved cognitive and behavioral competence, confidence, positive social connections, character, caring, and ability to contribute to society-the 6C's for positive youth development.

What can Teens Expect to get out of it in the Longer Term?

What are the long-term influences of teen science café participation on alumni's academic, professional, and personal trajectories? An in-depth study aimed at answering this question was carried out (J. Sickler Consulting, 2019) by engaging a substantial number of men and women who had been part of teen cafés between 2008 and 2012 in a survey and focus groups. Key results of this comprehensive investigation are best summarized by quotes from the resulting report, including quotes from participants in the focus groups.

Evaluator Note: Words in italics are from the students. Words not in italics are summaries from the evaluator.

Has the fundamental intent of the teen café program been fulfilled?

"Most alumni felt [the teen café program] had some influence on understanding of what scientists do; awareness of science careers; feeling part of a science-interested peer group; and knowledge of science."

"The mix of the authenticity of the scientists, the compelling topics, and the relaxed social environment that kept teens coming back.... In the focus groups and interviews, alumni helped interpret how these factors came together as the "secret sauce" of making [the program] so effective. They described that its value came from being a "casual" or "intimate" learning environment – a small group of friendly peers, food, and highly interactive and conversational with the scientist-presenters. Together, these factors created the casual learning environment that worked for teens. Some noted how distinct this was from most other STEM-related experiences available to teens."

Figure 5. Teens gain skills in leadership and confidence to reach out to science experts. (Photo credit Teen Science Café Network by permission.)

"For me it was the topics and the format and how it was very intimate, and everybody could kind of ask questions and have conversation, that was really nice."

"When alumni reflected on what they remembered liking most about [the program], three things rose to the top for more than one-quarter of respondents. Alumni enjoyed it because of the combination of real scientists (talking about real research), learning about interesting topics, and that it was a relaxed, social experience."

"Getting a glimpse at a variety of different speakers' interests and all the amazing things that are out there in this world to be explored or worked on. New information is invaluable, and exposure to something - even if the field is only a minor interest for me as an individual - is still one of my favorite things."

"Alumni tended to describe that [the teen café program] had made science very tangible or real for them. These comments often referred to the stark contrast between school science (a collection of facts and rote experiments) and [the teen café program], which revealed the complexity, applications, and passion of the people who pursue it."

"Coming into it, my only experience to science before then was sitting down in a classroom that I didn't necessarily want to be in. But then my choice, my option to be there and checking out all this cool stuff that's happening, it was just incredible for me. Being able to get my hands really on it rather than just the theory of it was a really big game changer for me."

"...really care about understanding the basics of science or staying literate or being able to read the news and know what they're talking about, or [that] being able to research things on my own are important, feasible, doable things.

The impact has been as great on regular participants as on teen leaders. "[The program] seemed to have stronger influence on those who were not involved in other STEM activities .... Alumni for whom [the program] was their only STEM-related activity reported that [it] was much more influential on their knowledge of science. They also were more likely to report that [the program] prompted them to be more interested in STEM-related life-long learning activities, including reading science articles, talking about science with friends and family, and watching science-related television programming.... it seems that [the program's] welcoming, casual, social environment has the potential to be particularly influential at shaping science knowledge and interest for the students who might not otherwise get involved in traditional out-of-school STEM activities and/or who may not be on a STEM career pathway."

Has the program, for some, been eye-opening in terms of the possibility of a life in science?

"About one in five of alumni who answered this question reported that the program had influenced their STEM career interests to some degree. This was sometimes sparking interest, sometimes reinforcing interest, and sometimes highlighting how these careers would impact the world around them."

"Café Scientifique definitely influenced my decision to pursue research opportunities in college, first working as a lab assistant and then conducting my own research for my thesis."

"It was not often that a Café inspired their passion for a particular topic, but that it made the work of scientists tangible, so that it suddenly seemed to be a real possibility for their lives."

"For me the casual atmosphere with a scientific professional, with someone who was an expert in their field made the science and the idea of actually becoming a scientist far more tangible for me. I didn't know a lot before [the program], and now I'm in grad school for chemistry and that definitely made it seem tangible…I think that definitely affected my career path….

"That exposure, it was life changing for me, honesty. So just knowing that these things are out there, and I can do them, huge .... Because, I mean, I came from a lower economic background and that just didn't seem possible...."

What has been the long-term impact on teen leaders in terms of skill development?

"Qualitative data suggested that the impacts of being a teen leader were less directly STEM-related, and were more in the development of "soft skills" of organization, leadership, planning, and confidence in one's own abilities."

"I did not end up in a scientific field, but when I reflect on my experiences, what I remember most are those planning sessions. And I think that that's pretty telling in terms of what I do now [in my job], [which] is a lot of project and people management, and a lot of strategy and planning and focusing on process. And so, I think I was realizing that through the leadership involvement with [the teen café program], that that was an interest to me, perhaps more than the subject matter itself."

What can Scientist Presenters Expect to get out of it?

While the initial focus was, naturally enough, on benefitting the teens, the authors soon realized that the program was proving to be of much benefit to the scientist-presenters as well, in particular in developing skills in communicating their science to the public, skills that extend beyond the teen audience to other public audiences, managers, and funders. One of the most significant benefits expressed by some presenters was a recognition that effectively communicating their science has fed back into their thinking about their own research. A sampling of scientist responses to a questionnaire:

"The Café experience was beneficial to me as a scientist in that preparing an interactive talk for an audience of young people helped me identify the critical issues in my work: why I was doing it, why it is challenging, what we are trying to accomplish (vs. being bogged down in technical details). This really focused my thoughts about my work."

"Professionally, I feel it improves my ability to teach to a broad range of people. Sometimes trying to explain science in more simplistic terms helps me see the questions and potential new means of investigating them in a fresh and helpful light."

"It helped me to think about the basics of my research and discipline, and to rephrase."

Figure 6. This word cloud gets across desired qualities in a teen science café presenter

CLOSING REMARKS

The most essential elements of the Cafe´ model are the people involved (program leaders, scientist-presenters, youth leaders and attendees). Teen science cafés have been launched by a myriad of organizations, some with adult leaders and others that are organized completely by teens. They have been implemented in rural areas, as well as New York City. It is the mix of social engagement and learning that has drawn over 160,000+ teens to attend these events.

The TSCN is a community of practice that is passionate about creating opportunities for teens and experts to explore through stories, conversation, and activities how science and technology innovations are changing the world. Teen science cafés are free and open to all interested teens.

This material is based upon work supported by the National Science Foundation under Grant Nos. 0714762 and1223830. *Any opinions, findings, and conclusions or recommendations expressed in this material are those of the author(s) and do not necessarily reflect the views of the National Science Foundation.*

REFERENCES

Bell, P. (2009). *Learning science in informal environments: People, places, and pursuits*. Committee on Learning Science in Informal Environments.

Cafe Scientifique. (2012). *Science for the price of a coffee*. www.Cafescientifique.org

Dallas, D. (2006). *Café Scientifique— Café Scientifique—déjà vu. Is the café scientifique a fashionable by-product of a comfortable age or an indicator of the changing relationship between science and society?* https://www.cell.com/fulltext/S0092-8674(06)00898-1

Davis, B. G. (1993). *Tools for teaching*. Jossey-Bass Publishers.

Farmelo, G. (2004). *The dana centre: Science for adults only, dimensions, biweekly news journal of the association of science-technology centers*. May/June.

Foutz, S., & Luke, J. (2010). *A youth-directed Café Scientifique summative evaluation (Report prepared for Science Education Solutions)*. Institute for Learning Innovation.

Gambone, M., & Connell, J. (2004). *The community action framework for youth*. Academic Press.

Gambone, M., Klem, A., & Connell, J. (2002). *Finding out what matters for youth: Testing key links in a community action framework for youth development*. Youth Development Strategies, Inc., and Institute for Research and Reform in Education.

Hall, M., Foutz, S., & Mayhew, M. (2010). Design and impacts of a youth directed café scientifique program. *International Journal of Science Education. Part B. Communication and Public Engagement*. Advance online publication. doi:10.1080/21548455.2012.715780

Kekelis, L., Heber, E., & Countryman, J. A. (2005, May). *Bridge to technology: Designing a program that attracts girls. ASTC Dimensions*.

Larson, R., Walker, K., & Pearce, N. (2005). A comparison of youth-driven and adult driven youth programs: Balancing inputs from youth and adults. *Journal of Community Psychology*, *33*(1), 57–74. doi:10.1002/jcop.20035

McLaughlin, M. W. (2000). *Community counts: How youth organizations matter for youth development*. Public Education Network.

Moore, A. J. (1987, October). Native Indian Learning Styles: A review for researchers and teachers. *Journal of American Indian Education*, *27*(1). https://jaie.asu.edu/v27/V27S1nat.htm

Nisbet, M. C., & Mooney, C. (2007a). Framing science. *Science*, *316*(5821), 56. doi:10.1126cience.1142030 PMID:17412941

Nisbet, M. C., & Mooney, C. (2007b). The risks and advantages of framing science - Response. *Science*, *317*, 1169–1170.

Sargent, J. F., Jr. (2017). *The U.S. science and engineering workforce: Recent, current, and projected employment, wages, and unemployment*. https://fas.org/sgp/crs/misc/R43061.pdf

Sickler, J. (2018) *Long-term impact of teen science cafés: Results of a pilot study with cafe scientifique New Mexico Evaluation Report*. https://www.informalscience.org/long-term-impact-teen-science-cafes-results-pilot-study-cafe-scientifique-new-mexico

Sullivan, P. & Reed, W. L. (2009). *Connecting youth to high-resource adults: Lessons from effective youth programs*. doi:10.1177/0743558409350505

Swisher, K. (1994). American Indian learning styles survey: An assessment of teachers' knowledge. *Journal of Educational Issues of Language Minority Students*, (13), 59-77. http://www.ncela.gwu.edu/pubs/jeilms/vol13/americ13.htm

Wenger, E. (2009). *Communities of practice: a brief introduction*. http://www.ewenger.com/theory/

KEY TERMS AND DEFINITIONS

Café Scientifique: Café Scientifique is a place where, for the price of a cup of coffee or a glass of wine, anyone can come to explore the latest ideas in science and technology. Meetings take place in cafes, bars, restaurants and even theatres, but always outside a traditional academic context.

Evaluation: Is a process that critically examines a program. It involves collecting and analyzing information about a program's activities, characteristics, and outcomes. Its purpose is to make judgments about a program, to improve its effectiveness, and/or to inform programming decisions.

Informal Learning: Informal learning is unstructured and takes place away from formal learning settings.

Science: The intellectual and practical activity encompassing the systematic study of the structure and behavior of the physical and natural world through observation and experiment.

Science Communication (SciCom): May be defined as the use of appropriate skills, media, activities, and dialogue to produce one or more of the following personal responses.

Self-Efficacy: Refers to an individual's belief in his or her capacity to execute behaviors necessary to produce specific performance attainments.

Teen: The years of a person's age from 13 to 19.

Teen Leadership: Is the practice of teens exercising authority over themselves or others. Youth leadership has been elaborated upon as a theory of youth development in which young people gain skills and knowledge necessary to lead civic engagement, education reform and community organizing activities.

Workforce Development: Trains individuals to be more productive and prosperous in the workplace, which benefits both the employer and the worker.

Chapter 6
Learning About the Authentic Nature of Scientific Work From Professional Scientists:
Preparation for Research Participation and Mentoring

Stephen Burgin
University of Arkansas, USA

ABSTRACT

In this chapter, it is argued that in order to be able to participate in authentic science through something like a research apprenticeship, students and teachers alike must be well versed in what professional scientists do as they conduct their research. Partnerships with working scientists can be leveraged in a number of ways to do just this. As a result, teachers and students can be made aware of the professional work of scientists and the ways in which they can participate in and mentor others to engage in authentic research both in and out of the classroom.

INTRODUCTION

The authenticity of scientific practices engaged in by secondary science students has taken on an increased import with the adoption of the Next Generation Science Standards in the United States and the reform documents supporting their development (Achieve, 2013; Furtak & Penuel, 2019). For example, it continues to be desirable and even ideal for students to learn science content simultaneously with their own involvement in scientific investigations. Problems arise when the differences between the practices of professional science conducted by working scientists and the scientific activities of secondary science students are overlooked (Chinn & Malhotra, 2002; Buxton, 2006; Hofstein & Lunetta, 2004; Sandoval, 2005). In the classroom, students are frequently asked to conduct rote investigations with known outcomes in a linear stepwise manner virtually void of creativity despite consistent evidence of the posi-

DOI: 10.4018/978-1-7998-4966-7.ch006

tive impacts of participation in more student-centered forms of open-inquiry where students generate research questions and the methods to investigate those questions (Dorfman et al., 2020). Many would look at the average laboratory activity as typifying authentic scientific inquiry because students are doing some sort of hands-on activity and subsequently turning in a report of their investigation. Authenticity though is not guaranteed as a result of students only utilizing scientific equipment within a laboratory environment. The nature of the question being asked and the data being analyzed are important factors to understand in order to make that determination. Perhaps a result of a lack of involvement in and explicit reflection upon what would make an inquiry experience authentic, secondary students tend to have a relatively naïve view of scientific inquiry a problem that these recent science education reforms is seeking to address (Lederman et al., 2019).

Along those same lines, many undergraduate science laboratory experiences fall short of truly resembling the work of professional scientists, and efforts have been made to reform them accordingly (Boyd-Kimball & Miller, 2018; Chatterjee et al., 2009; Lord & Orkwiszweski, 2006; Weaver et al, 2008; Wheeler, et al., 2017). Perhaps less-than-authentic undergraduate laboratory experiences are necessary for certain purposes like students learning how to manipulate scientific apparatus or how to prepare chemical solutions in a safe way. For example, in a typical secondary or undergraduate chemistry laboratory, students might be asked to perform an acid base titration and compare their calculated concentration of hydrochloric acid to the known concentration of that acid. In terms of authenticity, this sort of investigation (though valuable for the learning of science) falls far short. Authentic scientific work ultimately generates results that push the field forward rather than simply reproducing that which has already been accomplished.

AUTHENTICITITY DEFINED

What has just been described is vastly different from the understandings of the development of scientific knowledge both from sociological examinations of the history and development of scientific knowledge (Merton, 1973) and from professional scientists themselves reflecting on their own work (Wong & Hodson, 2009; 2010). Taking into account these perspectives as well as ideas from the science education community (e.g. Burgin, 2020; Buxton, 2006; Martin, Kass, & Brouwer, 1990; Roth, 1995) a relatively comprehensive definition of authenticity within science teaching and learning can be derived. That definition includes the following features. First, the practices that students engage in must closely resemble the practices that working scientists engage in (Roth, 1995). Second, learners, much like scientists, must recognize the value of the work that they do and therefore recognize its import and utility within specific contexts (Buxton, 2006). Finally, authentic science endeavors ought to be worthwhile to others beyond the those who are engaged in that work (Burgin, 2020). Combining these dimensions, results in a three-dimensional conceptualization of authenticity that can serve to situate the authenticity of a scientific endeavor (Burgin, 2020). Each science learning activity can be assigned a value for each dimension. A visual representation of this conceptualization is presented in Figure 1 below.

Teachers could theoretically take this conceptualization and use it as a guide to revise typical laboratory investigations in ways that are more authentic according to the three dimensions and that therefore more closely resemble the work, motivations, and significance of professional scientific research.

However, as has just been discussed, the typical secondary and undergraduate science learning experience is lacking in terms of authentic experiences along these lines. In North America at least, secondary

Figure 1. A 3-D conceptualization of authenticity. This is figure originally appeared in an article in International Journal of Science Education (Burgin, 2020)

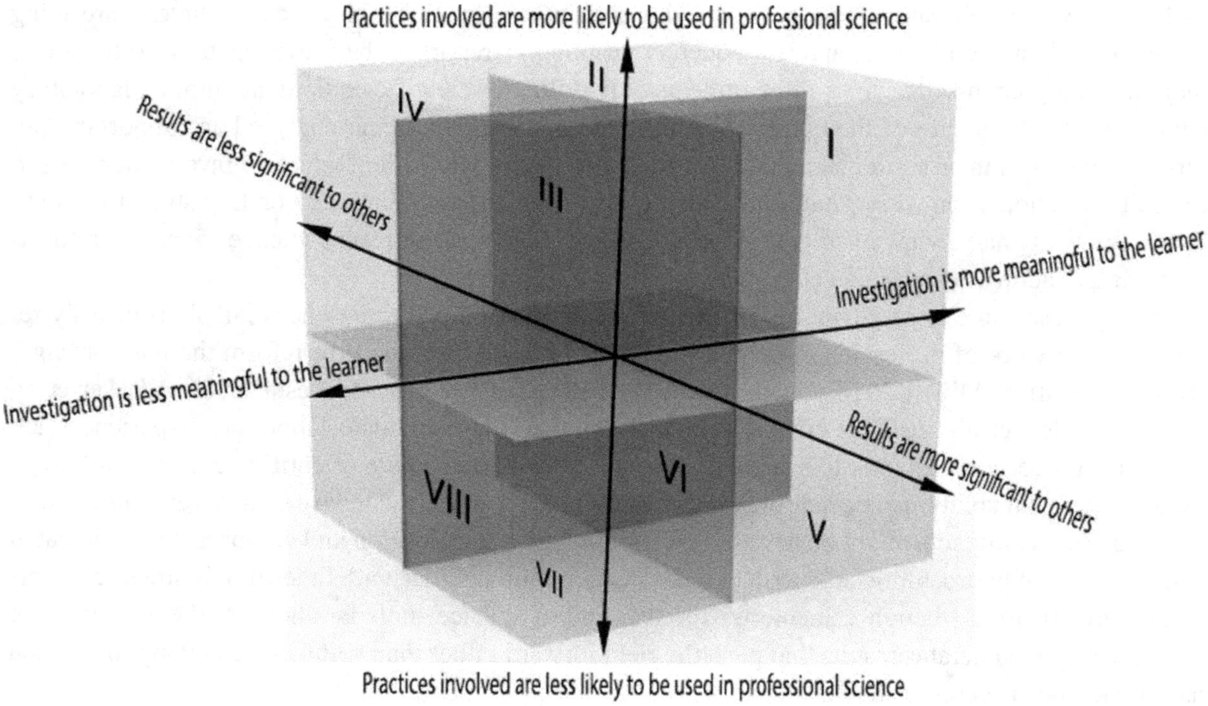

science teachers are for the most part required to have an undergraduate degree in their content area to obtain a teaching license with some exceptions (Olson et al., 2015). It follows that relatively low numbers of secondary science teachers would have a graduate degree in science specifically. Therefore, large numbers of secondary science teachers are unlikely to have first-hand experience in authentic professional scientific research that typically does not take place until a student enters a graduate degree program within their scientific discipline. If these science teachers lack experience conducting authentic scientific research, it would be hard for them to facilitate truly authentic opportunities for their own students to engage in scientific research. It is therefore important for many teachers to learn about authentic scientific research just as it is important for all students to do so.

Figure 2 illustrates a progression from learning about to mentoring within authentic scientific endeavors and specifies who among the stakeholders in science education has a role to play in that regard.

It is this progression that will be used to frame the purposes, players, and nature of student, teacher, scientist partnerships discussed in this chapter. Specifically, what will be argued is that these partnerships offer a wonderfully powerful opportunity for students and teachers to first learn about the nature of each of these three dimensions in the context of a specific scientist's professional research. This can happen within traditional classrooms and can involve virtually all science students. The stage can then be set for both students and teachers to engage in full participation in authentic scientific practices that meet all three of the dimensions of authenticity presented in Figure 1.

In summary, the basic premise for this chapter is that in order to engage in authentic scientific activities, students need to be explicitly taught the features of authenticity just described. This requires science teachers who are prepared to do just that. These teachers need to be well-versed in the authentic work

Figure 2. Progression from learning about to mentoring within authentic science by stakeholders within science education

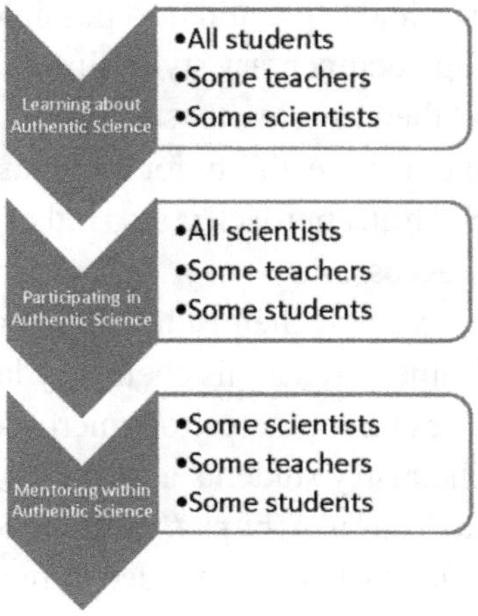

and research of professional scientists. Since many science teachers themselves have no background personally conducting professional scientific research, they specifically need to come into contact with working scientists to learn about and or participate in the work that they are doing (e.g. Enderle et al., 2014). Science educators then need to work with science teachers to help them translate this knowledge of authenticity into realistic authentic classroom experiences for their students. These experiences first would introduce students to the authentic work of professional scientists and would hopefully lead to opportunities for them to participate in that work.

LEARNING ABOUT AUTHENTIC SCIENCE FROM SCIENTISTS

Virtual Video-based Laboratory Field Trips

One effective tool to get students out of the classroom and physically observe professional settings is the traditional field trip (e.g., Braund & Reiss, 2006, Parvin & Stephenson, 2004). These professional settings would include zoos, chemical industrial sites, field-based research centers, and working laboratories within them. Field trips to laboratories on university campuses would be another option. On these trips, a visiting class (teacher and students) would have the opportunity to observe the interactions between groups of researchers (i.e. graduate students, post-doctoral students, laboratory technicians, principal investigators, professional scientists, etc.), the practices they employ, and the tools they use in their work. By so doing, the visitors might notice the collaborative nature of authentic scientific research as well as the humanity of those working in professional science. The teacher could then leverage the experience into subsequent school laboratory investigations in which the students participate, thereby increasing the authenticity of those activities.

Despite these benefits, visits to working laboratories come with a number of challenges. Safety concerns paired with space limitations make it hard for large groups of students to visit these and other similar contexts. For example, students would likely all need safety glasses to physically enter the workspace. The logistics of trying to bring entire secondary science classes on such field trips would likely make such an idea unfeasible.

In response to these issues, a virtual field trip could be a possible solution (Tuthill & Klemm, 2002). One way to do this would be to develop documentary style films featuring the work of professional scientists in their laboratory contexts and then to bring those films into the high school classroom (Smith, 2014). Advantages of this approach include the ability for large numbers of students to simultaneously view a working authentic laboratory. Such an approach is also rather inexpensive. If the video is uploaded to the internet, it would also be easily accessible.

Recently the author worked with a research chemist to document through video the procedures that he utilized to develop metal organic frameworks in his chemistry lab and then the methods his research group employed to identify the structures through electron microscopy (Burgin et al., 2020). This video was then shown to area high school chemistry students as a way to bring the research of the chemist to the classroom. We uploaded the video to YouTube (https://youtu.be/lqTQfnK9DEU) so that the chemistry teacher that we were working with could easily use the video with all of his chemistry classes. Through the use of surveys, we found that the students self-reported the development of new understandings related to how professional scientists go about their research and the significance of that work to the broader public (two of the three dimensions of the authenticity framework).

Classroom Visits by Scientists

Face-to-face presentations in high school science classrooms by professional scientists could be another way to teach about the three dimensions of authenticity. These visits are definitely more time consuming than the former approach in that they involve a greater time commitment on the part of the scientist and more intricate collaborative planning between the scientist and the science teacher. However, the advantage of the ability to have real time interactions between the students and the scientist may make these efforts worthwhile. They could be more effective in revealing the human qualities of the scientist related to their personal motivations to do the work that they do.

As part of the study described in the previous section, the chemist came and presented the same information both with and without the accompanying youtube video. We found that when the chemist was physically there, the interactions that took place as students asked questions related to the video/presentation were more substantial than when the chemist was not present. Specifically, the chemist was able to communicate the significance of his research to the general public (one of the three dimensions of authenticity). Figure 3 displays a photograph of this chemist interacting with a classroom of high school students.

PARTICIPATING IN AUTHENTIC SCIENCE

Traditional Student Teacher Scientist Partnerships

Typically, Student Teacher Scientist Partnerships (STSPs), involve students and teachers partnering with scientists to gather data which the scientist then analyzes and incorporates into their work (Lawless & Rock, 1998, Sadler et al., 2010). These experiences have been shown to be quite effective in how they have the potential to involve students in parts of the authentic research process. Despite some hesitation,

Figure 3. Research chemist presenting his work to a classroom of high school chemistry students.

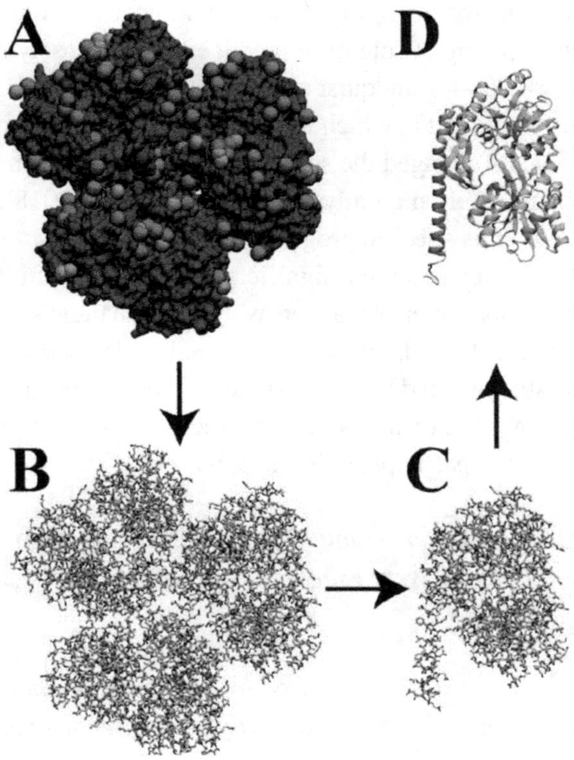

studies have demonstrated the quality and the utility of the data collected (e.g. bird populations, weather data, water quality, etc.) which has alleviated some of the concerns held by scientists about the value of such collaborations (Bonney et al., 2014; Cohn, 2008; Kosmala et al., 2016). However, these experiences often referred to as "citizen-science" have been documented to be frustrating for students who are not given the opportunity to participate fully in the scientific process and therefore do not develop robust views of scientific processes (Moss, Abrams, & Kull, 1998). As a result, call have been made to involve students and teachers more fully in the entire research process from asking questions to analyzing data and presenting findings within these partnerships (Trumbull et al., 2000; Wormstead et al., 2002). Maybe the lack of involving partners fully in the scientific research on the part of scientists is at least partly the result of the students and the teachers lacking a comprehensive understanding of the nature of the work that scientists are engaged in. For examples, without explicit conversations and examples from the partnering scientist, they likely will not know what is going to happen with the data that they provide. Perhaps other experiences are needed to lead up to the involvement of the K-12 learner and science teacher in fully authentic practices. As previously discussed, presentations by the scientist in a variety of forms are one way that this can be accomplished.

Utilization of Technological Tools

Another approach that has the potential to involve an entire class of science learners in authentic science participation would be to bring the actual research tools employed by research scientists directly into the high school classroom. For example, in chemistry the practice of modeling is one of the authentic tasks engaged in by research scientists. Involving students in molecular modeling then would be one way to specifically have them participating in scientific practices which closely mirror if not exactly model professional chemistry (Gilbert, 2004; Lundquist et al., 2016).

An example of this is a recent project in which we brought professional molecular modeling software into the high school classroom and engaged the students in the visualization of protein structures in the same way that a research chemist did on a daily basis (Burgin et al., 2018). Figure 4 displays sample protein structures as modeled by this specific program.

Students utilizing the software demonstrated significant gains in their understandings of the nature of models. Additionally, the high school science teachers working with them spoke highly of the experience in that it allowed for students to get hands-on with the actual tools used by professional scientists; an experience that none of these students had had. The students themselves told us that it was obvious that the experience was real in that it wasn't using some watered-down educational program. What follows are a few quotes from students who participated in the experience.

[Our teacher] explained it. That it was something that people on campus [professional research scientists on the University campus in town] were using too. . .It gives it more a sense of this is real. People are actually using it.

If I am able to understand. . .most parts that are used by professionals then I think it can be used in a high school setting. . . And it is really cool that we actually got to use it because it is like hey we did this, and we are only 9th and tenth graders in class.

They recognized that they were actually using real modeling techniques currently being employed by research biochemists.

Science Fair Projects

The previous examples either had students learning about the 3 dimensions of authentic science (Figure 1) or participating in scientific practices through the utilization of professional molecular modeling software. Attention is now turned to participating in scientific research that has the potential to additionally be both meaningful to the learner (teacher and/or student) and significant to others (potentially scientists). One classic example of this is a correctly implemented student science fair project (Bencze & Bowen, 2009; Koomen et al., 2018; McComas, 2011).

Students have the potential to investigate something of personal value to them through a science fair in much the same way that a professional scientist does. Additionally, through presenting their work, students may find themselves sharing their research with members of their community who find the information to be worthwhile.

Again, I wish to make the point, that science fair project quality could only be increased through interactions with professional scientists who are sharing the ins and outs of their own personal research with teachers and students in addition to serving as potential mentors and/or judges within the science fair.

Figure 4. Protein structures modeled by professional modeling software (Burgin et al. 2018).

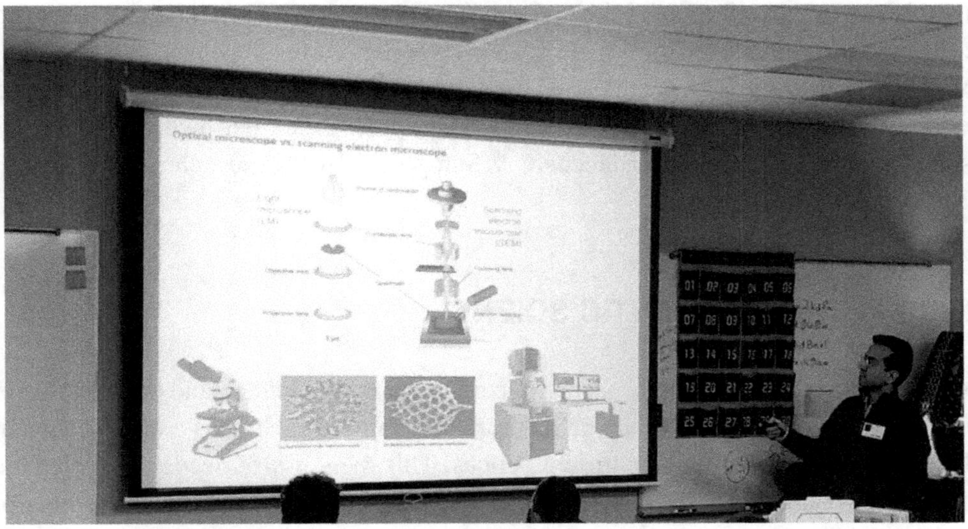

Research Apprenticeships for Students and Teachers

It could be argued that some of the most authentic contexts in which to perform scientific research are found outside of the classroom through internship experiences with professional scientists (Hsu & Venegas, 2018). Summer research experiences (internships/apprenticeships) for high school students where they are mentored by research scientists in authentic laboratory and field settings are a great example of this (Burgin & Sadler, 2016; Burgin, Sadler, & Koroly, 2012; Sadler et al., 2010). That these opportunities are highly authentic in terms of the similarities between the practices employed by the participant and a professional scientist is hard to deny (Bell et al., 2003; Hsu & Venegas, 2018; Richmond & Kurth, 1999). However, it is possible that the work students are involved in may not be personally relevant or meaningful to them. The mentor scientist might also choose to involve the student in work that is merely busy work and not of value to the professional scientific community. It is therefore imperative that mentor scientists are carefully chosen and intentionally paired with the students who will be interning with them.

The issue with these experiences compared with those outlined above is that they do not reach all K-12 science students but rather only a select few who already have developed an interest in science and science-related careers. Additionally, these students have the means to spend a summer devoted to scientific research and the access to programs that allow them to do so (Burgin & Sadler, 2013).

It is imperative therefore that the high school science teacher be involved both before and after research apprenticeship experiences for their students. From recruiting students and recommending students for these programs, to providing a venue for sharing the students' research upon their return, the science teacher is crucial (Burgin & Sadler, 2013). Developing a relationship with practicing scientists in the local community through field trips, visits, and the like is also an important component of bringing awareness of these opportunities to science teachers.

As mentioned earlier, these internship experiences are not only valuable for science learners, but also for science teachers. Many science teachers do not have experience personally conducting authentic science research and would therefore have trouble mentoring and engaging their own students through research that could be classified as authentic according to the three dimensions of authenticity (Figure 1). One valuable way to afford teachers the opportunity to partner with scientists on authentic research projects is through programs classified as Research Experiences for Teachers (RETs). These experiences allow teachers to serve as research scientists and have been shown to lead to increases in their confidence to perform scientific research and their ability to engage their own students in performing authentic science in their classrooms (Blanchard & Sampson, 2018; Cutucache et al., 2017; Enderle et al., 2014; Ruston & Reiss, 2019).

MENTORING WITHIN AUTHENTIC SCIENCE

In order for a K-12 student or a science teacher to participate in and understand the three dimensions of science authenticity (Figure 1) they must be mentored by someone who has a deep knowledge of and substantial experience engaging in scientific practices. This mentor also needs to be prepared to have explicit conversations with those partnering students and teachers about the nature of professional scientific work (e.g. the time it takes to reach meaningful conclusions, the ways in which research questions and procedures evolve, how results are disseminated, etc.), their personal motivations for participating in the research, and the significance of the work that they are doing to others. Potential mentors (scientists, teachers, and perhaps even a few students) therefore need some professional development related to how to effectively communicate these dimensions throughout the research process.

Of unique concerns to STSPs is the importance of the relationship between the scientist and the teacher to be reciprocal. The most effective teachers in these partnerships will value the expertise of the scientist and vice-versa (Shanahan & Bechtel, 2020). Indeed, the scientist needs to expertise of the teacher to think about the pedagogical realities of the traditional classroom (Ng & Fergusson, 2019). These are the sorts of issues that could be explicitly addressed by a science education researcher who might be facilitating a professional development (PD) for mentor scientists and teachers. Ultimately these PD opportunities would hopefully result in an increased understanding of and participating in the three dimensions of authentic science practice.

CONCLUSION

In summary, partnerships between scientists, teachers, students (and I would add science educators) are important in order to inform teacher and student understandings of just what makes for authentic science so that participation in scientific practices could be as authentic as possible given the contextual constraints. A progression (Figure 2) within these partnerships from learning about authentic science to participating in that research and ultimately mentoring others as they engage in the work is one that should ideally be followed from beginning to end in that order. It is argued most students will need to enter that progression at the learning stage compared to only some science teachers (likely those who have not had any opportunities conducting authentic scientific research). Some scientists also, might have some learning that needs to be done specifically along the lines of how to effectively mentor students and teachers as they communicate the three dimensions of authentic science (the practices employed, the personal meaningfulness of the work, and the significance of the results to others) to them. Once the learning has taken place and someone has been personally involved in authentic science practices, then I believe they are ready to mentor others through collaborative partnerships. It is hoped that the discussions within this article will serve to encourage all stakeholders in STSPs that engaging in the whole range of authentic scientific practices that are both meaningful and significant is possible with proper training and exposure to those practices.

REFERENCES

Achieve. (2013). *Next generation science standards.* Retrieved July 12, 2018, from http://www.nextgenscience.org

Bell, R. L., Blair, L. M., Crawford, B. A., & Lederman, N. G. (2003). Just do it? Impact of a science apprenticeship program on high school students' understandings of the nature of science and scientific inquiry. *Journal of Research in Science Teaching*, *40*(5), 487–509. doi:10.1002/tea.10086

Bencze, J. L., & Bowen, G. M. (2009). A national science fair: Exhibiting support for the knowledge economy. *International Journal of Science Education*, *31*(18), 2459–2483. doi:10.1080/09500690802398127

Blanchard, M. R., & Sampson, V. D. (2018). Fostering Impactful research experiences for teachers (RETs). *Eurasia Journal of Mathematics, Science and Technology Education*, *14*(1), 447–465.

Bonney, R., Shirk, J. L., Phillips, T. B., Wiggins, A., Ballard, H. L., Miller-Rushing, A. J., & Parrish, J. K. (2014). Next steps for citizen science. *Science*, *343*(6178), 1436–1437. doi:10.1126cience.1251554 PMID:24675940

Boyd-Kimball, D., & Miller, K. R. (2018). From cookbook to research: Redesigning an advanced biochemistry laboratory. *Journal of Chemical Education*, *95*(1), 62–67. doi:10.1021/acs.jchemed.6b00722

Braund, M., & Reiss, M. (2006). Towards a more authentic science curriculum: The contribution of out-of-school learning. *International Journal of Science Education*, *28*(12), 1373–1388. doi:10.1080/09500690500498419

Burgin, S. R. (2020). A Three-Dimensional Conceptualization of Authentic Inquiry-Based Practices: A Reflective Tool for Science Educators. *International Journal of Science Education*, *42*(9), 1465–1484. Advance online publication. doi:10.1080/09500693.2020.1766152

Burgin, S. R., Oramous, J., Kaminski, M., Stocker, L., & Moradi, M. (2018). High school biology students' use of VMD as an authentic tool for learning about modeling as a professional scientific practice. *Biochemistry and Molecular Biology Education*, *46*(3), 230–236. doi:10.1002/bmb.21113 PMID:29457336

Burgin, S. R., & Sadler, T. D. (2013). Science immersion: Summer programs that expose students to authentic scientific research. *Science Teacher (Normal, Ill.)*, *80*(9), 44–49. doi:10.2505/4/tst13_080_09_44

Burgin, S. R., & Sadler, T. D. (2016). Learning nature of science concepts through a research apprenticeship program: A comparative study of three approaches. *Journal of Research in Science Teaching*, *53*(1), 31–59. doi:10.1002/tea.21296

Burgin, S. R., Sadler, T. D., & Koroly, M. J. (2012). High school student participation in scientific research apprenticeships: Variation in and relationships among student experiences and outcomes. *Research in Science Education, 42*(3), 439–467. doi:10.100711165-010-9205-2

Burgin, S. R., Sakamaki, Y., Tsuji, M., Watson, O., Heidrick, Z., Childress, M., Chitwood, T., Benamara, M., Martin, E., & Beyzavi, M. H. (2020). Using a Faculty-Developed Documentary-Style Film to Communicate Authentic Chemistry Research to a High School Audience via YouTube. *Journal of Chemical Education, 97*(8), 2351–2355. Advance online publication. doi:10.1021/acs.jchemed.0c00376

Buxton, C. A. (2006). Creating contextually authentic science in a "low-performing" urban elementary school. *Journal of Research in Science Teaching, 43*(7), 695–721. doi:10.1002/tea.20105

Chatterjee, S., Williamson, V. M., McCann, K., & Peck, M. L. (2009). Surveying student attitudes and perceptions towards guided-inquiry and open-inquiry laboratories. *Journal of Chemical Education, 12*(12), 1427–1432. doi:10.1021/ed086p1427

Chinn, C. A., & Malhotra, B. A. (2002). Epistemologically authentic inquiry in schools: A theoretical framework for evaluating inquiry tasks. *Science Education, 86*(2), 175–218. doi:10.1002ce.10001

Cohn, J. P. (2008). Citizen science: Can volunteers do real research? *Bioscience, 58*(3), 192–197. doi:10.1641/B580303

Cutucache, C. E., Leas, H. D., Grandgenett, N. F., Nelson, K. L., Rodie, S., Shuster, R., Schaben, C., & Tapprich, W. E. (2017). Genuine faculty-mentored research experiences for In-Service science teachers: Increases in science knowledge, perception, and confidence levels. *Journal of Science Teacher Education, 28*(8), 724–744. doi:10.1080/1046560X.2017.1415615

Dorfman, B. S., Issachar, H., & Zion, M. (2020). Yesterday's students in today's world—Open and guided inquiry through the eyes of graduated high school biology students. *Research in Science Education, 50*(1), 123–149. doi:10.100711165-017-9683-6

Enderle, P., Dentzau, M., Roseler, K., Southerland, S., Granger, E., Hughes, R., Golden, B., & Saka, Y. (2014). Examining the influence of RETs on science teacher beliefs and practice. *Science Education, 98*(6), 1077–1108. doi:10.1002ce.21127

Furtak, E. M., & Penuel, W. R. (2019). Coming to terms: Addressing the persistence of "hands-on" and other reform terminology in the era of science as practice. *Science Education, 103*(1), 167–186. doi:10.1002ce.21488

Gilbert, J. K. (2004). Models and modeling: Routes to more authentic science education. *International Journal of Science and Mathematics Education, 2*(2), 115–130. doi:10.100710763-004-3186-4

Hofstein, A., & Lunetta, V. N. (2004). The laboratory in science education: Foundations for the twenty-first century. *Science Education, 88*(1), 28–54. doi:10.1002ce.10106

Hsu, P. L., & Venegas, L. (2018). Activity features of high school students' science learning in an open-inquiry-based internship programme. *International Journal of Science Education, 40*(12), 1391–1409. doi:10.1080/09500693.2018.1479801

Koomen, M. H., Rodriguez, E., Hoffman, A., Petersen, C., & Oberhauser, K. (2018). Authentic science with citizen science and student-driven science fair projects. *Science Education*, 102(3), Kosmala, M., Wiggins, A., Swanson, A., & Simmons, B. (2016). Assessing data quality in citizen science. *Frontiers in Ecology and the Environment*, *14*(10), 551–560.

Lawless, J. G., & Rock, B. N. (1998). Student scientist partnerships and data quality. *Journal of Science Education and Technology*, *7*(1), 5–13. doi:10.1023/A:1022575914118

Lederman. J., Lederman, N., Bartels, S., Jimenez, J., Akubo, M., Aly, S., Bao, C., Blanquet, E., Blonder, R., Bologna Soares de Andrade, M., Buntting, C., Cakir, M., El-Deghaidy, H., ElZorkani, A., Gaigher, E., Guo, S., Hakanen, A., Al-Lal, S. H., Han-Tosunoglu, C., Hattingh, A., … Zhhou, Q. (2019). An international collaborative investigation of beginning seventh grade students' understandings of scientific inquiry: Establishing a baseline. *Journal of Research in Science Teaching*, *56*(10), 486-515.

Lord, T., & Orkwiszewski, T. (2006). Moving from didactic to inquiry-based instruction in a science laboratory. *The American Biology Teacher*, *68*(6), 342–345. doi:10.2307/4452009

Lundquist, K., Herndon, C., Harty, T. H., & Gumbartt, J. C. (2016). Accelerating the use of molecular modeling in the high school classroom with VMD lite. *Biochemistry and Molecular Biology Education*, *44*(2), 124–129. doi:10.1002/bmb.20940 PMID:26751137

Martin, B., Kass, H., & Brouwer, W. (1990). Authentic science: A diversity of meaning. *Science Education*, *74*(5), 541–554. doi:10.1002ce.3730740505

McComas, W. F. (2011). A new look at an old tradition. *Science Teacher (Normal, Ill.)*, *78*(8), 34.

Merton, R. K. (1973). *The sociology of science: Theoretical and empirical investigations*. University of Chicago press.

Moss, D. M., Abrams, E. D., & Kull, J. A. (1998). Can we be scientists too? Secondary students' perceptions of scientific research from a project-based classroom. *Journal of Science Education and Technology*, *7*(2), 149–161. doi:10.1023/A:1022564507639

Ng, W., & Fergusson, J. (2019). Technology-enhanced science partnership initiative: Impact on secondary science teachers. *Research in Science Education*, *49*(1), 219–242. doi:10.100711165-017-9619-1

Olson, J. K., Tippett, C. D., Milford, T. M., Ohana, C., & Clough, M. P. (2015). Science teacher preparation in a North American context. *Journal of Science Teacher Education*, *26*(1), 7–28. doi:10.100710972-014-9417-9

Parvin, J., & Stephenson, M. (2004). Learning science at industrial sites. *Learning Science Outside the Classroom*, 129-149.

Richmond, G., & Kurth, L. A. (1999). Moving from outside to inside: High school students' use of apprenticeships as vehicles for entering the culture and practice of science. *Journal of Research in Science Teaching*, *36*(6), 677–697. doi:10.1002/(SICI)1098-2736(199908)36:6<677::AID-TEA6>3.0.CO;2-#

Roth, W. M. (1995). *Authentic school science: Knowing and learning in open-inquiry science laboratories*. Kluwer. doi:10.1007/978-94-011-0495-1

Rushton, E. A., & Reiss, M. J. (2019). From science teacher to 'teacher scientist': Exploring the experiences of research-active science teachers in the UK. *International Journal of Science Education, 41*(11), 1541–1561. doi:10.1080/09500693.2019.1615656

Sadler, T. D., Burgin, S., McKinney, L., & Ponjuan, L. (2010). Learning science through research apprenticeships: A critical review of the literature. *Journal of Research in Science Teaching, 47*, 235–256.

Sandoval, W. A. (2005). Understanding students' practical epistemologies and their influence on learning through inquiry. *Science Education, 89*(4), 634–656. doi:10.1002ce.20065

Shanahan, M. C., & Bechtel, R. (2020). "We're taking their brilliant minds": Science teacher expertize, meta-discourse, and the challenges of teacher–scientist collaboration. *Science Education, 104*(2), 354–387. doi:10.1002ce.21550

Smith, D. K. (2014). iTube, YouTube, WeTube: Social Media Videos in Chemistry Education and Outreach. *Journal of Chemical Education, 91*(10), 1594–1599. doi:10.1021/ed400715s

Trumbull, D. J., Bonney, R., Bascom, D., & Cabral, A. (2000). Thinking scientifically during participation in a citizen-science project. *Science education, 84*(2), 265–275. doi:10.1002/(SICI)1098-237X(200003)84:2<265::AID-SCE7>3.0.CO;2-5

Tuthill, G., & Klemm, E. (2002). Virtual field trips: Alternatives to actual field trips. *International Journal of Instructional Media, 29*(4), 453–467.

Weaver, G. C., Russell, C. B., & Wink, D. J. (2008). Inquiry-based and research-based laboratory pedagogies in undergraduate science. *Nature Chemical Biology, 4*(10), 577–580. doi:10.1038/nchembio1008-577 PMID:18800041

Wheeler, L. B., Maeng, J. L., Chiu, J. L., & Bell, R. L. (2017). Do teaching assistants matter? Investigating relationships between teaching assistants and student outcomes in undergraduate science laboratory classes. *Journal of Research in Science Teaching, 54*(4), 463–492. doi:10.1002/tea.21373

Wong, S. L., & Hodson, D. (2009). From the horse's mouth: What scientists say about scientific investigation and scientific knowledge. *Science Education, 93*(1), 109–130. doi:10.1002ce.20290

Wong, S. L., & Hodson, D. (2010). More from the horse's mouth: What scientists say about science as a social practice. *International Journal of Science Education, 32*(11), 1431–1463. doi:10.1080/09500690903104465

Wormstead, S. J., Becker, M. L., & Congalton, R. G. (2002). Tools for successful student–teacher–scientist partnerships. *Journal of Science Education and Technology, 11*(3), 277–287. doi:10.1023/A:1016076603759

ADDITIONAL READING

Burgin, S. R. (2020). A three-dimensional conceptualization of authentic inquiry-based practices: A reflective tool for science educators. *International Journal of Science Education*, *42*(9), 1465–1484. do i:10.1080/09500693.2020.1766152

Burgin, S. R., & Sadler, T. D. (2013). Consistency of practical and formal epistemologies of science held by participants of a research apprenticeship. *Research in Science Education*, *43*(6), 2179–2206. doi:10.100711165-013-9351-4

Burgin, S. R., Sadler, T. D., & Koroly, M. J. (2012). High school student participation in scientific research apprenticeships: Variation in and relationships among student experiences and outcomes. *Research in Science Education*, *42*(3), 439–467. doi:10.100711165-010-9205-2

DeBoer, G. (1991). *A history of ideas in science education*. Teachers College Press.

Luft, J., Bell, R. L., & Gess-Newsome, J. (Eds.). (2008). *Science as inquiry in the secondary setting*. NSTA Press.

Meadows, L. (2009). Change in secondary science settings: A voice from the field. In Reforming secondary science instruction. Arlington, VA: NSTA Press.

National Research Council. (1996). *The role of scientists in the professional development of science teachers*. National Academies Press.

Xie, Q., & Pallant, A. (2011) The molecular workbench software: an innovative dynamic modeling tool for nanoscience education. In Models and modeling: cognitive tools for scientific enquiry. Springer. doi:10.1007/978-94-007-0449-7_6

KEY TERMS AND DEFINITIONS

Authenticity: A measure of the similarities between an activity engaged in by a learner and the professional practice of the discipline being learned. It can be thought of as consisting of three dimensions; the similarities between school work and professional activity, the meaningfulness of the activity to the learner, and the significance of the work to others.

Computer-Based Tools: Software programs or websites that can be used by scientists and learners alike in their work to produce and/or learn about scientific knowledge. Molecular modeling software would be one example of a computer-based tool.

Research Apprenticeship: An experience in science that typically consists of a secondary or undergraduate learner who is learning under an experienced scientist in a professional setting as they participate in the activity of science.

Scientist Visits: A traditional presentation/lecture given by a scientist who is visiting a classroom of science learners.

Virtual Field Trips: A web-based video and/or simulated experience whereby a learner can experience a specific professional context remotely.

Section 3

Chapter 7
An Interdisciplinary Model to Diversify STEM Participation:
College, High School, and Industry Partnerships

Margaret R. Blanchard
North Carolina State University, USA

Richard A. Venditti
North Carolina State University, USA

Shana L. McAlexander
https://orcid.org/0000-0003-0075-7233
North Carolina State University, USA

Katherine R. McCance
https://orcid.org/0000-0002-6843-8207
North Carolina State University, USA

Karen M. Collier
North Carolina State University, USA

ABSTRACT

This chapter describes an interdisciplinary program between a College of Education and a College of Natural Resources and their partnerships with rural high schools, regional colleges, and bioeconomy industries. The overarching goal of the program was to provide engaging professional development and support for teachers and diverse undergraduate students to prepare and promote diverse students to consider STEM majors and careers related to bioproducts and bioenergy. A team of faculty and graduate students from a Forest Biomaterials department and a Science Education department developed online courses, workshops, and laboratory activities and internship placements for undergraduate students and high school science and CTE teachers. This chapter details the need for the program, its context, online course development, and laboratory activities. For each of the key partners in the program—the faculty and graduate students, the high school teachers, and the undergraduate students—key strategies, lessons learned, and recommendations are shared.

DOI: 10.4018/978-1-7998-4966-7.ch007

INTRODUCTION

There is a need to develop the workforce in bioenergy, bioproducts, and the bioeconomy (Energy Independence and Security Act, 2007). Biobased product employment (excluding biofuels) grew from 2014 to 2016 by 17% and continues to grow, contributing $459 billion to the U.S. economy (Golden et al., 2018). In addition to the technical/scientific knowledge specific to the bioproducts and bioenergy industry, there are technological skill sets required in 21st century manufacturing environments. These include skills in chemistry, biology, experimental design, process control, written and oral communication, data management, and working across disciplines (Glaser, 2013). Accomplished individuals in these areas are highly competitive in today's workforce. Jobs in these industries span the educational spectrum: from 2-year technical school to advanced, professional degrees (Glaser, 2013). Currently, many sectors of the STEM workforce lack the diverse perspectives of underrepresented minorities and females (National Science Foundation, 2017). Increasing the diversity of this workforce will help to provide new perspectives for innovation and sustainable solutions (Herring, 2009). Given this, there is a need to build these skills into high school curricula (NEA, 2012).

STEM experiences can vary greatly between students, with socio-economic factors, geographic region, racial background, and gender influencing course persistence and future careers (Alegria & Branch, 2015; Andersen & Ward, 2014; Brown et al., 2016). Approximately 25% of K-12 US students are enrolled in rural schools that primarily enroll Black and Brown students, 14% of which are designated as high-poverty schools (Lavalley, 2018). Schools designated as high poverty tend to have fewer teachers who are certified in their content areas, fewer advanced courses offered in STEM subjects, and higher turnover than in wealthier schools (Darling-Hammond & Hammond, 2015).

North Carolina Context

This program was focused in North Carolina, which is a tremendous agricultural and forest based biomass region, having the seventh highest degree of bioproducts industry concentration in 2013 (Golden et al. 2018). North Carolina's 18.6 million acres of forest land cover 60% of the state, and timber receipts at the first point of delivery totaled $807 million in 2015 (McConnell, Jeuck, Bardon, & Hazel, 2016). In total, the forest products industry directly supports over 70,000 employees at 2,000 businesses located in ***all*** counties of the state, making it the seventh highest bioproducts industry concentration in the U.S., in 2013 (Golden et al., 2018). The total value added that is returned to North Carolina by forestry and forest products activities is $11 billion (McConnell et al. 2016). Much of this biomass based industry is concentrated in Eastern NC (see Figure 1). Farming is equally important in NC, supporting about 700,000 jobs on more than 8 million acres of land. North Carolina leads the nation in sweet potato, egg and poultry production, and ranks in the top three for Christmas trees, pork, trout, and turkeys (Adams, 2019).

Eastern North Carolina is a more rural region that has lagged in several socioeconomic indicators (U.S. Census Bureau, 2016). Therefore, the ability to leverage the human resources of the state with the needs of an emerging bioproducts and bioenergy industry made this region uniquely positioned for the program. Discussions with stakeholders have indicated the combination of this science/technology knowledge and skills is highly valued, requiring more preparation and skill development. One way to build these skills and promote interest to advance the opportunities for rural youth was to add these skills to high school curricula.

Figure 1. Distribution of farm and timber receipts per county
Note. Darker shading indicates higher values. The red box indicates the area the program serves. Figure reused with permission from Venditti & Blanchard (2016) and created based on data available from the North Carolina Department of Agriculture and Consumer Services (2015) and McConnell et al. (2016).

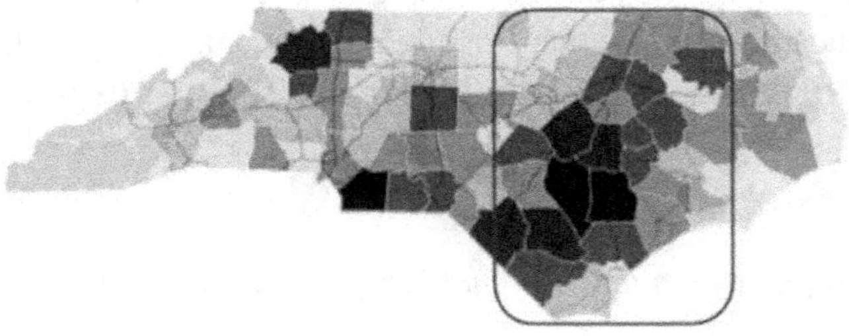

In rural, eastern North Carolina, school districts tend to be small in population and spread out over a larger geographic area. Most of the schools serve low income families, with approximately 70% receiving Free and Reduced Price lunch (F&RL) and about 54% of whom are minorities, with the majority of high schools designated as high poverty (North Carolina School Report Cards, n.d.; School Digger, n.d.). The charter schools tend to be an exception to this, as they pull from a larger area and families who attend are able to transport their children to school. In spite of the difficult conditions of the region, physical conditions within schools have improved over the last ten years, with wireless infrastructure, computers, and instructional technologies brought in from grant funds (e.g., Blanchard, 2010; Blanchard et al., 2014) and federal *Race to the Top* funds (US DOE, 2012). The superintendents from these rural districts are eager to provide professional development opportunities for their teachers, and enhance the students' exposure to STEM content, experiences, and to support their pathways to potential careers.

Pillars of Sustainability

To integrate bioproducts and bioenergy concepts into existing high school science and Career and Technical Education (CTE) curricula, the program drew from the field of sustainable development, broadly defined as using innovation and engineering approaches in development to meet the needs of the present without compromising the ability of future generations to meet their own needs (Keeble, 1988).

Central to the program's approach, the Sustainability conceptual framework Venn diagram (Purvis, Mao & Robinson, 2019) includes environmental, social, and economic components as the three pillars of sustainability. In Figure 2, we have situated the program in the center, although the program tended to emphasize environmental considerations more than economic or social elements. For example, students could weigh the costs and benefits of traditional plastic versus bioplastics for bottles in terms of biodegradable properties of the materials, water requirements for manufacturing and pollution (Environmental), cost of the materials and processes (Economic), and what workers are paid and working conditions in the US versus other countries for manufacturing these products (Social).

PROGRAM DEVELOPMENT

The program *Preparing Diverse and Rural Students and Teachers to Meet the Challenges in the Bio-energy and Bioproducts Industry* was designed to address challenges of shortages in Bioeconomy industries through an educational innovation (Figure 3). It was funded by a United States Department of Agriculture (USDA) NIFA grant for $2.75 million from 2017-2021 (#2017-67009-26771) (Venditti & Blanchard, 2016).

Figure 2. Situating the project within the Sustainability framework
Adapted from Purvis et al. (2019)

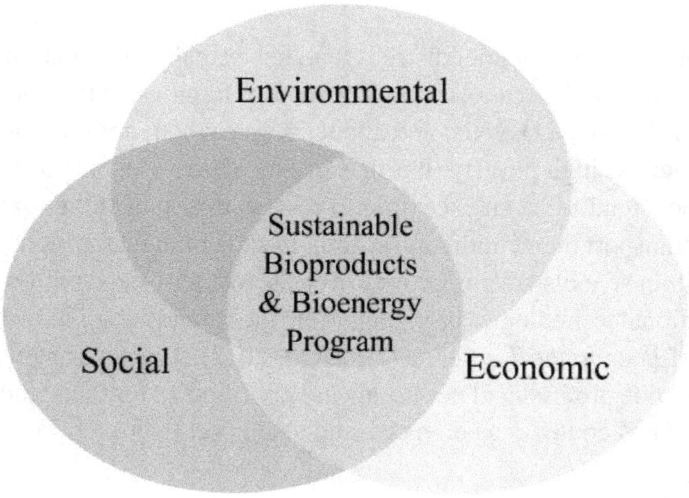

The program was an interdisciplinary collaboration between a College of Education and a College of Natural Resources at a university with very high research activity. The design tapped into: 1) the expertise of researchers in bioproducts and bioenergy and in science education at a major land grant institution, NC State University, 2) the experience of industry stakeholders understanding the needs of the workforce, 3) high school teachers in rural areas of the state who had a strong interest in the Bio-economy and were committed to the success of their students and communities, and 4) the talent pool of college undergraduates in several regional universities and colleges with a diverse student population.

The program was designed to show diverse high school students and undergraduates the opportunities and rewards that exist in STEM fields, motivate them to attain degrees, and promote their awareness and interest for a range of careers in the bioproducts and bioenergy industry. The primary activities of the program included:

1. Developing engaging distance courses about the bioeconomy for rural high school science teachers and undergraduate students attending regional community colleges, colleges, and universities.
2. Co-developing and testing laboratory-based activities with faculty and graduate students.
3. Preparing the high school teachers during summer professional development workshops so that they can implement the activities in their classrooms.

Figure 3. Program model of participants, activities, and impacts

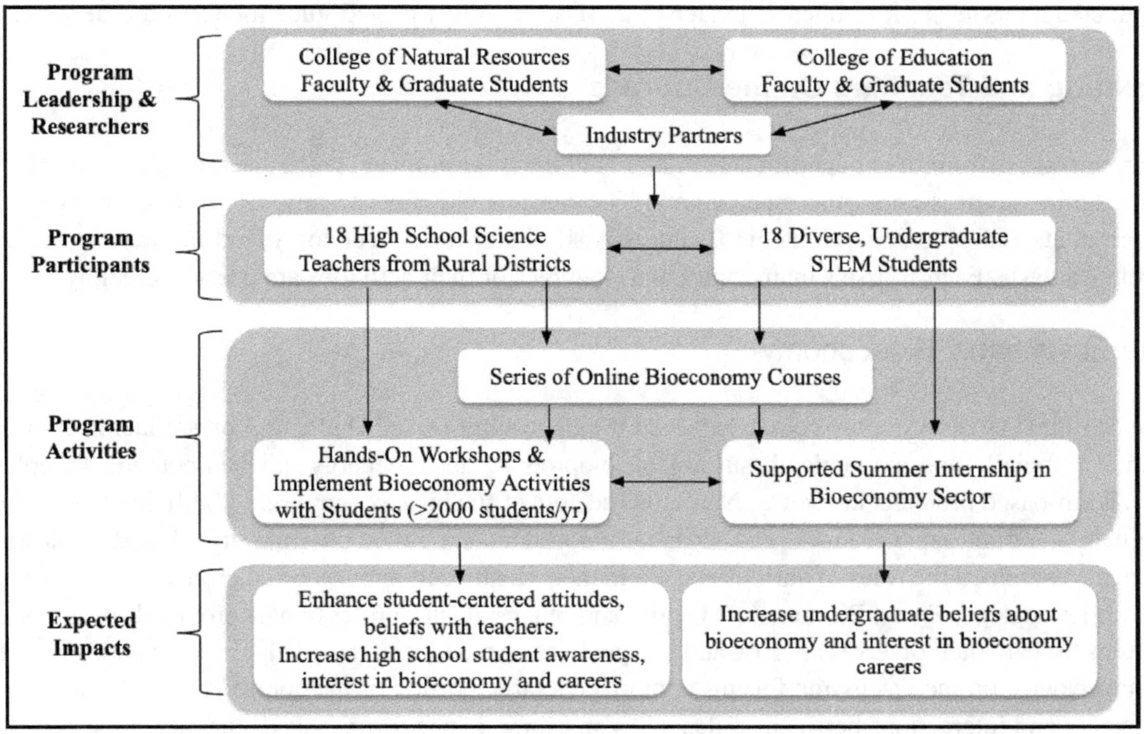

4. Providing all laboratory supplies and equipment, and supporting teachers as they implemented bioproducts and bioenergy activities and shared relevant careers in their classrooms.
5. Arranging for stimulating summer internships for the undergraduate college students at bioproduct and bioenergy companies and organizations.

Developing Industry Partnerships

Industry partnerships were developed through existing departmental contacts and newly recruited organizations. These partnerships helped to inform the program curriculum, highlight various careers and the knowledge and skills required for specific roles, and to provide summer internship placements.

The program team arranged travel to industry sites (ethanol plant and paper mill) and captured video footage to demonstrate operations in North Carolina. Video-recorded interviews with people in various job roles and from different educational backgrounds were incorporated into online courses. At a summer workshop with high school teachers, a panel of young professionals who actively worked in the bioeconomy sector shared their educational and career experiences.

Internship recruitment efforts were conducted via announcements at a bioproducts trade conference, an article in a bioenergy trade journal, social media posts, and through existing departmental contacts in industry, government, and research institutions. Participating undergraduate students were encouraged to identify additional organizations which might serve as internship hosts. Over 100 organizations were contacted by email with program details and an invitation to participate. Forty companies and organizations expressed interest in hosting internships with the program. Ultimately, seventeen perspective hosts conducted interviews with students, and ten of these successfully hosted student interns for the summer.

Creating and Teaching Online Courses

Team members from both departments collaborated to co-develop and teach a series of four asynchronous, online courses about the bioeconomy: The Sustainable Bioeconomy, Biomass Conversion for Bioproducts and Bioenergy, Strategic Business Analysis for the Bioeconomy, and Environmental Life Cycle Analysis. Each course's main topics and examples of their activities are described below.

The Sustainable Bioeconomy

The Sustainable Bioeconomy course explored the opportunities and challenges of producing environmentally, socially, and economically sustainable bioproducts and bioenergy and their potential to replace petroleum-based products in society. Numerous sectors of the bioeconomy were highlighted, including biofuels, wood and paper products, and advanced bioproducts such as biopharmaceuticals and bioplastics. Students examined the roles of various stakeholders in the bioeconomy, such as producers, consumers, and policy-makers. Forest Biomaterials faculty and other industry professionals provided several guest lectures to share their diverse expertise and perspectives on the bioeconomy (Figure 4). In addition to the lecture videos, online discussion forums comprised a large portion of the coursework and encouraged reflection and interactions between students. As an example, one forum asked students to use the U.S. Department of Energy's Bioenergy Career Map (Office of Energy Efficiency & Renewable Energy, n.d.) to explore entry-level to advanced-level careers in different sectors (e.g., engineering and manufacturing, operations and management, and education, communications, and outreach), select a career of interest, describe why they selected it, and reply to two other students. The forums and assignments allowed students' interests to guide their learning and exploration of the bioeconomy. A modified version of the course is now freely available through a MOOC (https://campus.extension.org/enrol/index.php?id=1641).

Example assignments from the Sustainable Bioeconomy course include:

- SDGs (Sustainable Development Goals): Students completed research on SDGs from provided links, focusing on one goal. In an online forum, they discuss how their goal connected to Zero Hunger and created a list of organizations and initiatives in their area that actively address SDGs and beyond (nation, world).
- Amazon Effect Poll: After reviewing the "Amazon Effect" article, students consider market demands and desires for sustainability. Based on this information, they created a 3-5 question poll to gather information about consumer behavior and preferences about packaging, gathered data from at least 3 people, and summarized their findings.

Biomass Conversion for Bioproducts and Bioenergy

The Biomass Conversion course was based on the fundamental science and engineering concepts, processes, and technologies involved in converting biomass resources (e.g., trees, crops, agricultural residues, and municipal solid waste) to usable chemicals, energy, and other bioproducts and biomaterials. Through the lecture videos, students learned about topics such as biorefineries (facilities that process

Figure 4. Example of a lecture video from the Sustainable Bioeconomy online course.

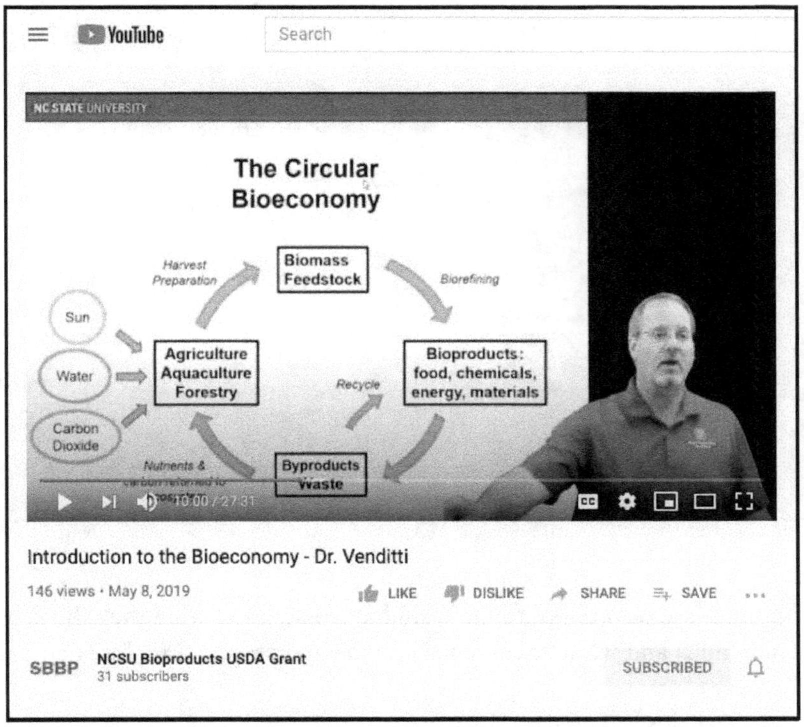

biomass into a range of higher-value products and byproducts) and the chemical composition of biomass, in addition to specific biomass conversion processes, including pretreatment, hydrolysis, fermentation, distillation, and anaerobic digestion. The lecture videos featured pop-up quiz questions and served as formative assessments for students and the course instructors. A majority of the course assignments focused on mathematical, chemical, and engineering concepts and calculations. Examples of calculations are comparing U.S. energy consumption and sources over time (e.g., solar, wind, biomass, and petroleum) and how much material is flowing through each unit operation in a conversion process (see Figure 5). A modified version of the course is now freely available through a MOOC (https://campus.extension.org/enrol/index.php?id=1642).

Example assignments from the Biomass Conversion course include:

- Circular Bioeconomy = No Waste: Students read the Ethanol Producer Announcement about Gevo's work on isoprene. From this article and class discussions, students chose one co-product from the biomass conversion to discuss where it was made, the conversion process used to turn this waste into a valuable product, how the co-product was marketed, its uses, and who would buy it.
- Engineering Challenges: xPrize – Transforming CO_2: In the xPrize competition, teams around the world compete for 20 million dollars by engineering the best process and product for turning CO_2 into a valuable product(s). After exploring the website, students selected which finalists they thought were more promising and why, then elaborated on how this project connected to biore-

Figure 5. Example of a unit operations problem for the Biomass Conversion online course.

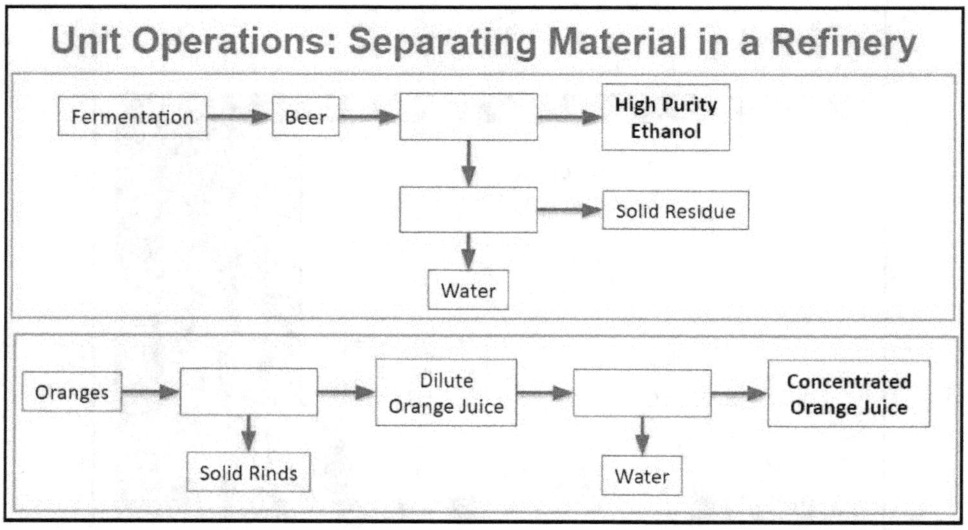

fineries and/or biomass conversion. Finally, students discussed what they thought of incentivizing engineering challenges and what advantages and disadvantages exist for such competitions.

Strategic Business Analysis for the Bioeconomy

The Strategic Business Analysis course guided students to examine the major drivers behind the growth of the bioeconomy through a series of lecture videos, quizzes, and assignments. Course topics included a survey of macro and micro economics, financial concepts and statements, cost structures, marketing and pricing theory, value chain, and economic challenges for commercialization of biobased products. Students practiced basic conversion economics and financial analysis. A modified version of the course is now freely available through a MOOC (https://campus.extension.org/enrol/index.php?id=1777).

Example Assignment:

- Students built upon their prior knowledge from Course 1 and Course 2 along with new financial content to develop a brief (2 page) research plan proposal for the following items:
    - How did consumers perceive sustainability?
    - What environmental impacts were associated with different fibers and technologies?
    - What fiber blend could provide a better performance, achieve the minimum sustainability acceptance without substantial increases in manufacturing cost?
    - What is the supply chain of the alternative fibers?
    - Provide excel with calculations.

Environmental Life Cycle Analysis

In the Environmental Life Cycle Analysis (LCA) course, students explored various methods used to systematically determine the sustainability of products and processes. One case study compared the raw material and energy inputs and outputs of organic milk production compared to conventional milk production. Students surveyed aspects of conducting and interpreting an environmental life cycle analysis through video lectures, software practice sessions, and projects, all emphasizing systems thinking. In-depth case studies utilizing OpenLCA software, allowed students to construct a life cycle inventory, a life cycle assessment, and interpret their results. A modified version of the course is now freely available through a MOOC (https://campus.extension.org/enrol/index.php?id=1778).

Example assignment

- Assignments were differentiated for the graduate and undergraduate sections of the course. Participating high school teachers and science and engineering graduate students conducted LCA for a product/technology/serve related to their area of interest. They first conducted a research literature search and review. Students delivered 1-2 page project proposals comprised of:
  - Introduction and Background discussing the topic and why a LCA for the product/technology/service was needed.
  - The plan for the LCA study defining the goal and scope of the LCA, determining the functional unit and system boundary, expected sources of LCI data and LCIA categories
  - Expected Results
- For their final projects, students created video presentations about LCAs they conducted. Each video included:
  - Four phases of LCA and all mandatory components
  - A process flow diagram showing the system boundary of their study
  - LCA results with necessary hotspot analysis
  - Conclusions and insights from the LCA

Course Staffing

The Program Manager and an Instructional Designer in the Department of Forest Biomaterials assisted with the course development and implementation process. The four courses were taught by four Forest Biomaterials faculty members who were experts in that particular course content. Although they were the primary course instructors for their one course, a total of 17 faculty in the College of Natural Resources and College of Agriculture recorded guest lectures to contribute to the online courses. A Forest Biomaterials or Science Education graduate student served as a teaching assistant for each of the online courses. The teaching assistants were responsible for grading coursework, co-developing assignments, communicating with students, and facilitating online discussion forums on Moodle. The Forest Biomaterials faculty primarily recorded lecture videos (most were 15-30 minutes) and designed the assignments and assessments.

The courses were dual-listed at undergraduate and graduate levels. Homework assignments were differentiated for the two groups. This dual-listing allowed the high school teachers and undergraduate student participants in the grant program, as well as other students at the university, to develop a learning community with diverse perspectives and backgrounds. All of the program participants' course credits were paid for by the grant. Teachers who were interested in pursuing graduate programs in Science Education or Forest Biomaterials could use the courses toward their degree programs. Undergraduates could transfer the course credits back to their home institution. The online courses were also converted to Massive Open Online Courses (MOOCs, see Figure 6) to provide course content at no cost to anyone interested in learning more about the bioeconomy (https://research.cnr.ncsu.edu/sites/sustainablebio-products/online-courses/).

Figure 6. Example of the Sustainable Bioeconomy MOOC content and structure

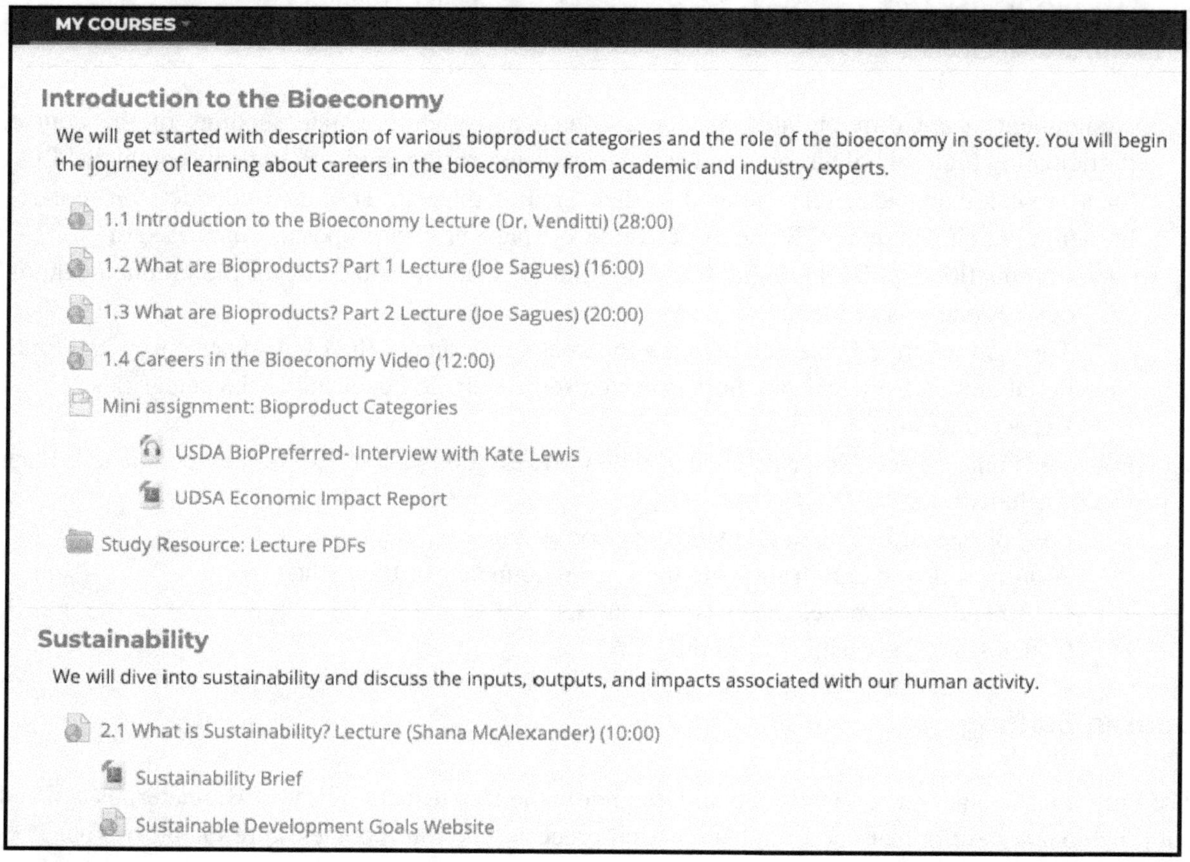

Co-Developing and Piloting the Laboratory and Classroom Activities

The interdisciplinary team of faculty and graduate students initiated the development and testing of bioeconomy-related laboratory and classroom activities that were refined and carried out during the teacher professional development (PD) summer workshops. The activities were implemented by science teachers with their high school students. The activity development process was enhanced by the interactions of the interdisciplinary team members from both departments, drawing on technical and scientific expertise from Forest Biomaterials faculty and graduate students and pedagogical and classroom experience of the Science Education members. Led by the graduate student team, fourteen hands-on laboratory and classroom activities were co-developed and tested for alignment with Next Generation

Science Standards (NGSS) and NC State Essential standards. Programs such as this one, that partners teachers and scientists together, are considered a productive way to help teachers develop the skills and knowledge needed to design and carry out lessons that provide students opportunities to participate in the practices of science (e.g., Blanchard & Sampson, 2018; Enderle et al., 2014).

During the first summer, the program hosted orientation days when teachers, undergraduate students, and program leaders met for the first time and connected throughout the day over their shared experiences and interest in the bioeconomy. For instance, in the Year 1 summer orientation, the high school teachers (and the undergraduates) completed an ink flotation paper lab (Figure 7) and an enzyme optimization lab and were introduced to the program goals, expectations, and research participation.

Figure 7. During the Year 1 Summer Orientation, high school teachers and undergraduate students worked with graduate students and faculty in an ink flotation paper lab.

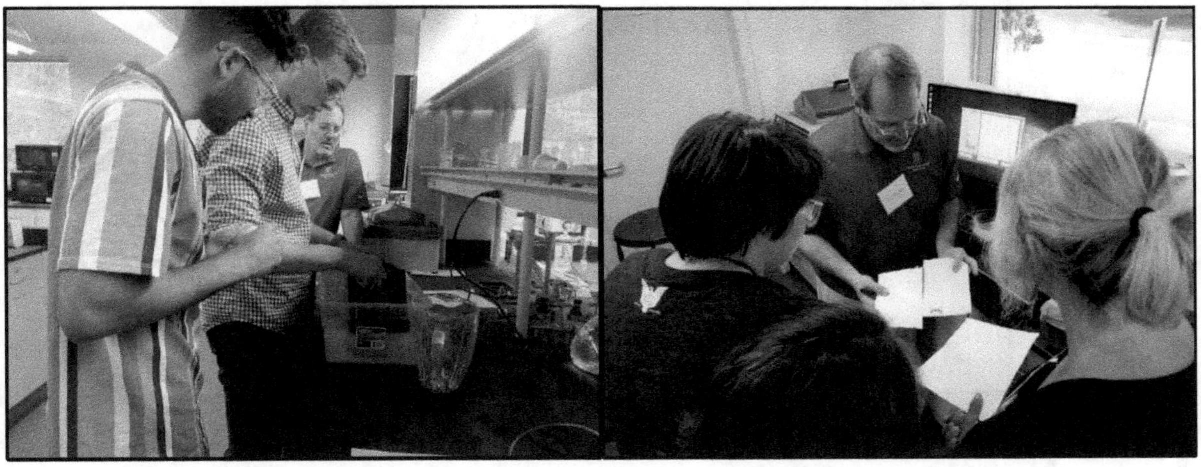

Hands-on laboratory activities developed for the Year 2 workshop included: 1) biorefinery modeling with pineapple biomass to ethanol and paper, 2) modeling of greenhouse gases and global warming potential, 3) production and characterization of biopolymers, 4) yeast bioreactor design and fermentation optimization, 5) photosynthesis lesson with algae, 6) paper recycling, 7) enzymes for biofuels, and 8) life cycle analysis comparison of three different types of shoes. For Year 3, six more activities were developed, including: 1) anaerobically digesting organic waste to generate biogas, 2) assessing biomass availability with big data analytics, 3) mixed waste recycling, 4) renewable packing peanuts, 5) soap-making factory, and 6) wastewater treatment simulation. Graduate students led the development effort of these labs, and mentored undergraduate students in the Forest Biomaterials department who assisted in developing and/or pilot testing the lab activities.

The labs were constructed to follow the 5E Instruction Model: Engagement, Exploration, Explanation, Elaboration, and Evaluation (BSCS Science Learning, n.d.), and included a teacher lesson guide and student version. For each activity, short videos were created and posted on YouTube and the program website as seen in Figure 8. Access lessons at https://research.cnr.ncsu.edu/sites/sustainablebioproducts/.

These videos were to be utilized by teachers to introduce their high school students to the concepts of the activities from the perspective of the science curriculum and the bioeconomy. The videos provided

Figure 8. Sample of Laboratory and Classroom Activities available on the program website.

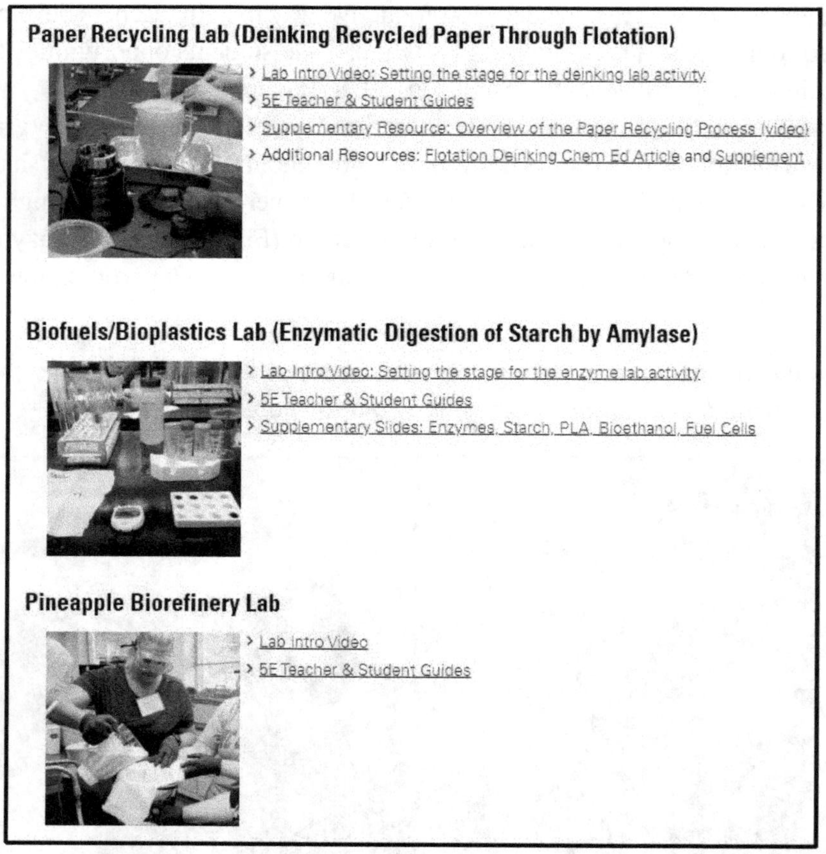

career connections for the students and also featured real-world applications of the activity content. These included footage of industries located in eastern North Carolina that are involved in the bioeconomy and their operations connected to the objectives of the activity, such as recycling. Some of the labs were modified from existing published methods or those previously used in the department, and others were newly created (Table 1). The graduate students were responsible for testing and piloting the activities in the lab. The Forest Biomaterials and Science Education team members worked together to ensure the activities emphasized connections to the bioeconomy, bioproducts, and/or bioenergy, potential careers, and the national and state science standards.

In the next section of the chapter, we describe the activities of the Graduate Students, Teacher Participants, and Undergraduate Students. In each of the sections, we include the strategies, research focus & theoretical frameworks, lessons learned, and recommendations.

GRADUATE STUDENT INVOLVEMENT IN THE INTERDISCIPLINARY TEAM

The interdisciplinary program team was formed by a Forest Biomaterials faculty member and Science Education faculty member who wrote and were awarded the USDA grant that funded the four-year program. A Program Manager, who was a full-time staff member in Forest Biomaterials and a part-time

Table 1. Details for select Laboratory and Classroom Activities

Name of Activity	Technologies & Tools	Description	Real World Applications
Assessing Biomass Availability with Big Data Analytics (2 x 45-minute class period)	Microsoft Excel or Google Sheets	Using Microsoft Excel or Google Sheets, students will create Pivot Tables to assess biomass resource availability in the United States. Students will learn to quickly and efficiently retrieve data from a large, master data set with nearly 80,000 data points.	Data Analysis
Bio-based Polymer Films (2 x 45-minute class period)	Vernier Go Direct® Force and Acceleration probeware, Opacity scale, General Laboratory Equipment	Students will create three bio-based polymer films and study the mechanical, optical and barrier properties. Students will compare the features of bio-based polymer films with petroleum-based polymer films to establish performance differences and the challenges of transitioning to bioplastics from petroleum-based plastics.	Bioplastics
Designing the Greener Shoe: A Lesson in Sustainability and Life Cycle Analysis (1 x 45-minute class period)	Microsoft Excel or Google Sheets	Students will make observations of the materials that make up their shoes. They will consider inputs and outputs of those materials and map out where and how those materials are made. The students will review current media and literature sources about the environmental impacts of various materials. Next, students will explore the making of mathematical models to apply quantitative measures to compare the sustainability of different shoes/materials using Excel or Google Sheets.	Bioproducts for Consumer Goods
Renewable Packing Peanuts (2 x 45-minute class period)	Corn Starch, Baking Powder, Glycerol, Iodine, Paper Cups, Microwave, General Laboratory Equipment	Students will produce a more environmentally friendly option using starch to produce starch-based packaging peanuts and compare its performance to polystyrene-based packaging peanuts.	Environmentally Friendly Packaging Material

Science Education doctoral student, was hired to manage all aspects of the grant and program activities. The interdisciplinary team also funded three Science Education graduate students and nine Forest Biomaterials graduate students, who were assigned to work a portion of their time with the program. Several undergraduate students, both in Forest Biomaterials and in Science Education, were hired to assist with program activities as needed during the academic year, as well as summer.

Activities: Graduate Students

The graduate students played a critical role in the program development and implementation. There were full-time graduate students from Forest Biomaterials and from Science Education, and their involvement in the program differed. The Forest Biomaterials graduate students contributed to the grant and program activities in numerous ways, including serving as teaching assistants (TAs) for the online courses, developing high school laboratory activities, and facilitating lab sessions during the teacher PD (Figure 9).

However, their primary responsibility as graduate students was conducting bioproduct and bioenergy-related laboratory research for their degree programs, and their laboratory-based research was not directly tied to the program's educational goals. Additionally, some of the graduate students were funded by and

Figure 9. Graduate students served as teaching assistants for the online courses (left), and developed and piloted laboratory activities for the high school teachers and students (right).

worked for the grant program over multiple years, while others flowed on and off of the project for a semester or two at a time, serving as TAs and helping with lab activity development as needed.

The responsibilities of the Science Education graduate students included working with the Forest Biomaterials graduate students in the development of the courses and the laboratory activities. In addition, and in contrast to the Forest Biomaterials students, the Science Education graduate students' research and grant program commitments were closely aligned with the program's educational activities. The Science Education graduate students were typically funded on the grant program for multiple consecutive years. This allowed the research and dissertation work of the Science Education graduate students to be directly related to the grant program, and they were responsible for designing and conducting research on the grant program participants as well as studying interactions of the interdisciplinary team, itself. The Science Education graduate students also worked closely with the Program Manager to coordinate various aspects of the program, such as planning for the summer teacher professional development workshops, particularly making the laboratories consistent with state course objectives and effective pedagogical practices. The Science Education graduate students also maintained communication with the program participants and dealt with a wide range of logistical aspects, from ordering equipment to going out to schools to video record program lessons.

Strategies: Graduate Students

Develop Clear Agendas and Objectives for Team Meetings

The interdisciplinary team held frequent in-person team meetings, lasting 60 to 90 minutes each, throughout the process of co-developing the high school laboratory activities and online courses. Typically, the Program Manager or a Science Education graduate student created an agenda to guide each meeting. Feedback from the graduate student members of the team indicated they felt that having clear agendas and objectives promoted more productive meetings. These meetings consisted of discussions about the development of the high school lab activities. The graduate students took turns sharing about their progress, and other graduate students and faculty provided feedback, offered suggestions, and asked questions to collectively enhance the lab activities. To close out each meeting, the team identified "next steps" and assigned each member tasks to complete by the next meeting.

Incorporate One-on-One Sessions with Graduate Students

In addition to the large-group team meetings, the Program Manager and the Science Education graduate students also offered one-on-one support to the Forest Biomaterials graduate students during the lab activity co-development process. These meetings were scheduled early on when beginning to conceptualize the lab activities, during the lab activity development and pilot process, and also toward the end of the lab activity development. At this point, the graduate students created formal lesson plans and prepared to use best teaching practices to facilitate the lab activities during the summer workshops. This was an opportunity to provide specific, targeted feedback and enhance the collaborative interactions between team members.

Hire a Program Manager who has a Background in Science and in Education

The Program Manager played a critical, supportive role for the team, working closely with team members from both the Forest Biomaterials and STEM Education departments. The Program Manager was employed by the Department of Forest Biomaterials and was also a graduate student in Science Education. Importantly, the Program Manager had experience in both fields and was well-connected with both departments, which helped to bridge the two departments' efforts on the grant and merge the science, engineering, and pedagogical expertise from both departments.

Provide Opportunities to Connect with High School Teachers and Students

The interdisciplinary program team collaborated to coordinate STEM and Bioproduct/Bioenergy events that specifically focused on career exposure for the high school students. The team helped organize field trips to local industry sites and sent graduate student team members to Curriculum Nights at the high schools where they spoke with students and parents about college programs and careers related to the bioeconomy (Figure 10).

Figure 10. The Program Manager and a graduate student visited a high school Curriculum Night to discuss career opportunities and college majors.

These events educated and gave students exposure to possibilities for future careers they personally could seek out. Following the Year 2 summer workshop, graduate students also were available to visit schools and assist with lab activities. Many of the teachers did not utilize the support that was offered, but they acknowledged struggling to carry out labs, and many failed to complete the quantity of labs they originally planned.

Research Focus and Theoretical Framework

The graduate students participated in a research study to enable us to better understand their experiences and interactions within the project. A Science Education graduate student and interdisciplinary team member conducted interviews with the Forest Biomaterials graduate students and the Program Manager during the second and third years of the grant program. Interviews were transcribed and qualitatively analyzed using constructs from the Community of Practice framework (Wenger, 2010). Detailed research findings are communicated in a research paper (McCance et al., 2020). The interviews served as data for the science education research study. They also provided findings back to the participating graduate students, and formative feedback to the program team on how to modify and improve grant program activities.

Lessons Learned: Graduate Students

The 'Team Meeting' Model Enhanced Accountability and Team Relationships

The iterative process of conducting frequent whole-team meetings, assigning work to complete in between meetings, and providing substantive feedback during meetings was a model that seemed to enhance accountability for the graduate student team members and allowed the members to more effectively utilize their time together. Additionally, during the earlier team meetings, some of the graduate students seemed less confident in sharing their ideas. However, as the team members spent more time together and the graduate students continued to make progress on their lab activities, they developed confidence in contributing to team discussions and felt more agency in their work.

Diversity of Expertise with Shared Goals Strengthened the Program

The amount of time, nature of the tasks, and level of involvement and commitment on the grant varied among the program team members. During interviews, some of the graduate students expressed that they felt like the two departments had different levels of commitment with the grant program; it seemed as though the Science Education team members felt more invested in the program than did the Forest Biomaterials members. One factor that could have contributed to this was that the graduate students were assigned to work on specific tasks as needed; the grant program was multifaceted, as mentioned earlier, and required the graduate students to participate in various ways. Because of this, some of the graduate students had multiple roles, and they were sometimes asked to take on new responsibilities on short notice such as working as a course teaching assistant, serving as an undergraduate resume coach, organizing classroom kits, or attending classroom visits. Understandably, this was stressful for the graduate students. Despite the Forest Biomaterials and Science Education team members having somewhat different levels of commitment and motivations for working on the grant as well as bringing different perspectives to their work, all team members clearly cared about producing a high-quality program, and this diversity of expertise was necessary for carrying out the program activities.

The Grant Program was a Challenging Yet Valuable Experience for the Graduate Students.

The TA experience and the development of laboratory activities were influential experiences in giving the Forest Biomaterials graduate students exposure to reform-based teaching practices, improving their teaching and communication skills, and enhancing their understanding of their own field. However, learning to balance their own science laboratory research and grant program expectations was challenging for several of the students in Forest Biomaterials. This was particularly the case when they first joined the grant team and were learning about - and expected to practice - reform-based pedagogies, often for the first time.

For instance, when developing the lab activities based on the 5E learning model, the Forest Biomaterials graduate students struggled the most with designing the 'engage' portion that was intended to grab high school students' attention and increase their interest and excitement about the lab activities and learning about the bioeconomy. During multiple team meetings, graduate students presented for three to five minutes to practice how they would pique students' interest and engage them in the lab activities. After each graduate student's presentation, the team members provided constructive feedback for improvement, and students said they liked and appreciated the feedback. Their learning process about these pedagogical practices was gradual, but clearly was enhanced by the interdisciplinary team's collaboration. When the graduate students presented the lab activities and their 'engage' presentations during the online PD workshop in the third summer, almost all of the teacher participants felt that the graduate students' presentations were effective and that the laboratory activities were better than in previous years. (This was in spite of the need for the professional development to be held online due to COVID-19.)

The Science Education Research Components were not Clear to all Team Members

During the interviews with the Forest Biomaterials graduate students, it was apparent that several of them felt they did not fully understand the nature of the Science Education team members' research and that it was not sufficiently explained to the team. The Forest Biomaterials graduate students were not directly involved in conducting the educational research activities for the grant program (e.g., designing the research studies, collecting and analyzing data, and writing up results), but they mentioned they would have been interested in receiving occasional updates about the Science Education team members' research projects throughout the duration of the grant.

Interviews with Program Team Members Provided Opportunities for Reflection and Feedback

The interviews were a productive way for team members to reflect on the ways in which they felt they personally benefited from working on the grant program. The interviews were also informative for the project team to gain an in-depth understanding of what was working well and what needed improvement from the perspective of the graduate students.

Recommendations: Graduate Students

1. Increase the graduate students' engagement with the high school teachers and their students through formal, expected class visits and video calls with the students where they could share about science and engineering majors and careers and assist in conducting the laboratory activities.
2. Increase the transparency of the Science Education team members' work with the graduate students by mapping out the progress of the different research projects, explaining how and why data was being collected, and sharing preliminary results.
3. Improve communication of roles and expectations with the graduate students well in advance of duty transitions to reduce stress and enhance results.
4. Build graduate students' agency, investment, and motivation by matching students' interests to tasks and asking for their ideas.

HIGH SCHOOL TEACHER BIOECONOMY PROFESSIONAL DEVELOPMENT

Recruitment and Selection

Recruitment for high school teachers was done by emailing all of the science and science-focused CTE teachers in the high schools of participating districts, with an attached program flyer, which included program requirements, benefits, a link to the application, and deadlines. Additionally, over 200 mailers were sent to eastern North Carolina high school principals and administrative assistants leading to more than 80 high school teacher applicants, predominantly from schools designated as Title I. Nineteen high school teachers from ten counties in Eastern North Carolina were selected, based on their expressed interest in the bioeconomy, demographic information, and their teaching experience. Selection also favored 'paired' teachers who applied from the same school to enhance teacher support and student impacts. Six of the 20 teacher participants were paired teachers from three high schools. Five of the original 19 teachers dropped out early in the program, due to feeling too overwhelmed to participate fully. Subsequently, two additional teachers from schools with similar demographics joined the program.

There were two main parts of the program for the teachers: two online graduate-level courses and workshops over three summers, with the option for teachers to enroll in one additional online course. Additionally, teachers selected from various additional bioproducts and bioenergy career experiences for their students; field trips to an industry or the university, a career night, or graduate students visiting class for an activity and providing role models. The team also regularly emailed teachers about STEM activities and events available for participation, such as a STEM day at a science museum in the region.

Activities: High School Teachers

Onlines Courses

During School Year 2, the high school teachers enrolled in Fall and Spring online courses (with the undergraduate students), the Sustainable Bioeconomy and Biomass Conversion. In the Fall of Year 3, teachers had the option to take a third course. Throughout several of the courses, teachers' assignments included creating lesson plans - some individually and others in small groups - to incorporate content from the online courses into immediately transferable high school lessons and activities. Teachers were encouraged to share bioeconomy-related assignments they created throughout the program, using a shared Google Drive folder. Teachers also were able to take assignments from the graduate courses and utilize them within their classroom, modifying them to be simpler or more appropriate for their high school students. The classes also had a strong career exploration element for the teachers. This piece was extremely useful for it provided resources, via websites and a plethora of YouTube videos, to share with students about careers in the bioeconomy.

In addition to career connections, the courses provided excellent resources to be incorporated into science classrooms, such as videos for enzyme modeling with computers, green fluorescent dye as an indicator in cells, and tree harvesting. Through the workshops and the online courses, teachers developed a learning community to share ideas, resources, and support throughout the program. The benefit of this support is further exemplified by the positive comments from teacher pairs who worked together at the same high schools. Two of the pairs were separated after Year 1 due to changes in assignments, and all teachers involved in the separation acknowledged the greater difficulty in completing tasks, such as labs or courses, without their partner teacher. The development of comradery between teachers within this program through summer workshops, online courses, and teacher pairs supported teachers in the successful completion of the program. A high school teacher commented about the influence of the courses: *"I have changed my curriculum to include more focus on bioproducts, bioproduct production, and bioindustry careers."*

Summer Professional Development Workshops

For the Year 1 Orientation & Year 2 Summer Workshops, the high school teachers practiced and offered refinements for the laboratory activities (Figure 11). One high school teacher wrote: *"This workshop was one of the best professional development courses that I have attended in a long time. The labs were well written and the debriefing sessions were helpful for classroom implementation."*

Figure 11. High school science teachers working with university faculty and graduate students to model greenhouse gases and global warming potential at the Year 2 Summer Workshop.

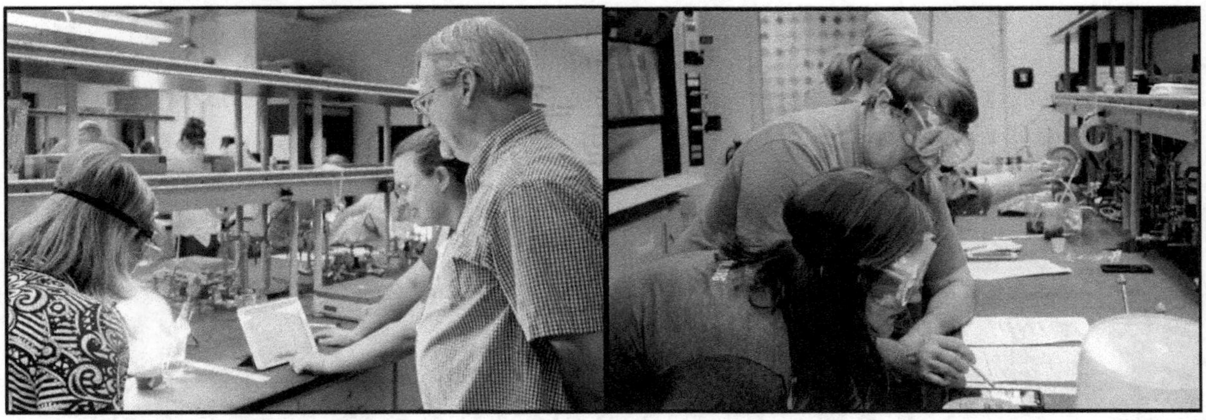

Teachers learned about the bioeconomy through industry panels and presentations. In Year 2, a panel of five young professionals actively working in the bioeconomy - including a Senior Research Associate, a Process Engineer, and a Bioanalytics Technician - shared their experiences with teachers (Figure 12). They discussed their educational backgrounds, job responsibilities, and career expectations. In addition to the industry panel, three professors within the College of Natural Resources conducted individual presentations throughout the four-day workshop about their role in the bioeconomy and what they were researching. Both of these facets, the industry panel and professors' presentations, provided research updates and information about career opportunities that teachers could share with their students. During the Year 3 summer virtual workshop, a professional in the industrial hemp sector also conducted a presentation about hemp bioproducts and the role of hemp in the state's economy.

Figure 12. Young professionals actively working in the bioeconomy shared their experiences with teachers.

Also during the Year 2 summer workshop, teachers ventured on a field trip to a forestry research area near campus that studies breeding in trees and spoke to the individuals who oversee the research (Figure 13). Teachers also visited a facility on campus that made paper from various fiber sources. On the day they toured, the facility was making paper out of banana leaves. Similar to the presentations by the industry panel and professors, the field trips provided memorable experiences for teachers to share with students about ongoing activities in the bioeconomy and to encourage them to consider career possibilities in this sector.

Figure 13. Teachers learn about new technologies in silviculture and tree cultivation at a research forest near the university's campus (left) and about paper production at a facility on campus (right) during the Year 2 summer workshop.

Due to constraints of social distancing, the Year 3 workshop was hosted online via a virtual platform over two days (Figure 14). Despite the physical barriers, graduate students were able to share background information regarding their new lab activities through slide presentations and share videos and pictures of the labs, still in development. Teachers displayed high interest in the new activities and had many conversations about possibly converting several of the activities to remote learning lab activities during the period of social distancing. Also during the workshop, each teacher shared which labs and information they had incorporated into their classrooms, along with successes and challenges that arose with each. These presentations allowed teachers to reflect on activities they conducted in their classrooms and provided motivation for other teachers to incorporate more activities in the upcoming school year.

Figure 14. The Year 3 workshop was held online during COVID-19 required social distancing. Teachers shared about their experiences implementing bioproduct lessons with their students.

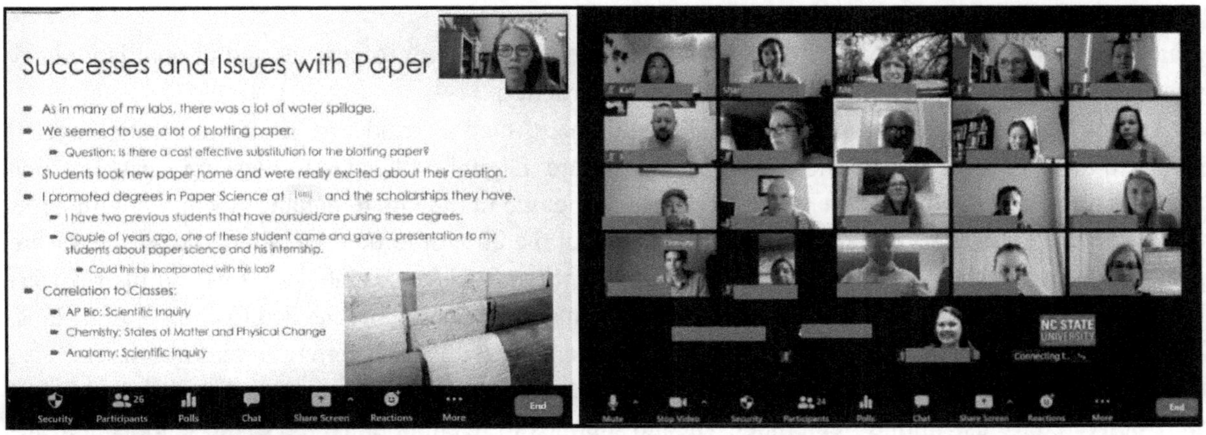

During the summer workshops, strategies and activities were modeled by the research group during pedagogy sessions that encouraged hands-on, inquiry based learning by the teachers, which tends to foster more reformed based teaching methods (Blanchard et al., 2010; Blanchard et al., 2009). Sessions included information on conducting class discussion on science topics, such as a Marine Plastics Case Study, conducting better labs, facilitating career exploration, and utilizing technology and data in a science classroom.

Teachers also completed a variety of surveys for the program. Teachers completed a post-lesson reflection describing how it went and what could be done differently. This information provided feedback for the research group about possible modifications that may be needed for the activities. The other portion of the surveys related to teacher practices with a Teacher Beliefs Interview (Luft & Roehrig, 2007). The surveys further encouraged these reform-based practices by providing insightful reflection on teachers' beliefs and practices. Teachers completed these surveys at several points throughout the first year. The workshops and surveys connected content knowledge with pedagogical methods to facilitate inquiry-based teaching methods.

Following the workshops, teachers selected science equipment and supplies to use for all of the bioproduct and bioenergy lab activities they anticipated implementing in their classroom, funded by the

Figure 15. Examples of supplies and lab equipment used during the summer PD and available for inclusion in teachers' supply kits.

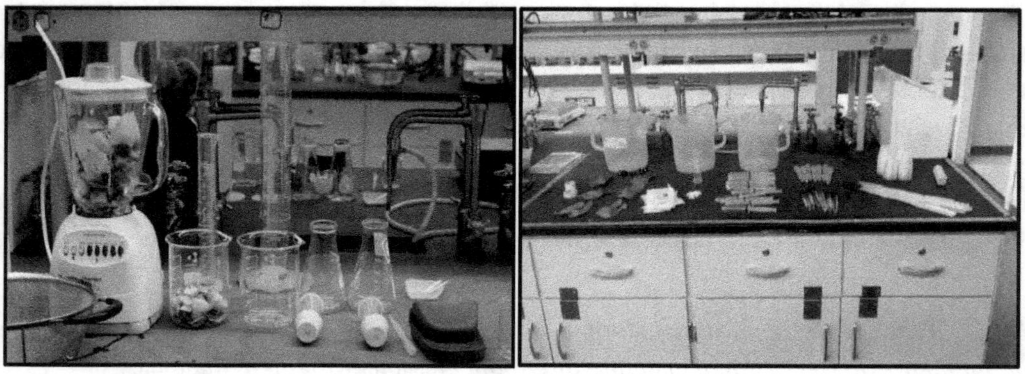

grant program (Figure 15). Many teachers had sufficient grant funds to purchase additional laboratory materials or equipment to benefit their classroom instruction.

The supplies ranged from activity-specific equipment, such as handsheet molds to create recycled paper, to more general lab supplies, including glassware. Digital probeware to measure variables such as pH, temperature, and force were also available to include in the kits to enhance technology integration and collect and analyze data. During the last day of the 2019 summer workshop, the teachers were provided with a spreadsheet detailing all the equipment required for the lab activities and their prices. From this spreadsheet, teachers were able to select items they specifically needed to complete the labs. This allowed teachers to avoid duplicating lab equipment they already had in their classrooms and also tailor their equipment request for the lab activities the teachers were most interested in. These materials were delivered for use in mid-September. The lab supplies were either delivered to the schools or available for pick-up if teachers were close to Raleigh.

Through participation in the workshops, teachers also were able to earn continuing education credits (CEU's) for participation in the workshops at a rate of 1 hour of contact time for 0.1 CEU. The teacher professional development workshops were designed as collaborative sessions for teachers, faculty, graduate students, and industry partners to share experiences and gain experience with engaging bioeconomy-focused lessons for high school science classes.

High School Supports

Lesson Implementation and School Support

Assessment

For Year 2, eight activities included pre- and post-quizzes for the high school students, with questions created by the graduate students. The graduate students worked with each teacher to create unique quiz question sets pertaining to the laboratory activities each teacher planned to carry out during the school year. The quizzes could be completed on an array of electronic devices (e.g., computers, tablets, or smartphones). High school students took one inclusive pre-lab quiz at the beginning of the semester that included questions from all the activities their teacher planned to carry out. After each lab activity was complete, students took an immediate post-quiz related to that specific lesson. The university team converted class quiz results (collected in Qualtrics) into graphs and data tables the teachers could access via a google drive. This provided insightful feedback for the teachers relating to what the students already knew (inclusive pre-lab quiz), learned from the activity (specific post-lab quiz), and areas needing additional reinforcement. Some teachers used the results to prompt a discussion with their students regarding their lab experience and the bioeconomy. Results also provided formative feedback for the research team to consider quiz item revisions.

School Year PD Support

As additional support, there were two video calls between one of the Science Education graduate students and the teachers. The teachers shared strategies they had for completing the lab(s), issues they may have run into, such as the pineapple fibers binding up the blenders or long drying times for biofilms, and to ask questions. Teachers reported good rapport with the Science Education graduate students working on the program and felt able to contact them with any concerns about the activities via email or by phone throughout the program. On several occasions, graduate students visited high schools to assist teachers in carrying out the activities. Not only was this beneficial for the teachers in completing a new lab activity, but the visits also provided college role models for high school students. During these visits, the graduate students gave an informal presentation about attending college, their graduate program of study, and careers in the bioeconomy.

Graduate students led conference presentations to disseminate the grant work and members' experiences. On several occasions, high school teachers co-presented with Science Education graduate students on a range of activities from laboratories and online course assignments (Figure 16). The conference allowed this activity to be shared with other professionals in the field of K-16 STEM education.

Figure 16. Teacher participants share the Energy Diagram activity with their peers at a local K-12 education conference (McAlexander et al., 2019).

Career Nights, University Visits and Field Trips

The interdisciplinary program team and graduate students coordinated STEM events to introduce the high school students to the bioeconomy, potential careers, and role models. Tabling events were organized for school curriculum nights. To highlight one outreach event held at the university, fifteen biology and chemistry high school students visited campus to learn about natural resources majors and careers. Their teacher wrote: *"Our students learned a lot of science as well as realizing that there are other fields of science that are equally satisfying career-wise and financially."*

Funds were made available to school for organizing Bioproducts/Bioenergy in STEM events. The Program manager helped coordinate several class field trips to industry facilities. Unfortunately, a number of the events were cancelled due to the COVID-19 pandemic and resulting shut-down.

Strategies: High School Teachers

Recruitment focused on teachers interested in the Bioeconomy, pair teachers were recruited from the same schools, and schools serving underrepresented students had priority.

On the application, teachers wrote an essay describing their interest in taking part in the program. An explicit part of the recruitment process was to make sure there was an expressed interest in sustainability, and a clear desire to translate their learning into their classrooms. Their school site was checked to see if it was serving underrepresented students (first generation, minority, low socio-economic status, rural). From prior research, we knew that having multiple teachers in the same school could hold greater benefits for student learning and teacher support (Blanchard et al., 2016), so we indicated in the application that it would enhance their chance of being selected if they had a partner teacher (three sets of partner teachers were selected).

Design of online course curricula was linked to the skills and knowledge teachers could use in their high school classes.

The high school teachers were able to take up to three online graduate-level courses during the program. Topics ranged from the parameters of sustainability, life cycle analysis, biomass conversion to business concepts of the bioeconomy. Many of these topics were easily woven into high school classrooms regardless of the subject. The online courses provided a large array of supplemental videos, forum discussion prompts, and homework assignments for teachers to utilize, many of which were directly transferable to the high school classroom. In addition, the information from course lecture videos could be incorporated and connected to concepts in most science high school courses. Some courses, like environmental science, had a higher degree of transferability, but many concepts were transferable to other courses, such as biomass conversion problems in chemistry and utilizing enzymes utilized to digest wood in biology.

Design activities that engage students.

The lab activities were jointly created between the Department of Forest Biomaterials and STEM Education. With the talents and skills of both departments, high quality inquiry-based activities were developed for the students. The lab activities were developed with the 5E Learning Model: Engage, Explore, Explain, Elaborate, and Evaluate. The labs also incorporated concepts from a variety of curricula (e.g., biology, environmental science, and chemistry), so they could be incorporated into various classrooms with ease. Design and engineering were a central focus for many of the inquiry-based labs. In many of the labs, students designed modifications to the initial experimental design for consecutive trials. For students, the ability to create part of the experiment and observe the results produced was highly engaging. Also, the labs correlated to topics of the bioeconomy, of interest to many students. By the labs having real-life connections, such as paper recycling or fermentation using pineapples, students' engagement was high.

Provide laboratory equipment and supplies to teachers to support bioeconomy lab activities and other hands-on lessons in the classroom.

One area of difficulty for rural science teachers is a lack of supplies. Even if teachers want to incorporate labs and utilize inquiry-based learning, many of the classrooms lack basic laboratory supplies and glassware and have a huge deficit in more advanced technology. The grant funded all of the supplies needed, all of which were selected by the teachers and ordered, organized, and delivered (as needed) by the program team.

Connect teachers to regional industry professionals

For all online courses, there were many connections to careers in the bioeconomy. Several of the assignments were career exploration tasks that provided teachers with resources they could easily use with high school students. Likewise, many of the supplemental videos utilized in the courses focused specifically on companies located in eastern, rural North Carolina near the school sites. In the summer workshops, teachers participated in industry panels of young professionals working in regional bioeconomy positions and listened to presentations from professors about their locally-based research in the bioeconomy. Ongoing regional career connections were made with field trips and guest speakers. As one teacher commented on the workshop evaluation, *"The guest speakers and the discussion topics were relatable for me as a teacher and will impact my classroom for years to come."*

Provide implementation support during the school year

Graduate students and program staff were available to teachers for help with lesson planning, obtaining additional materials, and classroom visits. Graduate students also were able to visit classrooms on several occasions to help facilitate the laboratory activities and assist teachers in collecting student data for ongoing education research projects.

Provide access to materials through online resources

The program activities and high school lessons were made publically available with written teacher and student manuals and introduction videos supplied freely on the program's website and YouTube Channel (Figure 17). Modified versions of the courses taken by undergraduates and high school teachers are available at no cost via MOOC.

Research Focus and Theoretical Framework

Figure 17. Video Introductions were created and made public for each lesson to help draw connections between global challenges, bioproducts industries, and science curricula.

Seventeen high school teachers participated in a research study to help us to better understand their motivation to add new Bioeconomy-focused laboratories, information, and career connections to their instruction. Informed by Expectancy-Value Theory (Wigfield & Eccles, 2000), online coursework posts, interviews, summer workshop presentations and classroom implementation of Bioeconomy topics and laboratories were analyzed. Research findings are detailed in a research paper (Blanchard et al., 2020), based on qualitative analyses of value and cost constructs paired with teacher implementation data. Findings suggest that teachers' level of implementation of new bioeconomy-focused lessons is influenced by their focus on its value for their students.

Lesson Learned: High School Teachers

Teachers do better with a support person, including reminders for data collection, pre-quizzes, and other accountability measures.

After the Year 2 summer workshop, many teachers had high motivation and detailed plans of how they were going to incorporate many of the new activities in the upcoming school year. However, for many teachers, those goals were not met and some teachers failed to implement any new activities. Possible solutions to assist teachers in achieving goals of lab implementation include connecting stipends to labs implemented and data collected. Furthermore, increased use of partner teachers could also be used as an accountability factor in addition to providing a resource for lab implementation and problem solving.

Teachers and their students enjoyed authentic and relevant laboratory experiences.

Feedback from teachers indicated their students displayed high engagement and interest in labs that were authentic and relevant to their lives, such as the paper recycling and fermentation of pineapple to ethanol. From Summer 3's workshops, teachers had the highest interest in completing the packing peanuts and soap factory activities in the upcoming year, which can be attributed to the authentic and relevant nature of the activities.

Teachers valued online course materials that they found relevant to their teaching, and reactions were mixed about the quality and nature of assignments.

The feedback from teachers on the online courses presented two different viewpoints. Some teachers found the material intriguing and enjoyed the assignments. Another teacher group found the work more tedious, considered responding to Moodle posts as 'busy work,' and did not appreciate the material that was not immediately implementable into their classrooms (e.g., chemistry problems are difficult to implement into environmental science). Also, some teachers found the online lecture videos 'boring' and as an antiquated teaching method while others expressed a more positive opinion of the lectures.

There was a wide range of criticalness on the part of the teachers about the program.

Much like the two groups of opinions for the courses, the same pattern was repeated regarding teachers' perspective of the program. Some teachers expressed their excitement about being back in school, and several teachers have left the classroom to pursue graduate degrees while others are pursuing graduate degrees while also maintaining a teaching position. Other teachers expressed the program was not 'cutting edge' or different enough.

The asynchronous nature of the courses during the year delayed the development of the learning community of the program participants, for teachers who didn't have other involvement.

For the online courses, the lectures were pre-recorded and videos and assignments were assigned to respective weeks. All tasks had due dates, but timing of completion was up to the teachers' discretion. Though this model was convenient, it seemed to contribute to feelings of isolation for some participants. The teaching cohort displayed a delayed comradery, which is attributed to the asynchronous course design. For other teachers, a greater sense of community developed earlier in the program, and this is attributed to those teachers being more involved in the program (e.g., they utilized graduate student class visits and participated in conference presentations). By the Year 3 summer workshop, a stronger sense of community was evident between the teachers. Therefore, the learning community was capable of developing but possibly delayed due to the asynchronous courses.

More time is needed for teachers to figure out where new lessons can fit into their curriculum.

At the end of the Year 2 summer workshop, some time was dedicated to teachers planning where they would incorporate the labs into their curriculum. However, many teachers failed to carry through with their plans, and during interviews it seemed clear that not enough time had been dedicated to planning where the labs would work best for each of the teachers' intended courses. More time could be spent by the research team and teachers to find places the labs would be best utilized and students could gain the most from their experience.

Recommendations: Teachers

1. Focus more on teacher pairs or triads in participant selection.
2. Utilize accountability pieces for lab implementation and data collection, e.g., connecting it with stipend funds or receiving laboratory supplies for future activities.
3. Facilitate the creation of learning communities with the participant group through biweekly virtual meetings and/or synchronous course design.
4. Prioritize lab activity placement in high school curricula to greater assist teachers in the ease and frequency of classroom implementation.

UNDERGRADUATE STUDENT BIOECONOMY TRAINING PROGRAM

Recruitment and Selection

Participant selection was guided by the grant's goal to provide opportunities for individuals who are historically underrepresented in science and engineering careers (i.e., females, minorities, first generation college students). A unique aspect of the program was the recruitment of undergraduate students from other institutions of higher education - with the goal of diversifying participation. Program staff recruited students at community colleges and primarily sophomores and juniors from Historically Black Colleges and Universities, a women's college, and regional universities. A number of strategies were used: administrative contacts at the partnering institutions assisted with forwarding emails to faculty and student groups with a program flyer, which included program requirements, benefits, a link to the application, and deadlines. Tables were set up in student centers, and some short class and science club presentations were delivered. Over 80 undergraduate students applied for the program. The program team selected participants based on students' expressed interest in the bioeconomy, demographic information, and their academic background. Twenty-one students, representing 8 institutions, were selected to participate in the program.

Activities: Undergraduate Students

Online Courses

The undergraduate students participated in an on-campus orientation event, three university courses, and a 10-12 week summer internship. Students joined the university's Non-degree Studies program where they could enroll in courses without being admitted to a degree program. Grant funds paid for tuition and fees. Using a cohort model, the students enrolled in The Sustainable Bioeconomy course in the Fall and the Biomass Conversion course in the Spring. The Fall of their second year, students chose to take a third course based on their interest area; either Environmental Life Cycle Analysis or Strategic Business Analysis. Each of the courses could be transferred for credit hours at the student's home institution, for up to 9 credit hours. Students who completed the first two courses with a C+ or better were eligible for the summer internship program.

Internships

The internship program gave most of the students their first science/engineering work experience. Each student engaged in a resume review and coaching session with the program staff. Students were asked which of the organizations they were interested in working with and each company was sent a curated packet of student resumes. Companies responded with a short list of students they wanted to meet, and the Program Manager introduced the students to company representatives via email for scheduling online and in-person interviews. Twelve students who successfully completed the required courses were matched to internships.

Each internship was funded with a base stipend of $8000 for 10-12 weeks. Additional funding provided support for students moving within the state and for students traveling outside the state. Each internship included an independent project in which students engaged in research, project development, and/or operations. The Forest Biomaterials Department benefited from the new and further established relationships with internship hosts (Figure 18) as they also may serve as future student and graduate employers and research collaborators.

Figure 18. Undergraduate students were matched with bioproduct and bioenergy companies and research organizations within North Carolina and across the United States.

Internship Hosts

Algaen (NC)
Arbiom (NC)
California Polytechnic State University (CA)
Domtar Plymouth (NC)
Lignetics R&D (CO)
North American Plant Production Office (NC)
North Carolina State University (NC)
Novozymes Biofuels R&D (NC)
RTI International (NC)
Velocys (TX)

Students identified the summer internship as their most influential career development experience to date. A female student shared "I learned more in 10 weeks than I would have in an entire semester...I am so grateful for this amazing experience and program."

Strategies: Undergraduate Students

Gain Entry Into Institutions of Higher Education Through a Liaison

While many regional institutions were contacted to encourage their students to participate in the program, the most successful recruiting efforts occurred at institutions when there were existing relationships between program host faculty or administrators and instructors at the institutions. This facilitated access to classes, encouraged students to read emails and posted flyers, and allowed for campus recruitment events.

Holistic Criteria for Acceptance

Contrary to other special programs in the Forest Biomaterials departments and many science and engineering programs, student applicants GPA's were not the first criteria for selection. In an effort to increase the participation of underrepresented students, a more holistic review process was used, including the student expressing passion for sustainability and career interest related to the bioeconomy.

Industry Career Panels and Video Tours

Industry professionals shared about their career experiences through videos in the course. On-site footage of local and national research and development and production sites highlighted various bioproduct and bioenergy settings and potential jobs. Video interviews and tours at Tyton Biofuels and Domtar Paper demonstrated the scale of biorefineries in North Carolina. An undergraduate, first generation college student shared his experience: *"I have really enjoyed learning about the bioeconomy, bioproducts, and bioenergy* [class]... *I learned a lot about the various jobs available in this industry. From education program specialist, to engineer, to communication."*

Resume, Interview, and Internship Matching Support

Professional development elements were embedded within the curriculum of the coursework. Program staff provided resume coaching sessions and mock interviews in an effort to match all students with an internship in their area of interest. Some students had multiple internship offers and some companies didn't find a compatible match. The Program Manager helped with negotiating matches to satisfy both the needs of the students and the companies.

Research Focus and Theoretical Frameworks

The first study was the development and validation of two quantitative surveys for use with the program participants to understand student perceptions and beliefs about bioproducts and bioenergy (McAlexander, Noble, et al., 2021). Over 200 undergraduate students who were enrolled in natural resources courses completed the surveys for the validation. The items of the Beliefs about Bioproducts/Bioenergy (BABB) highlighted personal and global beliefs based on task values of Expectancy Value Theory (Wigfield & Eccles, 2000). The Career Interest about Bioproducts/Bioenergy (CIBB) measured overall intention to pursue a career in Bioproduct/Bioenergy, with items addressing specific job settings. These validated surveys serve to assess changes in student beliefs and interest over the course of their program experience.

In a second study, reflective writings and interviews that focused on career development experiences were analyzed through the lens of Social Cognitive Career Theory (Lent et al., 1994, 2000). This study focused on six of the undergraduates who were transferring from the community college to a four-year institution. Reflective prompts embedded in online course forums asked students to write about their expectations for the internship prior to starting; then, following the internship, students compared their initial expectations to their experiences. In reflective interviews a few months after the internships, students were asked to review all of their writing reflections and discuss what they learned about themselves. Each student created a career narrative timeline and reflected on barriers and supports to their career development. Student materials were analyzed based on the contextual and intrapersonal supports and barriers they experienced. These led to recommendations for supporting community college transfer students support for internship placements (McAlexander, Blanchard, et al., 2020a).

A third study compared the motivations and perceived outcomes of students and industry partners who participated in the internship program (McAlexander, Blanchard, et al. 2020b). Industry partners were motivated to host program interns for a number of reasons that included: a desire to give back or mentor, building their industry or recruiting employees, and to diversify their workplace. Two internship supervisors attributed their decision to accept a student intern due to the extra technical knowledge that students acquired from the bioeconomy-focused program. Students were motivated to participate in the internship primarily to try out a particular industry or setting, either to reinforce or eliminate a potential career interest. Ten of the twelve students focused on positions where they could gain more laboratory experience.

Lessons Learned: Undergraduate Students

The two instruments that were developed were found to be valid and reliable with a population of undergraduate students.

The new survey instruments, the CIBB and the BABB, may be useful for educators and bioproduct/bioenergy professionals interested in: establishing a baseline of beliefs for a certain group; examining changes due to a course, program, or other intervention; or comparing beliefs between demographic groups (McAlexander, Noble, et al., 2021). The four Likert-type CIBB items measure career interest for bioproducts and bioenergy of a given population. The fifteen Likert-type BABB items measure beliefs and values about bioproducts and bioenergy; eight items reference personal beliefs and seven items reference global beliefs and these items may be used together or separately.

Students honed their career ideas through online coursework and internships.

The students in the program showed consistently positive beliefs about bioproducts and bioenergy. Their interest in bioproducts and bioenergy careers also remained high, although their career goals changed as they became more aware of various STEM jobs.

Extra support was needed for online courses.

Students entered the program with varying science backgrounds, experience with university-level course work, and comfort with asynchronous online courses. Though faculty and teaching assistants provide regular office hours via online conferencing tools, this population of students needed more support than the instructors were accustomed to. Most students balanced full-time course work, part-time jobs, and family responsibilities. The Program Manager served as a liaison between students and instructors, helping students better understand the course expectations, relaying student feedback to instructors, and helping manage requests for extensions or modifications to coursework.

Externally-funded internships incentivized internship hosting.

The companies and organizations were more interested in hosting interns due to the available funding. Because companies were competing for the pool of interns, some offered additional compensation including housing stipends, relocation reimbursement, and extra stipend amounts.

For smaller organizations, many were hosting interns for the first time as it had previously been cost prohibitive. With no precedent for mentoring interns or designing an internship project, students at these organizations had less structured experiences. More communication was required between the Program Manager and internship host to coordinate projects and expectations. For larger organizations, or those with established internship programs, students joined the ranks of other summer interns and had significantly more mentorship support for their projects and their professional development. Some unexpected challenges arose due to the external-funding and had to be resolved with the company's Human Resource departments including liability insurance, time tracking, and taxes.

Student interviews provided more information about supports and barriers for career development.

During reflective writing and interviews, students described their career journeys including events and people that influenced their career direction. Students also described personal and contextual barriers and supports that they faced during their internship experience. Contextual supports included family, program staff, internship host staff and supervisors, stipend funding, program courses, and resume support (McAlexander, Blanchard, et al. 2020a). Although the internships were fully funded, students shared that the financial logistics, housing, travel, transportation, fear of the unknown, and lack of confidence were the greatest challenges. Following the internships, students expressed a new-found confidence in their career journeys. One female student shared about the experience: *"I learned more in 10 weeks than I would have in an entire semester [of school]...I am so grateful for this amazing experience and program."*

Program attrition occurred for many reasons.

Of the twenty-one students who started the program, ten completed all program activities. Students left the program for multiple reasons and at multiple points of the program, as seen in Figure 19. As students prioritized the many demands for their time and energy, their participation in the program was understandably low on their list, particularly given that program participation was completely voluntary and separate from their degree requirements at their home institution. Credit for the courses was transferable to home institutions, but these credits were not necessarily counted toward their requirements for their degree program. Working more closely with academic advisors, we would have liked to ensure the alignment of the online course with the student's major curriculum. Alternatively, offering an official 12-credit undergraduate certificate was originally in the project plan, but the year-long process to make

Figure 19. Undergraduate student persistence throughout the program.

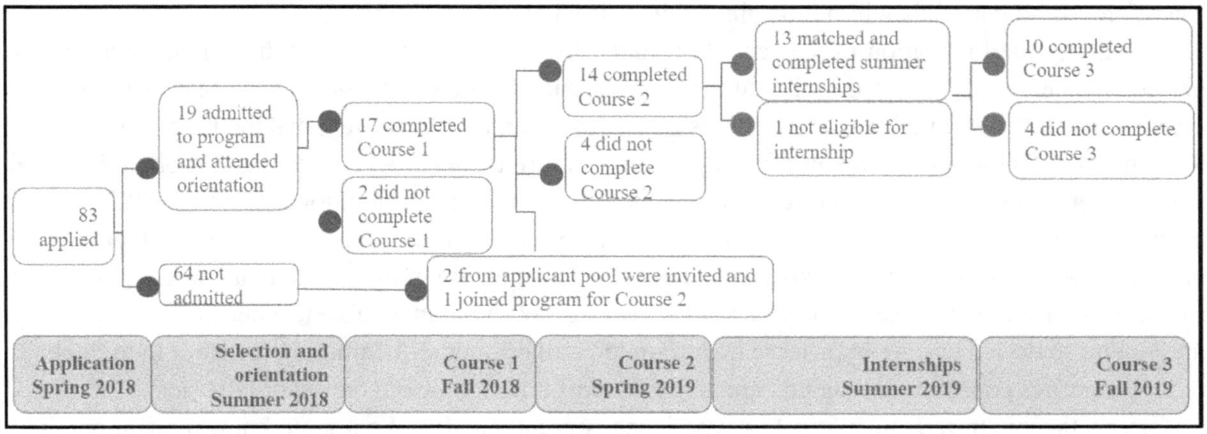

the certificate official and the lack of historical data for the courses created a delay between the students' participation and the availability of the official certificate.

Recommendations: Undergraduate Students

1. Provide extra academic support for students and prepare faculty for added flexibility in course expectations.
2. Boost communications with students' home institutions. Work with students' academic advisors and career advisors to align student academic goals with the program.
3. Create a formal program of study certificate to encourage student persistence and ensure the sustainability of the program activities.
4. Create an internship guide and engage in summer-long discussion with internship hosts especially for companies new to hosting interns. This will include a proposed schedule of events, expectations for interns, supervisors/mentors, and project guidelines.
5. Providing funding to students is not enough. Engage students in discussions about housing, travel and transportation, and budgeting so they are better prepared to take on the responsibilities of moving and starting a professional role.

REFLECTIONS AND NEXT STEPS

As we reflect back on the project, we realized that many of our participants have advanced professionally. Two of the teachers were honored as their school district's Teacher of the Year. One of the teachers became the State Teacher of the Year for North Carolina. One of the graduate students from Forest Bioproducts was hired as a faculty member in the College of Agriculture and Life Science at NC State University. Four of the teachers became graduate students: two master's students in the College of Natural Resources and two in the College of Education (one Master's and one Doctoral).

We also noted the active use of the online materials through the College websites, not only by the teachers, but also by the broader community. The MOOCs and online courses will be sustained for the future, and the lessons are becoming available on the website and in relevant practitioner publications. Therefore, these materials will be sustained with fairly low maintenance.

Sustaining the innovation in the schools is more difficult. We noticed that the teachers who had a partner teacher or who worked in the same school district tended to support one another. Although we are continuing to be available to the teachers for materials and technical support, the long-term support for continuing and the recruitment of new teachers is more tenuous. At two of the schools, we invited new teachers to replace one who is retiring and one who is out of the school for a year as the NC Teacher of the Year. As we think about future funding opportunities, we have submitted proposals that harness the expertise of teachers who are experienced with the Bioeconomy program content and activities, who can serve in a support capacity for new teacher recruits. With the inevitable teacher turnover, we hope these more seasoned teachers can serve in a 'train the trainer' model. Ultimately, turnover in high needs schools requires constant tending and creative solutions with help from the school districts.

Clearly the undergraduate internship experiences were impactful. We want to know more about their career pathways, and will work to stay in contact with these students. They hold the potential of being the next role models for high school students in the program teachers' classrooms.

Finally, we have learned a lot about our interdisciplinary work. Ongoing research projects study how to work in meaningful ways with industry professionals, science and education university faculty and graduate students, diverse undergraduates, and high school teachers. We hope that this program will inspire and inform others who want to build positive and productive partnerships.

ACKNOWLEDGMENT

We are grateful for the participation of all of the teachers, industry professionals, graduate students, university staff, and professors who participated in the program. Funding from the United States Department of Agriculture NIFA program supported this work and its related research (#2017-67009-26771). We are grateful for the support of the College of Natural Resources and the College of Education at NC State University.

REFERENCES

Adams, R. (2019, June 12). *N.C. Agriculture: Feeding Families, Growing the Economy*. NC State Extension. https://ipm.ces.ncsu.edu/2019/06/n-c-agriculture-feeding-families-growing-the-economy/

Alegria, S. N., & Branch, E. H. (2015). Causes and consequences of inequality in the STEM: Diversity and its discontents. *International Journal of Gender, Science and Technology*, *7*(3), 321–342.

Andersen, L., & Ward, T. J. (2014). Expectancy-value models for the STEM persistence plans of ninth-grade, high-ability students: A comparison between Black, Hispanic, and White students. *Science Education*, *98*(2), 216–242. doi:10.1002ce.21092

Blanchard, M. R. (2010). *STEM teams: Promoting science, technology, engineering, and mathematics (STEM) career interest, skills, and knowledge through strategic teaming*. ITEST Strategies, National Science Foundation.

Blanchard, M. R., Collier, K. M., Rajwade, A. M., McCance, K. R., Mcalexander, S. L., Venditti, R. A. (2020). *Analyzing whether teachers' task values influenced their implementation of bioeconomy-focused lessons: A pilot study.* Manuscript in Progress.

Blanchard, M. R., LePrevost, C. E., Tolin, A. D., & Gutierrez, K. S. (2016). Investigating technology-enhanced teacher professional development in rural, high poverty middle schools. *Educational Researcher*, *43*(3), 207–220. doi:10.3102/0013189X16644602

Blanchard, M. R., Osborne, J. W., Wallwork, C., & Harris, B. S. (2013). Progress on implementing inquiry in North Carolina: Nearly 1,000 elementary, middle and high school science teachers weigh in. *Science Educator*, *22*(1), 37–49.

Blanchard, M. R., Painter, J., Gray, D., Alsbury, T. L., Denson, C., Williams, B., Ragan, S., & Johnson, J. (2014). *STEM career clubs: Enhancing the potential of underrepresented students in STEM careers through a strategic teaming model.* ITEST Strategies, National Science Foundation.

Blanchard, M. R., & Sampson, V. D. (2018). Fostering impactful research experiences for teachers (RETs). *Eurasia Journal of Mathematics, Science and Technology Education*, *14*(1), 447–465.

Blanchard, M. R., Southerland, S. A., & Granger, D. E. (2009). No silver bullet for inquiry: Making sense of teacher change following an inquiry-based research experience for teachers. *Science Education*, *93*(2), 322–360. doi:10.1002ce.20298

Brown, B. A., Henderson, J. B., Gray, S., Donovan, B., Sullivan, S., Patterson, A., & Waggstaff, W. (2016). From description to explanation: An empirical exploration of the African-American pipeline problem in STEM. *Journal of Research in Science Teaching*, *53*(1), 146–177. doi:10.1002/tea.21249

Darling-Hammond, L., & Hammond, R. (2015). *Teaching in the flat world: Learning from high-performing systems.* Teachers College Press.

Enderle, P., Dentzau, M., Roseler, K., Southerland, S. A., Granger, E., Hughes, R., Golden, B., & Saka, Y. (2014). Examining the Influence of RET's on Science Teachers' Beliefs and Practice. *Science Education*, *98*(6), 1077–1108. doi:10.1002ce.21127

Glaser, V. (2013). Meeting current needs and assessing future opportunities to drive the global bio-economy. *Industrial Biotechnology (New Rochelle, N.Y.)*, *9*(5), 271–274. doi:10.1089/ind.2013.1594

Golden, J.S., Handfield, R. B., Daystar, J. S., & McConnell, T. E. (2018). *An economic impact analysis of the U.S. biobased products industry: A Report to the Congress of the United States of America Update.* A Joint Publication of the Duke Center for Sustainability & Commerce and the Supply Chain Resource Cooperative at North Carolina State University.

Herring, C. (2009). Does diversity pay? Race, gender, and the business case for diversity. *American Sociological Review*, *74*(2), 208–224. doi:10.1177/000312240907400203

Independence, E., & Act, S. (2007). *Public Law*, 110–140. https://www.govinfo.gov/content/pkg/BILLS-110hr6enr/pdf/BILLS-110hr6enr.pdf

Keeble, B. R. (1988). The Brundtland Report: 'Our common future'. *Medicine and War*, *4*(1), 17–25. doi:10.1080/07488008808408783

Lavalley, M. (2018). *Out of the loop: Rural schools are largely left out of research and policy discussions, exacerbating poverty, inequity and isolation*. Center for Public Education.

Lent, R. W., Brown, S. D., & Hackett, G. (1994). Toward a unifying social cognitive theory of career and academic interest, choice, and performance. *Journal of Vocational Behavior*, *45*(1), 79–122. doi:10.1006/jvbe.1994.1027

Lent, R. W., Brown, S. D., & Hackett, G. (2000). Contextual supports and barriers to career choice: A social cognitive analysis. *Journal of Counseling Psychology*, *47*(1), 36–49. doi:10.1037/0022-0167.47.1.36

Luft, J. A., & Roehrig, G. H. (2007). Capturing science teachers' epistemological beliefs: The development of the teacher beliefs interview. *The Electronic Journal of Science Education*, *11*(2), 39–63.

McAlexander, S., Topliceanu, A., Sharma, S., McCance, K., Starkey, H., Blanchard, M. R., & Venditti, R. (2019, October 12). *Building a sustainable future with bioproducts & bioenergy in science curricula* [Conference presentation]. Bridging the Gap 2019: Uniting North Carolina STEM Education, Raleigh, NC.

McAlexander, S. L., Blanchard, M. R., & Venditti, R. (2020a) *Supporting career development and internship experiences of historically underrepresented science and engineering community college to university transfer students.* Manuscript in preparation.

McAlexander, S. L., Blanchard, M. R., & Venditti, R. (2020b). *Investigating internship motivations and outcomes: Undergraduate student and industry perspectives*. Manuscript in preparation.

McAlexander, S. L., McCance, K., Noble, S., Blanchard, M. R., & Venditti, R. (2020). *The development and validation of the bioproduct and bioenergy beliefs (BABB) and career interest (CIBB) surveys.* Manuscript in preparation.

McCance, K. R., Blanchard, M. R., & Venditti, R. (2020). *Graduate Students' Perceptions of an interdisciplinary collaboration between science, engineering, and education: Developing community and navigating boundaries. Participating in a Science/Engineering and Education Interdisciplinary Collaboration: Developing Community and Navigating Boundaries*. Manuscript in preparation.

McConnell, T. E., Jeuck, J., Bardon, R., & Hazel, D. (2016). *North Carolina's forests and forest products industry by the numbers, 2013.* North Carolina Cooperative Extension Service Bulletin AG-817. 11

National Education Association. (2012). *Preparing 21st century students for a global society: An educator's guide to "the four Cs."* Washington, DC: Author. Available: https://www.nea.org/assets/docs/A-Guide-to-Four-Cs.pdf

National Science Foundation, National Center for Science and Engineering Statistics. (2017). *Women, minorities, and persons with disabilities in science and engineering: 2017.* Special Report NSF 17-310. www.nsf.gov/statistics/wmpd/

North Carolina Department of Agriculture and Consumer Services. (2015). *North Carolina agricultural statistics. Publication No. 216.* Author.

North Carolina School Report Cards. (n.d.). *School Report Cards.* North Carolina Department of Public Instruction. https://ncreportcards.ondemand.sas.com/src

Office of Energy Efficiency & Renewable Energy. (n.d.). *Bioenergy career map.* https://www.energy.gov/eere/bioenergy/bioenergy-career-map

Purvis, B., Mao, Y., & Robinson, D. (2019). Three pillars of sustainability: In search of conceptual origins. *Sustainability Science, 14*(3), 681–695. doi:10.100711625-018-0627-5

School Digger. (n.d.). https://www.schooldigger.com

Science Learning, B. S. C. S. (n.d.). *BSCS 5E instructional model.* https://bscs.org/bscs-5e-instructional-model/

United States Department of Commerce Census Bureau. (2016). *American factfinder.* http://factfinder.census.gov/faces/nav/jsf/pages/index.xhtml

United States Department of Education. (2012, January 10). *Race to the top: North Carolina Report. Year 1 (2010-2011).* Washington, DC: Author.

Venditti, R. & Blanchard, M. R. (2016). *Preparing Diverse and Rural Students and Teachers to Meet the Challenges in the Bioenergy and Bioproducts Industry.* USDA NIFA, $2,750,000/4 years.

Wenger, E. (2010). Communities of practice and social learning systems: The career of a concept. In C. Blackmore (Ed.), *Social learning systems and communities of practice* (pp. 179–198). Springer. doi:10.1007/978-1-84996-133-2_11

Wigfield, A., & Eccles, J. S. (2000). Expectancy–value theory of achievement motivation. *Contemporary Educational Psychology, 25*(1), 68–81. doi:10.1006/ceps.1999.1015 PMID:10620382

KEY TERMS AND DEFINITIONS

Bioeconomy: Using renewable biological resources and waste streams to produce value-added bioproducts, such as energy, fuel, chemicals, and consumer goods; an alternative to a fossil fuel-based economy and manufacturing processes.

Bioenergy: A form of renewable energy that is produced from biological materials including wood, agricultural residues, municipal solid waste, and algae.

Bioproducts: Products, such as biofuels, bioplastics, biopharmaceuticals, and other biomaterials, that are made from renewable biological resources and waste streams.

Community College Transfer: A student who moves from a community college (two-year postsecondary institution) to a four-year college or university.

High School: A secondary school, usually grades 9 through 12, that students attend after primary and middle school.

Interdisciplinary: Involving two or more academic disciplines.

Internships: A short-term position (paid or unpaid) with a company or organization to gain work experience in and exposure to a career or field of interest.

Online Courses: A course that is conducted in an online format rather than face-to-face.

Rural Education: Schools in rural areas that face unique challenges such as economic distress and high teacher turnover.

STEM Careers: Any career in science, technology, engineering, or mathematics fields.

Undergraduate: A student attending a postsecondary institution and who has not yet obtained an associate's or bachelor's degree.

Chapter 8
University Students:
Schoolteacher Partnership With Newly Developed Technologies

Shigeru Ikuta
https://orcid.org/0000-0001-5065-190X
Otsuma Women's University, Japan

Emi Sakurai
Otsuma Women's University, Japan

Misa Takayanagi
Otsuma Women's University, Japan

Nao Suzuki
Otsuma Women's University, Japan

Momoka Horiuchi
Otsuma Women's University, Japan

Honami Nakano
Otsuma Women's University, Japan

Yui Tamai
Otsuma Women's University, Japan

Kana Sugibayashi
Otsuma Women's University, Japan

Moeka Yoshida
Otsuma Women's University, Japan

Chikako Kaneko
Honmachidahigashi Elementary School, Japan

Yuko Nakazawa
Aiwa Elemnetary School, Japan

Satsuki Yamashita
Fuchu Keyakinomori Gakuen Tokyo Metropolitan School for the Physically Disabled and the Intellectually Disabled, Japan

Mariko Oshima
Hikawa Elementary School, Japan

Shinya Abe
Gridmark Inc., Japan

ABSTRACT

University students go to school several times a week, as voluntary supporters, and help teachers not only conduct lessons, but also create self-made teaching materials with newly developed technologies such as multimedia-enabled dot codes and EPUB 3 eBooks with Read-Aloud. The university students-schoolteachers partnership enables the younger students to learn the required subjects with newly educational technologies. For university students, especially those in the pre-service teacher program,

DOI: 10.4018/978-1-7998-4966-7.ch008

such a collaboration is really effective to not only know the class lessons, but also to encourage them to become teachers. The collaborative research works of the university professor with the ICT business companies are crucial to produce newly creative and gifted software and tools for students with various difficulties. This chapter briefly presents newly developed software and tools used for class lessons, and then it describes fruitful collaboration with schoolteachers in creating self-made teaching materials and conducting activities.

INTRODUCTION

Japanese students have acquired relatively high scores in international, well-known, competency-achievement tests such as the PISA (Organisation for Economic Cooperation and Development, 2018) and TIMSS (International Study Center, n.d.). As a result, it is generally believed that the Japanese educational system is effective in allowing students to acquire basic competencies. These wonderful outcomes in Japanese education, however, come from teachers' long working hours and extreme care they take with the students. The Teaching and Learning International Survey (TALIS) clearly indicated that while the teachers in Japan have success in creating wonderful classes and a wonderful teaching environment in schools, they also have the longest working hours in all the countries surveyed and always feel stressed (National Institute for Educational Policy Research, n.d.; Schleicher, 2020).

There exist 1,141 (separated) special-needs schools for students with disabilities in Japan (Ministry of Education, Culture, Sports, Science and Technology, Japan, 2019a). Even in the public schools, more than 6.3% of total students have significant learning and behavioral difficulties (Ministry of Education, Culture, Sports, Science and Technology, 2003). The number of these students has increased more than 2 times in these 10 years (Ministry of Education, Culture, Sports, Science and Technology, Japan, 2019b). The burden of homeroom teachers has increased, and their class managements have also become difficult. Also, the claims by parents and guardians to schoolteachers is increasing. The teachers, therefore, always feel stressed and are very busy every day at school, and do not have enough time to develop teaching materials for students in their classes.

The discussion about how to improve teachers' long working hours has started, but it is not certain it will improve the situation drastically and promptly. Consequently, schoolteachers and boards of education are now considering that each school might be supported by regional volunteers, including the nearest university's students. Every year, the university students in one of the present authors' (Shigeru Ikuta) laboratory have volunteered to help younger students learn in the class, and help schoolteachers develop self-made teaching materials, by using the recently developed information communication technologies (ICT) such as the multimedia-enabled dot codes (Gridmark, n.d.), EPUB 3 eBooks with media overlays (International Digital Publishing Forum, 2011a, 2011b), and Augmented Reality (HP Reveal, n.d.). The university students, especially those who are in the pre-service teacher program, will then be able to not only know the present status of schools, but also acquire wonderful and plentiful know-how through the in-service teachers' classes (Ikuta et al., 2020a).

Schools in Japan used to be closed and reluctant to accept people from outside the school for a long period of time. But now they are not saying such things; every school is accepting school volunteers positively. Of course, accepting such volunteers is not so easy for schoolteachers because it leads to more work for them. However, many schoolteachers are now willing to accept them, saying "School volunteers can provide supporting the students with significant learning and behavioral difficulties and the students who feel having no place in the class." Younger students also welcome university students being close to their own ages as sisters and brothers.

This paper briefly outlines the recent development of the state-of-the-art Grid Onput dot-code technology used in creating original self-made teaching materials for the younger students. Then, the paper describes the ongoing collaboration between university students and schoolteachers to help the younger students deepen their learning with the self-made teaching materials the university students created.

BACKGROUND

One of the authors, Shigeru Ikuta, has organized many communities using the Internet in his residential area (Tama district, Tokyo) for more than 30 years and had engaged in various activities like "Imaging the Future of the Tama district," "Learning Support for Children in Tama District," and "Activating Newly Developed Tama Monorail Connecting Tama District Area"; all the participants had thought the future of the Tama district in which they live. And, he had also organized a community to connect Miyakejima and Tama, when all the Miyakejima islanders had evacuated due to a big volcanic eruption there; he had managed the community as a representative. At the center of these voluntary communities, many schoolteachers were always there with one of the authors, Shigeru Ikuta. They act in the firm belief even now that school and schoolteachers should play a central role in the area around school.

More than 14 years ago, in order to support K-12 students with various difficulties, Shigeru Ikuta started using Scan Talk dot codes, which was developed by Olympus Co. (1999). Such codes transform voices and sounds into two-dimensional dot codes directly outputted on ordinary paper. However, students with severe hand, finger, or mental challenges could not correctly trace fairly long Scan Talk codes using the Scan Talk Reader. Therefore, he used new dot codes (i.e., Grid Onput) that Gridmark Inc. (the company of Shinya Abe, a coauthor of this paper) had originally developed. Shigeru Ikuta has now organized the collaborative research group with more than 200 schoolteachers all over the world (Gallup et al., 2021; Ikuta et al., 2015; Ikuta et al., 2017; Ikuta et al., 2019a; Ikuta et al., 2019b; Ikuta et al., 2020a; Ikuta et al., 2020b).

Shigeru Ikuta is now involved in the steering committee members at one of the public community schools in Tama district. In addition, he is also engaged in an external advisor at one of the largest special-needs education schools in Tokyo, where he and schoolteachers have been collaborating to create self-made teaching materials using the newly developed software and tools and conducting various school activities to support the students' learning in the classes. These activities are just a part of the collaborative research project described above.

Funds for the Ikuta's research project were provided by the Ministry of Education, Culture, Sports, Science, and Technology, Japan, and by Otsuma Women's University, which has enabled the purchase of the necessary tools and software for the collaborative research work. Gridmark Inc. also supported the research project by providing the software free of charge and discounting tools such as the speaking-pen and dot-code readers noted hereafter.

MAIN FOCUS OF THE CHAPTER

Students with Reading and Speaking Difficulties

Moats and Tolman (2009) described "reading-impaired" children as all those who score below the 30th percentile in basic reading skills. Among all poor readers, about 70–80 percent of them have trouble with accurate and fluent word recognition that originates with weakness in phonological processing, often in combination with fluency and comprehension problems. These students have obvious trouble learning sound–symbol correspondence, sounding out words, and spelling words. The term dyslexic is most often applied to this group. Another 10–15 percent of poor readers appear to be accurate readers but are too slow in word recognition and text reading. They have specific weaknesses with the speed of word recognition and the automatic recall of word spellings, although they do relatively well on tests of phonemic awareness and other phonological skills. These students have trouble developing automatic recognition of words by sight and tend to spell phonetically but not accurately. Yet another 10–15 percent of poor readers appear to decode words better than they can comprehend the meanings of passages. These poor readers are distinguished from dyslexic poor readers because they can read words accurately and quickly, and because they can spell (Moats & Tolman, 2009; National Institutes of Health, 2020). About 30 percent of all children with dyslexia also have Attention Deficit Hyperactivity Disorder (ADHD). Researchers have identified three kinds of developmental reading disabilities that often overlap but can be separate and distinct: (1) phonological deficit, (2) processing speed/orthographic processing deficit, and (3) comprehension deficit.

Autism Spectrum Disorders (ASDs) affect three areas of a child's life: social interaction, communication (both verbal and nonverbal), and behavior and interests. Each ASD child has a unique pattern of autism, however, all autistic people have some problematic core verbal and nonverbal communication symptoms, such as a delay in learning to talk, or having a complete lack of verbal ability. Most autistic students have some trouble relating to others (National Institutes of Health, 2018; Alli, 2018).

The Importance of Individual, Self-Made Teaching Materials and Tools for Each Student

Every year, especially the teachers of special-needs classes and schools consider whether the teaching materials and tools that had been effective for previous students are appropriate for new students and new classes, because the needs, desires, personality, and character of each student with disabilities tend to be unique to that student. Therefore, Shigeru Ikuta, always recommends that teachers create individual teaching materials for each student.

The importance of self-made original teaching materials and tools has been pointed out by many researchers, schoolteachers, and persons of great insight, but only a few persons have made their own in Japan. This is because educators need an easy-to-handle and less expensive software to create their own teaching materials with multimedia-enabled dot codes and EPUB 3 eBooks containing media overlays. The present author, Shigeru Ikuta, confirms that the collaborative research works of the university professors with the ICT business companies are crucial to produce newly creative and gifted software and tools for students with various difficulties.

Partnerships between University Students and Schoolteachers

There exist 786 universities in Japan (Ministry of Education, Culture, Sports, Science and Technology, Japan, 2019a); about 80% of them are private universities. There still exist 78 women's only colleges; Otsuma Women's University, where Shigeru Ikuta has been working, is one of the largest women's university in Japan, with nearly 8,000 students. All the students participate in a seminar for two years under a professor of their choices. In a weekly seminar of the first year, each student learns the professor's specialized knowledge and discusses together her graduation research theme that is tackled in the coming second year. In the second year of the seminar, each student performs her graduation study relating nearly to the professor's research field, and then, submits her graduation thesis after summarizing the research results and issues.

Every year, more than 10 university students join the Ikuta's seminar, since they are inspired by the challenges with newly developed software and tools, to realize what the younger students had not been able to achieve in the past. Many students in the Ikuta's laboratory participate in volunteers at schools near the university. They jump into the classes and help the younger students with disabilities and/or significant learning and behavioral difficulties. The university students also help schoolteachers create their original self-made teaching materials by using newly developed software and tools. Some of the university students prepare a lesson plan under the guidance of homeroom teacher and conduct actual class lessons using the teaching materials they created; and then, they can compile their graduation theses including the status of their efforts.

The university students taking a pre-service teacher program are full of hopes to become a teacher and the students hoping to continue research are full of hopes to go on to graduate school. Even, for other students, the collaborations with the schoolteachers really help them build a happy and worthy life in a society. The university students, greatly influenced by the schoolteachers who helped them during their compulsory education period, participate in the Ikuta's seminar with the hope of being involved with the younger students themselves, like their teachers at that time.

Scope of the Chapter

Author Shigeru Ikuta has been collaborating with a Japanese venture business company, Gridmark Inc., which developed Grid Onput dot codes. Shinya Abe is a staff member at Gridmark Inc. who has been developing original software that can handle multimedia-enabled dot codes. Authors Emi Sakurai, Misa Takayanagi, Nao Suzuki, Momoka Horiuchi, Honami Nakano, Yui Tamai, Kana Sugibayashi and Moeka Yoshida were university students who belonged to the Ikuta's laboratory. They voluntarily joined the classes at the schools. Chikako Kaneko, Yuko Nakazawa, Satsuki Yamashita, and Mariko Oshima are schoolteachers who welcomed the university students and collaborated with them affectionately. In this chapter, the authors first describe the newly developed software and tools briefly, and then, describe the fruitful activities that the schoolteachers performed for younger students in their classes, with the help of the university students.

SOLUTIONS AND RECOMMENDATIONS

Technology Used (1): Multimedia-Enabled Grid Onput Dot-Codes

Grid Onput dot-codes (Gridmark, n.d.) is a set of novel two-dimensional codes consisting of extremely small dots. Such dot codes can invisibly overlay any graphically printed letters, photos, and illustrations with no impact on the designed images. Author Shigeru Ikuta has created two types of original sheets (called "Magical sheets") that can be overlaid with the Grid Onput dot codes, as shown in Figure 1. Each sheet has 117 or 130 "Post-it-like" sticker icons and paper controller icons on the lower part that can change a mode and record and delete the voice/sound linked to each icon, which can be taken off and pasted on real items. Each sticker icon is linked with up to four multimedia mediums (i.e., four modes), such as movies, in addition to up to four voices/sounds. Teachers are able to obtain these Magical sheets from author S. I. free of charge. Recently, Gridmark Inc. created new sheets with "Post-it-like" sticker icons, with the maximum number of icons being 1,000 and the maximum available modes being 10, recently putting them into the market.

Figure 1. Original two types of "Magical sheets" with "Post-it-like" sticker icons

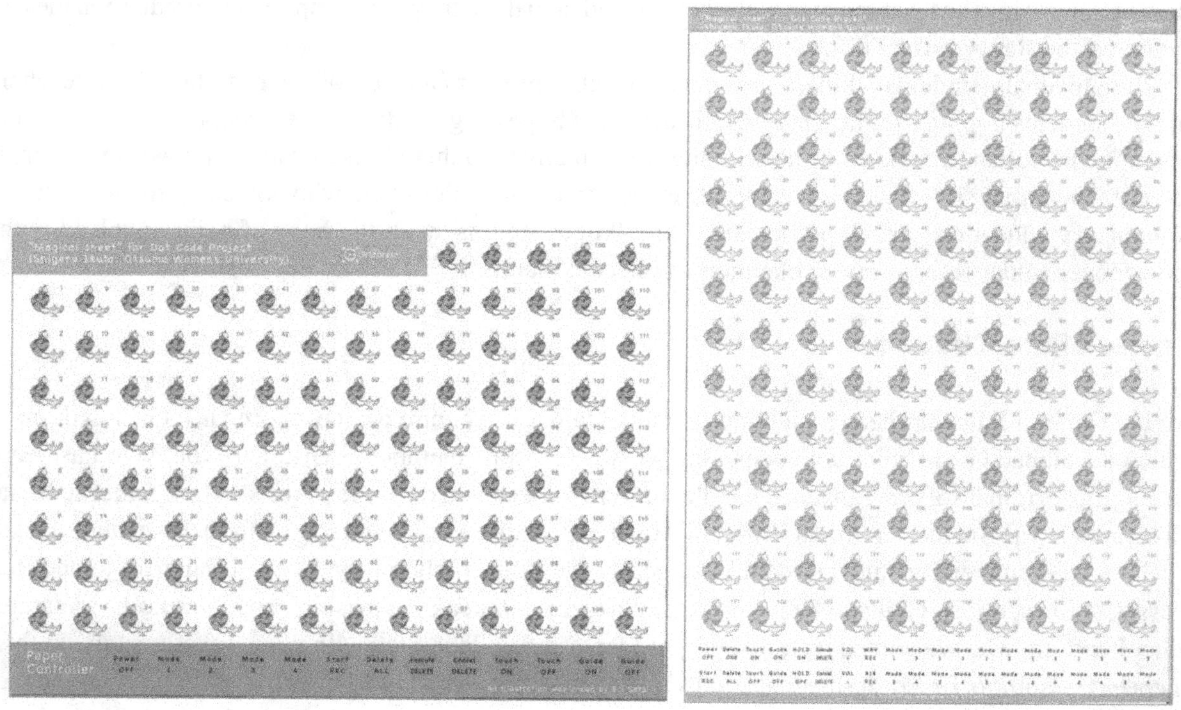

Gridmark Inc. has been developing the following four software programs that can handle the multimedia-enabled Grid Onput dot codes, referring to the suggestions made by Ikuta's research group (Ikuta et al., 2020b).

GM Authoring Tool

With the software GM Authoring Tool (Figure 2a), users can put the multimedia-enabled dot codes on their designed document at any place, in any size, and in any number they like. After completing the dot-code overlapping on all pages, users can output each document page by using Adobe Reader with a printer having more than 1200 dpi quality. The documents overlaid with dot codes can be used exactly as the "Post-it-like" sticker icons that the author Shigeru Ikuta created; a maximum of four audio clips and multimedia-like movies can be linked to each dot code overlaid.

Figure 2. a. GM Authoring Tool to overlap dot codes; b. Sound Linker to create content for a speaking-pen; c. File Linker to create a standalone application to replay multimedia, and d. GCV to create content for an iOS device.

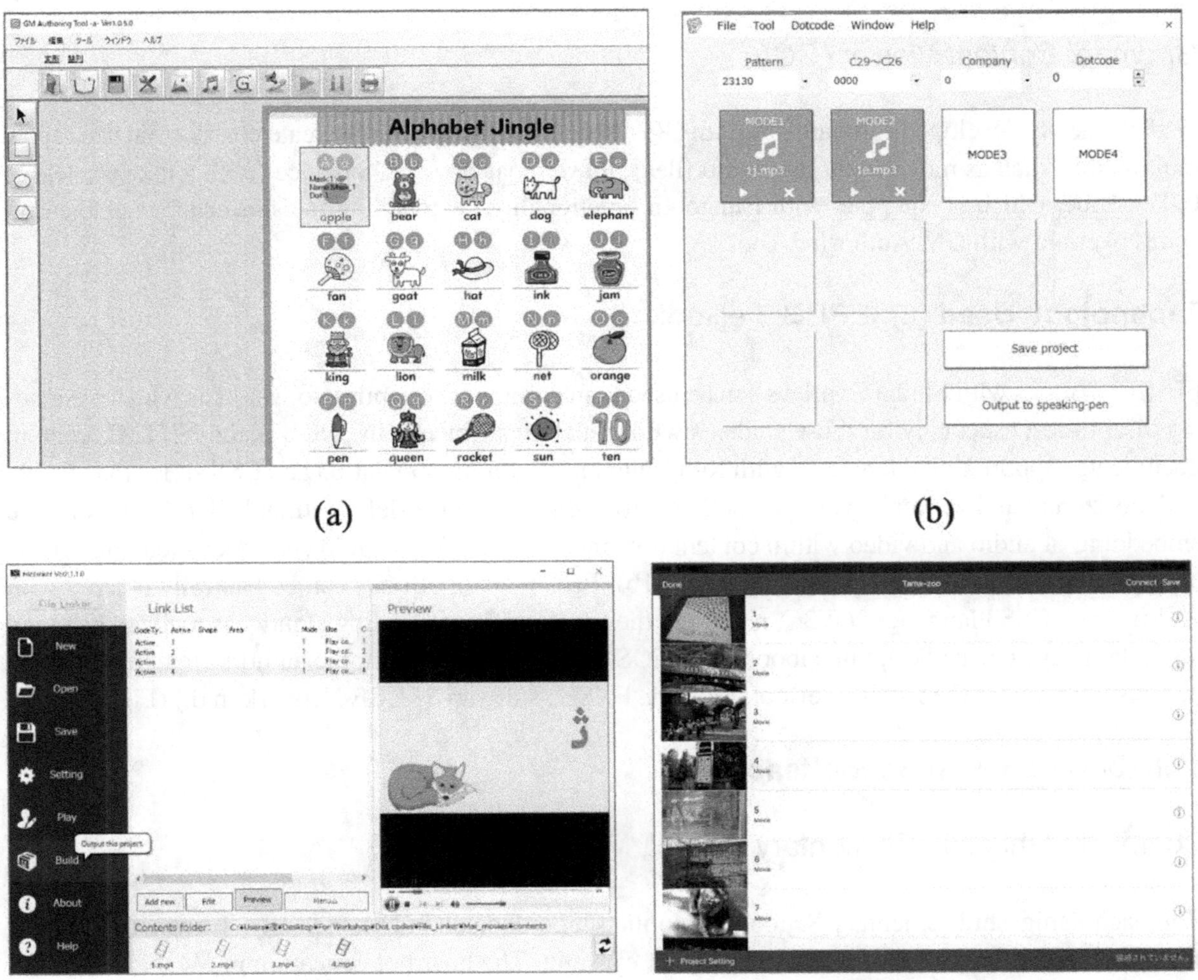

(a) (b)

(c) (d)

Sound Linker

With the software *Sound Linker* (Figure 2b), users can now easily create content for a speaking-pen. In order to create original content for a speaking-pen, they can just drag a maximum of audio clips into Mode sections, as in Figure 2b. After dragging the necessary audios, users can create a content and copy it into a MicroSD Card of a speaking-pen. Touching the sticker icons pasted on the real item or just beside the illustrations and texts on the users' designed document with a speaking-pen (G-Speak or G-Talk) plays back the original voices and sounds clearly.

File Linker

With the software File Linker (Figure 2c), a standalone application, an executable file that can reproduce multimedia such as movies can now be easily created. Users can now replay the multimedia by touching the "Post-it-like" sticker icons pasted on the real items with the same speaking-pen G-Speak connected with their Windows PC. The recently developed new dot-code reader G-Pen BT with Bluetooth functionality enables users to replay the linked multimedia without connection to their PC quite nicely.

Gridmark Content Viewer (*GCV*)

With the newly developed software program GCV (Figure 2d), users easily create content that can replay multimedia (such as movies, photos, audio files, and web pages) on iOS devices with a newly released G-Pen Blue, which is equipped with Bluetooth functionality, by touching the sticker icons or the dot-codes overlaid with GM Authoring Tool.

Technology Used (2): EPUB 3 eBooks

EPUB 3 eBooks with Media Overlays can be used to maximize accessibility for students who have reading disabilities, especially for those students with visual impairments. In EPUB 3, the XHTML content documents support global language with Ruby annotations for pronunciation and CSS styles that enable both horizontal and vertical layout and both left-to-right and right-to-left writing. EPUB 3 supports the embedding of audio and video within content documents via HTML5 audio and video elements (International Digital Publishing Forum, 2011a, 2011b). Each page in eBooks can be basically created with HTML5 and CSS3 languages (W3C, n.d.), and the functionality of reading aloud was realized by using SMIL language or JavaScript like ibooks.js on iOS devices. In this paper, the authors created self-made eBooks with one of the EPUB 3 eBooks editors, FUSEe software (FUSE Network, n.d.) (Figure 3).

Collaboration with Schoolteachers

Honmachidahigashi Elementary School, Machida, Tokyo

Honmachidahigashi Elementary School is a public school located in Machida city, Tokyo. This school has a special-needs class for students with weak eyesight. The board of education in Machida city created a side reader, *Our City Machida*, for third and fourth graders learning the required subject of Social Science. This paper supplement is composed of seven chapters: (1) Our city Machida, (2) Jobs of our

Figure 3. Screenshot of editing EPUB 3 eBook with FUSEe software

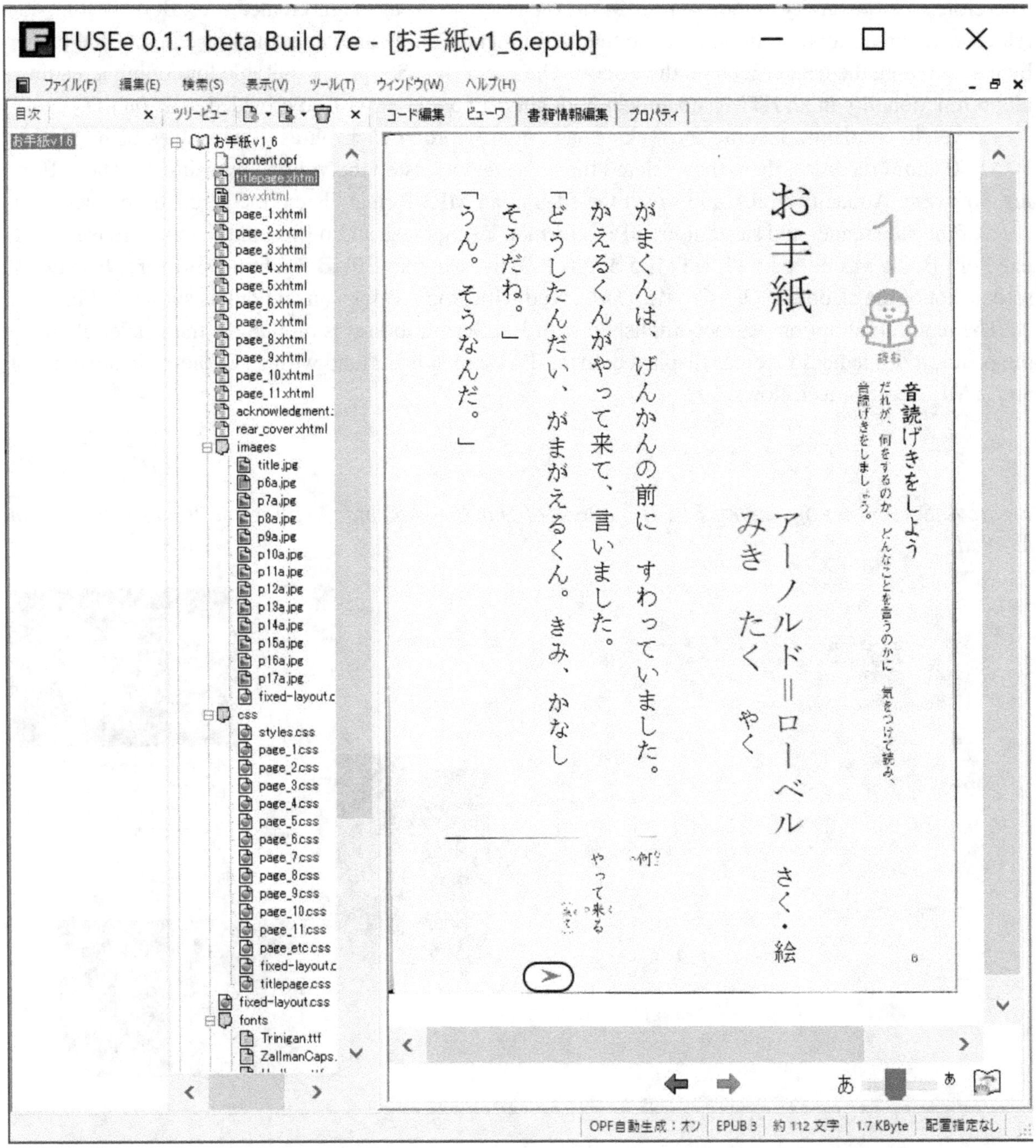

town folk, (3) Life of old people, (4) Live safely with peace of mind, (5) To protect our healthy lives, (6) People who had served our town, and (7) Our Tokyo government. This supplementary teaching material is required to be used at every public school in Machida City.

The students with weak eyesight could not read this supplementary teaching material, due to the small size of the text, and then, they could not understand it. The teacher, Chikako Kaneko, thought that an eBook with a read-aloud option might help such students read and understand it (Ikuta et al., 2019a). Therefore, the university students, Emi Sakurai and Misa Takayanagi, created the eBook of the side reader with a read-aloud functionality so that the students could confirm each phrase in the side reader by just touching the trigger icon on the eBook. The use of JavaScript was suitable for coding a reading-aloud functionality in an HTML document with almost no text or with so crowded texts on the page.

University students Sakurai and Takayanagi first recorded each phrase themselves using Sony's PCM-50 audio recorder; then, they deleted the noise and adjusted the volume by using an Audio Editing Software, Audacity (n.d.), and saved the files in an MP3 format. Then, the students scanned each page of the side reader and saved it in a JPEG format. Using these audio files and scanned images, each page of eBook was edited with HTML5 and CSS3, by using an EPUB 3 editing software, FUSEe. A screenshot of the eBook of *Our City Machida*, read with Apple Books on an iPad, is shown in Figure 4.

The read-aloud option was accomplished with JavaScript, ibooks.js on iOS. In the header element, the position of the audio icon as displayed on an iPad screen is defined with an inline-style attribute of an HTML element, as follows:

Figure 4. Screenshot of reading EPUB 3 eBook of Our City Machida, with Apple Books software, on an iPad

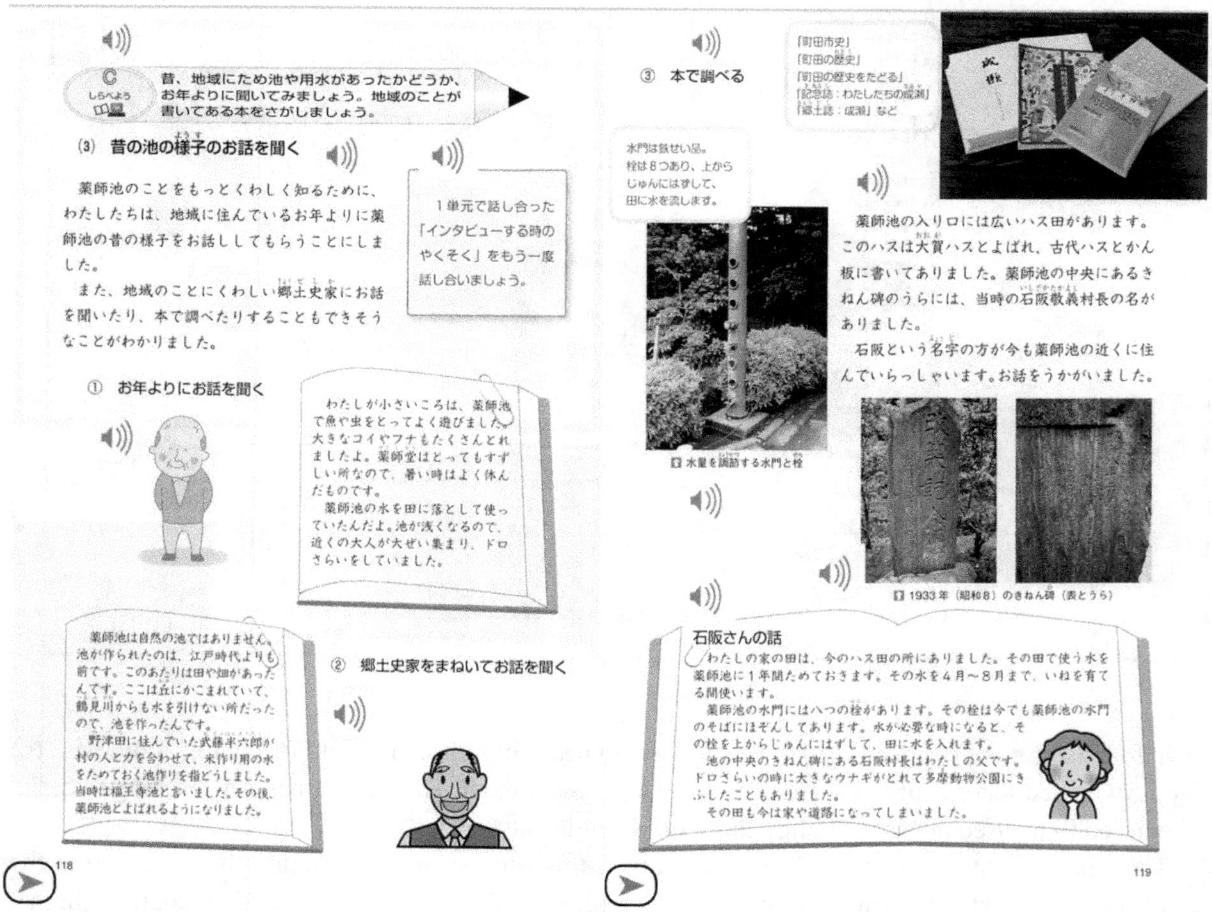

```
<head>
<style>#img1{position: absolute; display: block; top: 330px; left: 30px;}
</style>
<script src="ibooks.js" type="text/javascript"></script>
</head>
```

Ibooks.js, defined in the <head> element, is a collection of pre-written JavaScript code that makes it easier to develop books (for Apple Books) with JavaScript interactivity. Using this ibooks.js collection (Apple, 2019), audio playback in the eBook of *Our City Machida* works nicely. In the body element, the icon size (width="50"; height="50") and the audio file name (p90-1.mp3) are defined by using the <img> tag by assigning an "ibooks" class name and a data-ibooks-audio-src attribute, as follows:

```
<img id="img1" src="images/audio-icon.jpg" width="50" height="50"
class="ibooks-media-audio" data-ibooks-audio-src="audio/p90-1.mp3" />
```

If users want to start reading aloud eBooks, they can just touch the "audio-icon" illustration as shown in Figure 4; this illustration of the "audio-icon" works well as a trigger for starting, stopping, and restarting the reading-aloud functionality.

This eBook with a read-aloud option was used for a 4th grade student with premature retinopathy; he has 20/33 vision with glasses and joins a special class for people with weak eyesight. He has a problem with the visual function of eye movement, often skips over words while reading, has a problem reading and writing Kanji characters, and has a small working memory. He can barely read Kanji characters at the 2nd grade level, with Japanese Hiragana as Ruby annotations. His vocabulary is at the 1st grade level, since he has difficulty with reading textbooks and, as a result, he does not try to read any books. He usually uses multimedia DAISY textbooks with audio and an application for learning Kanji (The DAISY Consortium, n.d.). He often forgets to do his homework and to bring his notebook of parent–teacher correspondence. He is good-tempered in his class, and he has a watch-and-see attitude towards his classmates and follows their lead. The eBook with a read-aloud option, *Our City Machida*, was used in his individual lesson, with the Apple Books application on an iPad. During the lesson with this eBook, he followed the words and phrases with his eyes and added Ruby characters (Hiragana) to Kanji characters in the textbook. He had not been able to read a paper-based supplementary teaching material by himself before. He could really enjoy reading eBooks and get a sense of closeness, since the reading voices in the eBook were those of his homeroom teachers and the university-student volunteers. The teacher wants the students with reading disabilities to use this eBook in their individual lessons ahead of time, before the lesson in class begins.

The next year, the other university students, Nao Suzuki and Momoka Horiuchi, went to the school once a week as voluntary supporters and created self-made content of *Our City Machida* for the younger student with weak eyesight, by using "Post-it-like" sticker icons with multimedia-enabled dot codes. They first pasted a total of 374 dot-code icons near every sentence of the real side reader. Then, they recorded each audio clip corresponding to the sentence, and, then, the noises in the recorded audio files were deleted, and the volume was adjusted using the audio editor program Audacity. Some of the sentences were read by the school's principal to help the university students create the side reader with read-aloud. Finally, they created the content with a Sound Linker program and copied it to the MicroSD Card in the speaking-pen. Now, the younger student with weak eyesight became able to read each page of the side reader nicely, with the help of a speaking-pen (Figure 5a & 5b). This content was distributed to other schools' students with weak eyesight in Machida city.

Figure 5. a. Touching a dot-code icon with G-Speak to replay the audio in Our City Machida; b. School activity with the student with weak eyesight.

(a) (b)

Aiwa Elementary School, Tama, Tokyo

Aiwa Elementary School is a public school located in Tama city in Tokyo. A university student from Otsuma Women's University, Honami Nakano, created a multiplication table for second-grade students, in collaboration with a teacher, Yuko Nakazawa; the content includes each multiplication table, from 1 to 9, and a times table for each together with the song placed at the lower right.

According to the university student, Nakano, Grid Onput dot codes were first overlaid on the document sheet that all the second graders had, by using the GM Authoring Tool software (Figure 6a). Nakano prepared two types of the sheet: one had an answer for each multiplication question, and the other sheet did not have any answers. A total of 90 dot codes were overlaid on both documents and laminated.

The second-grade students are still using the multiplication table in both mathematics lessons and breaks, to get to know them by heart (Figure 6b). They can enjoy the content very much, since the voices were recorded by the volunteer student who had been with them for 10 months, Honami Nakano, and their schoolteacher. Especially, the original song of the times table is very attractive for the students, since the songs sung by a homeroom teacher are very rhythmical, and the content helps the students recite the multiplication from 1 times 1 to 9 times 9 very nicely. This content is now used at several schools, including a special-needs school.

Figure 6. a. Overlaying Grid Onput dot codes on a multiplication table with the software GM Authoring Tool; b. Activity with the student with a speaking-pen G-Speak.

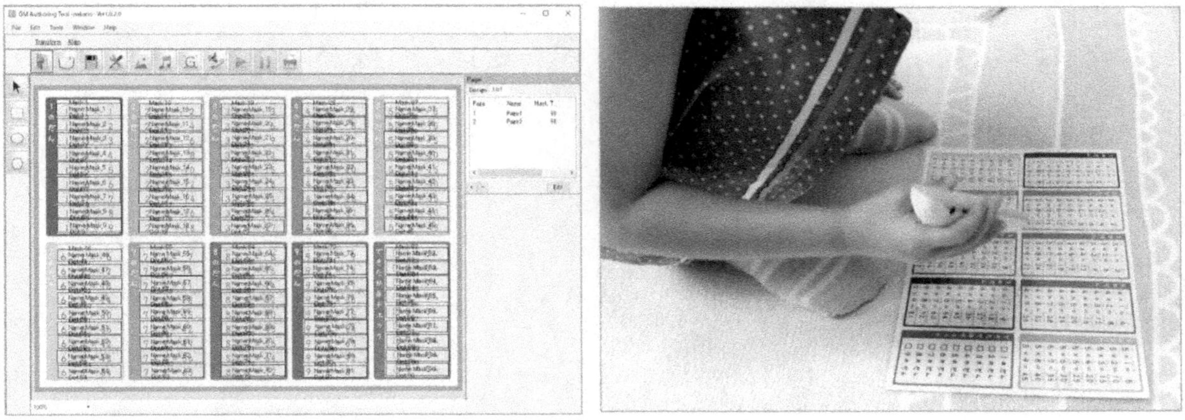

(a) (b)

The university student, Nakano, created one more piece of content for the second graders; it's an eBook, "お手紙 (in the book *Frog and Toad are Friends*, by Arnold Lobel)" with read-aloud functionality, created with the software FUSEe. In this eBook, to synchronize the highlighting of the text with the replaying voice, the SMIL language was used. The reading was done by the homeroom teacher Yuko Nakazawa. In the SMIL languages, the starting time (clipBegin) and the ending time (clipEnd) of each phrase's corresponding audio file should be inputted directly. In Figure 7, the screenshot of its eBook on the iPad screen is shown, and the SMIL language file corresponding to the page shown in the Figure is listed in the following:

```
<?xml version="1.0" encoding="UTF-8"?>
<smil xmlns="http://www.w3.org/ns/SMIL" version="3.0"
profile="http://www.idpf.org/epub/30/profile/content/">
<body>
<par id="par1">
<text src="page_2.xhtml#W1" />
<audio src="audio/nakano.mp3" clipBegin="50.8s" clipEnd="52.55s" />
</par>
…
<par id="par6">
<text src="page_2.xhtml#W6" />
```

Figure 7. Screenshot of reading the eBook, Frog and Toad are Friends, on an iPad, with the Apple Books application

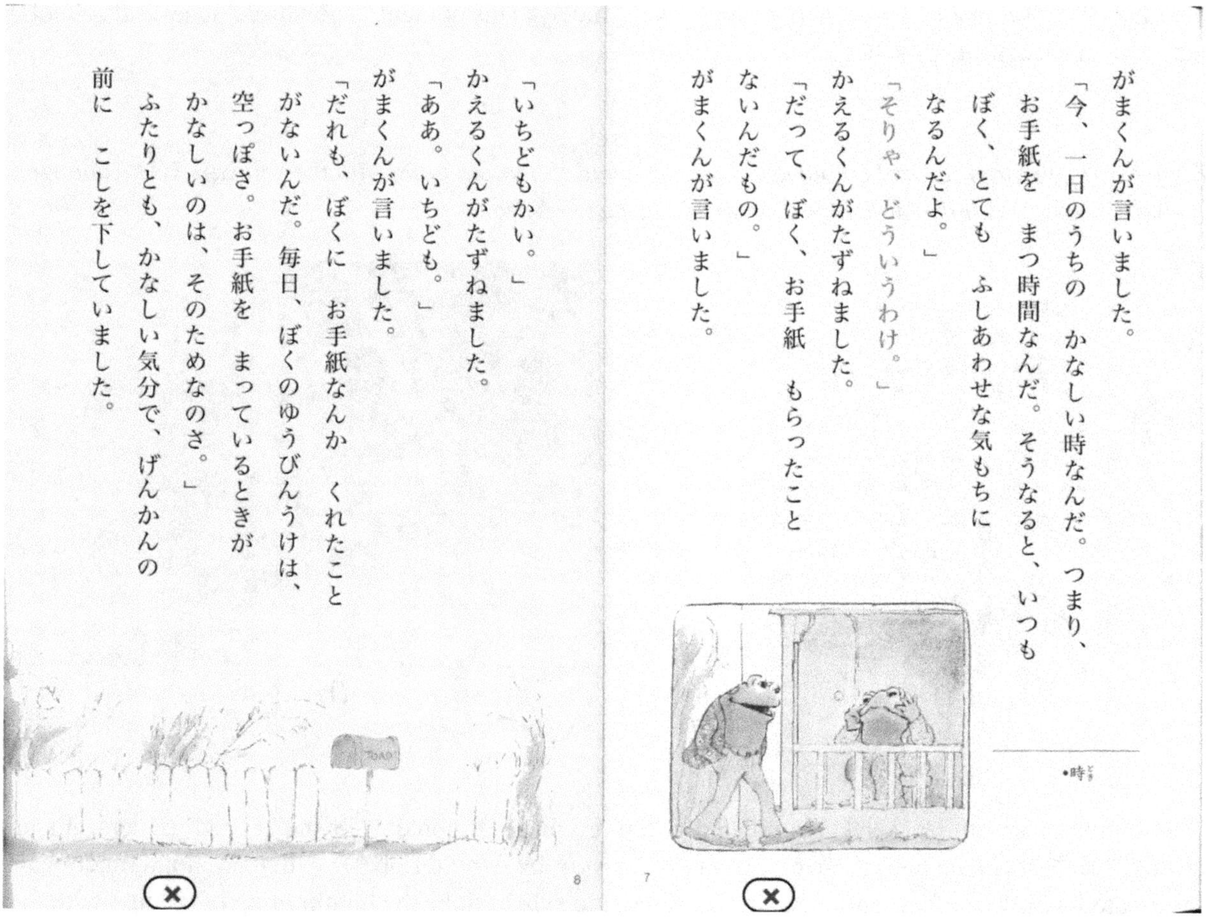
がまくんが言いました。
「今、一日のうちの　かなしい時なんだ。つまり、
お手紙を　まつ時間なんだ。そうなると、いつも
ぼく、とても　ふしあわせな気もちに
なるんだよ。」
「そりゃ、どういうわけ。」
かえるくんがたずねました。
「だって、ぼく、お手紙　もらったこと
ないんだもの。」
がまくんが言いました。

7

「いちどもかい。」
かえるくんがたずねました。
「ああ。いちども。」
がまくんが言いました。
「だれも、ぼくに　お手紙なんか　くれたこと
がないんだ。毎日、ぼくのゆうびんうけは、
空っぽさ。お手紙を　まっているときが
かなしいのは、そのためなのさ。」
ふたりとも、かなしい気分で、げんかんの
前に　こしを下していました。

8

Figure 8. a. Content for the description of the impression of books; b. Created content is placed at the school library.

(a) (b)

```
<audio src=”audio/nakano.mp3” clipBegin=”0:01:02.88” clipEnd=”0:01:06.77” />
</par>
…
</body>
</smil>
```

Reading aloud is nicely performed with the application program Apple Books, on iOS devices like the iPad and iPhone, as shown in Figure 7. All the students could read the eBook with joy, since the reading voice was of their homeroom teacher. Also, even the students with reading disabilities could nicely read it, with the aid of the synchronizing between the text highlighting and the replayed voice.

The other university student, Yui Tamai, went to the school once a week as a voluntary teacher. She created teaching material for a library that could encourage the younger student to read books. As homework during summer vacation, the sixth graders were assigned to provide a description of his/her impressions while reading books. Tamai first created each document composed of both the scanned book cover and the corresponding handwritten description of the student's impression of the book. Then, Grid Onput dot codes were overlaid on each document by using GM Authoring Tool software; and,

Figure 9. Replaying voices with a speaking-pen at the Aiwa school festival

all the documents with dot codes were outputted as a PDF file. Tamai printed out the PDF file with a high-quality color printer, where only carbon ink that absorbs infrared rays was used for dot-code printing. Tamai recorded each impression of reading a book, written by the younger students, by using each student's own voice, and linked it to the dot codes on the corresponding document sheet. Twelve sheets of students' handwritten documents were bundled with two rings and filed as a book (Figure 8a). This book is still now placed at the desk in the school library and anyone can read the six graders' description of their impressions while reading books with the help of a speaking-pen (Figure 8b). This book nicely promotes the activity for younger students to read more books.

Shigeru Ikuta and his university students collaborated with the teacher, Onizawa, to create content that was presented at the Aiwa School Festival, held every February. Each second grader spoke about what he/she made a great effort in regarding their school life in this academic year. The teacher Onizawa recorded every second grader's voice with a Sony PCM IC recorder and sent them all to Shigeru Ikuta to be created as content for a speaking-pen. Shigeru Ikuta and university students deleted the noises of the voices, adjusted the volume with the audio editing software Audacity, and created the content with Sound Linker software. The "Post-it-like" sticker icons were pasted near the students' photos printed on large paper. On the day of the festival, many students listened to what their classmates and friends made a great effort in regarding their school life in this academic year. The parents also confirmed it was their own child's voice by touching the dot-code icon with a speaking-pen, resulting in a big smile (Figure 9). Every year, this activity enjoys great popularity with the students and the parents.

Fuchu Keyakinomori Gakuen Tokyo Metropolitan School for the Physically Disabled and the Intellectually Disabled

Fuchu Keyakinomori Gakuen Tokyo Metropolitan School for the Physically Disabled and the Intellectually Disabled (Keyakinomori Gakuen) is a special-needs-education school founded by the Tokyo

Figure 10. a. Created content with the software GCV to replay the model performance; b. School activity at the subject of Physical Education

(a)

(b)

Government. Keyakinomori Gakuen is one of the biggest (separated) special-needs schools; more than 400 total students (143 students who were physically handicapped and 259 students with intellectual disabilities) were in this school, and more than 200 permanent teachers worked there, in the 2019 academic year. This school has three divisions, like an elementary, a junior high, and a senior high school division, and plays a leading role for special-needs education in Tokyo.

University students Kana Sugibayashi and Moeka Yoshida helped the teachers at Keyakinomori Gakuen create self-made teaching materials and conduct school activities. They first created content for the required subject of Physical Education to support the senior high school students in learning how to practice basketball.

They first took video clips of model performances on various forms, practiced by three schoolteachers with a license in Physical Education; then they deleted the noisy sounds, added the title and the name of each form of the model performance, and saved as new movie clips, using iMovie software. Then, these edited movie clips were linked to "Post-it-like" sticker icons pasted near the corresponding photo in the textbook by using a new software for iOS, Gridmark Content Viewer (GCV), with the help of a dot-code reader, G-Pen Blue (Figure 10a). The senior high school students usually could not grasp the images of the practicing exercises only from the photo and the explanation in the textbook. They could now replay each video clip nicely by touching the icon pasted near the photo in the textbook with the G-Pen Blue, confirm the model performance by the teachers, and, then, jumped into the court to practice the basketball form by themselves (Figure 10b). The senior high school students said to the teacher, "These video clips are so helpful, since we can check the model performance from various angles" and "These video clips are so fun, since our schoolteachers have performed the model performances." It has been difficult for the teachers to show model performances to the students during the real lessons, since the teachers have their hands full with paying attention to the students' behaviors. The teachers,

Figure 11. a. School activity in an English class with a speaking-pen; b. School activity in an English class by the teacher with the created teaching materials

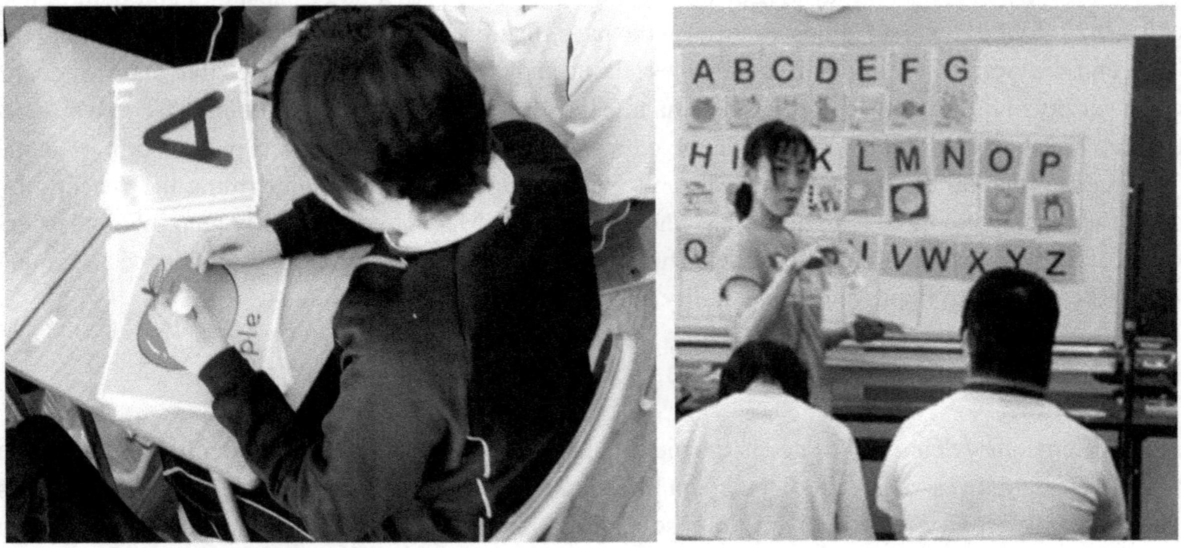

(a) (b)

however, have become to feel now that such instruction to the students might be possible, if they can use the present, newly developed system. The teaching material with video clips of model performances by the teachers really helps the high school students upgrade their Physical Education performances.

For students with intellectual disabilities at the senior high school division, the university students, Sugibayashi and Yoshida, created one more piece of content for English lessons. They created the following two types of content: (1) alphabetical cards and (2) illustration cards corresponding to each letter in the alphabet. The Grid Onput dot codes were overlapped on the whole surface of these cards, by using the software GM Authoring Tool, so that the younger students could easily touch with a speaking-pen. The magnetic rubber was also attached to the reverse side of each card, which could be pasted to the whiteboard in the class. All the cards were adjusted to be a size of a 15 cm square, for easy handling.

The English class was performed for students with intellectual disabilities, in the senior high school division. On the day the university students attended, 16 younger students were supported by six teachers in the class. The 16 students were divided into three groups, and thus one-third of all the cards with dot codes were distributed to each group. Each student has a different disability; for example, student K could not sit on the chair. However, all the students, including student K, could touch the cards with joy and mimic the replayed voices. Student K could not repeat the correct pronunciation reproduced with the speaking-pen, but he pronounced "fried eggs" when he touched the illustration card of an egg. And, some of the students could nicely understand the relation between the English replayed from the speaking-pen and the corresponding illustration, and they could also repeat the replayed English voices (Figure 11a). In one group, the teachers replayed the English voices only, while hiding the illustration, and then asked the students to look for the proper corresponding illustration. All the students could tackle the activity with joy, saying "Interesting," "Wonderful," and so forth. The schoolteachers said, "We can perform various activities with these cards by finding other instructional ways," "Usually the students have shame and do not speak willingly, but in the today's class, many students tried to speak voluntarily (Figure 11b)."

Shigeru Ikuta created EPUB 3 eBooks with Media overlays in collaboration with the university students and schoolteachers by using the software FUSEe. To realize the reading aloud, the SMIL language was applied to some eBooks, and JavaScript was applied to the books that had a substantial amount of text with a small font or that had almost no text on the page. In creating some eBooks (Figure 12a), both the SMIL and JavaScript were applied, since Shigeru Ikuta and the schoolteachers hoped that many students, with a variety of different disabilities, could handle the eBooks and read them efficiently. These eBooks (Figures 12a, 12b, and 12c) were used in the classes, where the iPad screen was projected into a large TV screen so that all the students could listen to the reading aloud, while confirming the synchronizing of the corresponding text highlighting. The eBooks with Media overlays really help schoolteachers conduct their lessons, and they help students not only to learn sound–symbol correspondence and sound out words, but also to improve reading ability and increase vocabulary skills.

Hikawa Elementary School, Okutama, Tokyo

Hikawa Elementary School is located in the mountains in Tokyo and has only 58 students in the school in the 2019 academic year. The school has six public classes and one special-needs class. This school does a book-reading activity very nicely in the whole school and many protectors are involved in reading a picture book to the students (Figure 13a). This school still uses the ScanTalk developed by Olympus Co. (1999) to create self-made teaching materials and perform various book-reading activities (Figure 13b).

Figure 12. a. eBook created with SMIL and JavaScript; b. eBook created with JavaScript; and c. eBook created with SMIL.

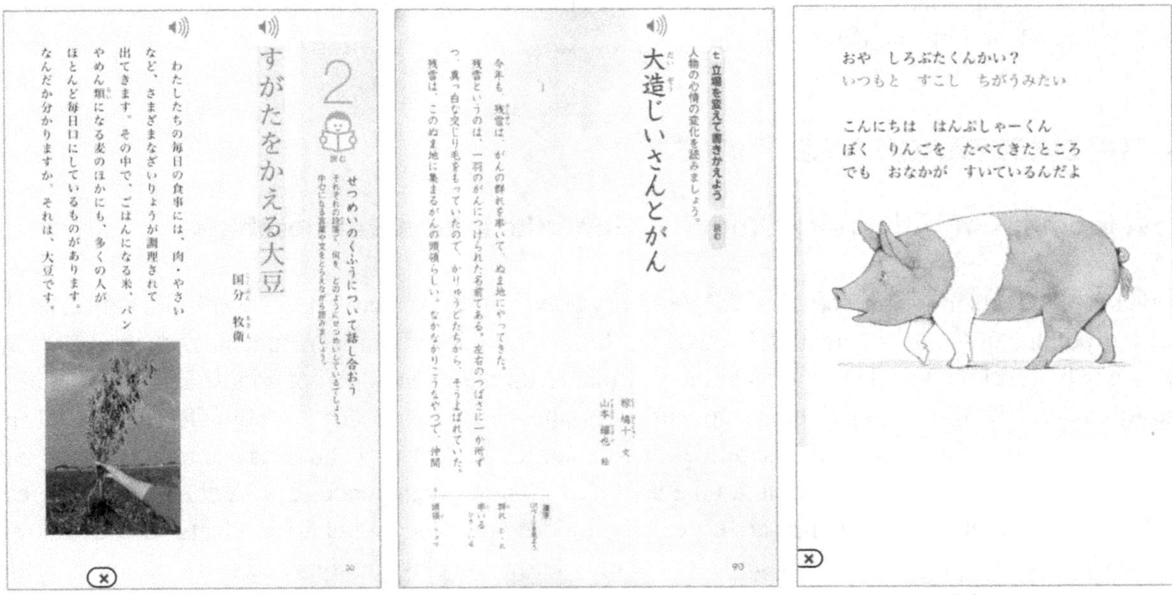

(a) (b) (c)

To help the schoolteachers perform English lessons, Shigeru Ikuta and the university students created several reading-aloud picture books with multimedia-enabled dot codes, in collaboration with Mariko Oshima and the assistant language teacher (ALT) for English lessons. The picture books created are *Where the Wild Things Are*, *999 Tadpoles*, *Brown Bear Brown Bear What Do You See*, *White Snow Bright Snow*, *Swimmy*, and *Fortunately*. The reading-aloud for each page was linked to the corresponding "Post-it-like" sticker icon pasted on the page, by using the software Sound Linker. These reading-aloud books with speaking-pens are usually placed at the school library so that all the students can try them (Figures 13c). Through the long-term activities with both ScanTalk and Grid Onput systems, all the students in this school are very familiar with the Information Communication Technologies and nicely use them.

Figure 13. a. Performing a reading activity by a parent; b. Self-made content corner with Sound Reader; and c. Self-made content corner with a speaking-pen.

(a) (b) (c)

In some of the books, Russian reading-aloud capabilities are included, since the ALT was born in Russia. The students compare the English reading to the Russian reading with some kind of wonder. The schoolteachers hope that this attempt will help the students know there are many countries and many languages spoken in the world.

FUTURE RESEARCH DIRECTIONS

Development of University Students-Schoolteachers Partnerships

University students in the Shigeru Ikuta's laboratory go to schools several times a week and help schoolteachers conduct lessons and create self-made teaching materials for younger students in their classes by using newly developed educational technologies like multimedia-enabled dot codes, EPUB 3 eBooks, and Augmented Reality. This university students–schoolteachers partnership is really effective not only to support the younger students with disabilities and/or significant learning and behavioral difficulties, but also provide other students who want to take lessons with newly developed, 21st century technologies.

These collaborative partnerships are especially important for the university students who are taking pre-service teacher programs. Such students not only know the class lessons and learn the most at the schools, but also are more encouraged to become a teacher. It may be crucial that university professors send university students to schools, after the students acquire some knowledge on newly developed educational technologies and know how to handle them. The author S. I. firmly believes that university professors, who know well the real education performed at schools, should collaborate with the developers who want to produce creative and invaluable educational software and tools. Needless to say, university professors have politely supported the university students' going to special-needs and public schools as voluntary supporters, and a careful analysis on the outcomes of such partnerships could be always conducted.

It is well-known that a child's relationship with their teacher is one of the most important fundamental factors to their success and happiness at school (Stade, 2018). John Hattie (2008, 2011) found that the relationship between a student and their teacher was more important than self-concept, class size, homework, and comparative age within the class. Student-faculty partnerships in higher education are defined as a collaborative, reciprocal process through which all participants have the opportunity to contribute equally, although not necessarily in the same ways, to curricular or pedagogical conceptualization, decision making, implementation, investigation, or analysis (Cook-Sather, et. al. 2014; Doktor, 2016). The partnerships between university students and schoolteachers described in this paper may be organized theoretically based on these invaluable concepts in more profound way.

Development of Software and Hardware

Recently, Gridmark Inc. has put an updated version of Sound Linker program into the market that can handle new sheets of the "Post-it-like" sticker icons, with the maximum number of icons being 1,000 and the maximum available modes being 10. One of the present authors, Shigeru Ikuta is also developing a new software for macOS users who want to create contents for a speaking-pen. By touching "Post-it-like" sticker icons with a speaking-pen (G-Speak or G-Talk pen), users can replay the audio files nicely. A standalone application, created with the File Linker program, can replay multimedia-like movies on

Windows PCs; touching "Post-it-like" sticker icons with a dot-code reader (G-Speak or G-Pen BT) can replay the linked multimedia perfectly. These software packages are very easy to use for teachers, even those who are not good at ICT. These software packages and new sheets are distributed to schoolteachers without cost through S. I.

Gridmark Inc. is still developing two software programs, (1) the GM Authoring Tool that can put dot-codes on your designed documents and (2) the Gridmark Content Viewer (GCV) that can create contents available for iOS devices, with the help of author Shigeru Ikuta's project. A continuing development of easy-to-handle and cheaper software and tools are crucial to continue and deepen fruitful partnerships between schoolteachers, university students, and university professors.

Development of Original, Self-Made Teaching Material

The author Shigeru Ikuta has been collaborating with schoolteachers for more than 14 years in the creation of "original," self-made teaching material and conducting school activities; over 200 schoolteachers all over the world have joined author Shigeru Ikuta's research project. University students also have been joining this project to create self-made teaching material and to conduct school activities for young students in collaboration with teachers at both special-needs and public schools (Ikuta et al., 2020a).

The content that is useful and effective for one student is not always useful for other students, as each student has a different set of needs and desires, degree of ability, and learning history. Therefore, teachers need to create independent content customized to each student; to do so, the authors deeply feel that developers should design the creative software and tools based on a profound understanding of education and real classes at schools. By doing so, the teaching materials created with such newly developed software and tools could be adapted effectively to each individual.

CONCLUSION

In collaboration with schoolteachers, Shigeru Ikuta and his university students have joined both general and special-needs classes and schools, creating plenty of original, self-made content, and helping the younger students deepen their learning and understanding. S. I. has created specially designed sheets with "Post-it-like" sticker icons overlaid with multimedia-enabled dot codes beforehand. Further, he has provided schoolteachers with speaking-pens, dot-code readers, and software applications in addition to such sticker icons free of charge. The schoolteachers can now create self-made content for the students in their classes without purchasing anything.

The fruitful collaboration between university students and schoolteachers provides the younger students with skillful class activities by using recently developed educational technologies such as multimedia-enabled dot codes, EPUB 3 eBooks with Media overlays, and Augmented Reality. The younger students feel friendly to the university students, since their ages are relatively nearer to the university students' age than to their teacher's age. The activities based on collaboration not only help schoolteachers equip classrooms with 21st century technologies, but also enhance their skills in educational technologies. For the university students, especially those taking a pre-service teacher program, such a collaboration is really effective to not only know the class lessons, but also to encourage them to become teachers.

The lessons on the university campus that are based on active learning and critical thinking are of course important for the university students, but the collaborative activities with schoolteachers outside the campus, as the authors presented in this paper, give them a real-life learning experience.

The study was approved by the Ethics Committee of Otsuma Women's University. Before the study, the schoolteachers received permission from their principals and parents to conduct the relevant school-based activities.

ACKNOWLEDGMENT

Author/researcher Shigeru Ikuta thanks "Grant-in-Aids for Scientific Research" (C) (JP16K04844) and "Otsuma Grant-in-Aid for Individual Exploratory Research" (Grant Numbers: s2810 and s2912). He also thanks the teacher Shinichi Onizawa at Aiwa Elementary School, Tama, Tokyo, for his helpful collaboration. The authors thank all the students and teachers at Honmachidaihigashi Elementary School, Machida, Tokyo; Aiwa Elementary School, Tama, Tokyo; Fuchu Keyakinomori Gakuen Tokyo Metropolitan School for the Physically Disabled and the Intellectually Disabled, Fuchu, Tokyo; and Hikawa Elementary School, Okutama, Tokyo. The authors would like to thank eContentPro International (www.econtentpro.com) for the English language review.

REFERENCES

Alli, R. A. (2018, November 12). *What are the types of autism spectrum disorders?* WebMD. https://www.webmd.com/brain/autism/autism-spectrum-disorders#1

Apple. (2019, October 28). Apple Books *Asset Guide 5.2.11*. https://help.apple.com/itc/booksassetguide/en.lproj/static.html#itcd0c2f7c59

Audacity. (n.d.). *Free, open source, cross-platform audio software*. https://www.audacityteam.org

Cook-Sather, A., Bovill, C., & Felten, P. (2014). *Engaging students as partners in learning and teaching: A guide for faculty*. Jossey-Bass.

Doktor, S. (2016, January 7). *5 Ideas to Shift Learning into a Co-created Teacher and Student Partnership.* https://www.pearsoned.com/5-ideas-to-shift-learning-into-a-co-created-teacher-and-student-partnership/

FUSE Network. (n.d.). *Software FUSEe*. https://www.facebook.com/FUSEeOFFICIAL/?ref=py_c

Gallup, J., Perihan, C., Tatsuma, Y., & Ikuta, S. (2021). Creating Inclusive Functional Content Using Dot-Codes: An Exploration of Multistep Recipes for Individuals With Autism in Post-Secondary Settings. In Y. Kats & S. Fabrizio (Eds.), *Education and Technology Support for Children and Young Adults With ASD and Learning Disabilities* (pp. 149–166). IGI Global. doi:10.4018/978-1-7998-7053-1.ch008

Gridmark. (n.d.). *Grid Onput*. http://gridmark.co.jp/technologies#section02

Hattie, J. (2008). *Visual learning: A synthesis of over 800 meta-analyses relating achievement*. Routledge. doi:10.4324/9780203887332

Hattie, J. (2011). *Visual learning for teachers: Maximizing the impact of learning*. Routledge.

Ikuta, S., Ishitobi, R., Nemoto, F., Urushihata, C., Yamaguchi, K., & Nakui, H. (2017). Handmade content and school activities for autistic children with expressive language disabilities. In Y. Kats (Ed.), *Supporting the education of children with autism spectrum disorders* (pp. 85–115). IGI Global. doi:10.4018/978-1-5225-0816-8.ch006

Ikuta, S., Morton, D., Kasai, M., Nemoto, F., Ohtaka, M., & Horiuchi, M. (2015). School activities with new dot-code handling multimedia. In L. Lennex & K. Nettleton (Eds.), *Cases on instructional technology in gifted and talented education* (pp. 311–338). IGI Global. doi:10.4018/978-1-4666-6489-0.ch015

Ikuta, S., Nagano, S., Sato, S. E., Kasai, M., Ezoe, T., Mori, K., & Kaneko, C. (2019a). Original teaching materials and school activities with e-books containing media overlays. In S. Ikuta (Ed.), *Handmade teaching materials for students with disabilities* (pp. 76–110). IGI Global. doi:10.4018/978-1-5225-6240-5.ch004

Ikuta, S., Tamai, Y., Odaka, M., Nakazawa, Y., Sato, M., Mitamura, R., Akiyoshi, T., & Abe, S. (2020a). University students–schoolteachers partnership with newly developed technologies. In D. Schmidt-Crawford (Ed.), *Proceedings of Society for Information Technology & Teacher Education International Conference* (pp. 1746–1750). https://www.learntechlib.org/primary/p/215951/

Ikuta, S., Yamashita, S., Higo, H., Tomiyama, J., Saotome, N., Sudo, S., Hoshi, S., Endo, T., Narushima, T., Suzuki, K., & Watanuki, M. (2019b). Original teaching materials and school activities with multimedia-enabled dot codes. In S. Ikuta (Ed.), *Handmade teaching materials for students with disabilities* (pp. 50–75). IGI Global. doi:10.4018/978-1-5225-6240-5.ch003

Ikuta, S., Yoshida, A., Ishitobi, R., Kudo, M., Sekine, M., Yamashita, S., Edagawa, Y., Edagawa, T., & Abe, S. (2020b). Software handling multimedia-enabled dot codes and gifted school activities for students with disabilities. In S. Ikuta (Ed.), *Handbook of research on software for gifted and talented school activities in K-12 classrooms* (pp. 217–242). IGI Global. doi:10.4018/978-1-7998-1400-9.ch010

International Digital Publishing Forum (IDPF). (2011a). *EPUB 3.0*. http://idpf.org/epub/30/

International Digital Publishing Forum (IDPF). (2011b). *EPUB Media Overlays 3.0*. http://www.idpf.org/epub/30/spec/epub30-mediaoverlays.html

International Study Center. (n.d.). *About TIMSS 2019*. https://timssandpirls.bc.edu/timss2019/

Ministry of Education, Culture, Sports, Science and Technology, Japan. (2003). *Students having significant learning and behavioral difficulties*. https://www.mext.go.jp/b_menu/shingi/chousa/shotou/054/shiryo/attach/1361231.htm

Ministry of Education, Culture, Sports, Science and Technology, Japan. (2019a, September 24). *Report on School Basic Survey*. https://www.mext.go.jp/kaigisiryo/2019/09/__icsFiles/afieldfile/2019/09/24/1421554_3_1.pdf

Ministry of Education, Culture, Sports, Science and Technology, Japan. (2019b, February 22). *Present status on special support services in resource rooms*. https://www.mext.go.jp/component/a_menu/education/micro_detail/__icsFiles/afieldfile/2019/03/06/1414032_09.pdf

Moats, L., & Tolman, C. (2009). *Excerpted from Language Essentials for Teachers of Reading and Spelling (LETRS): The Challenge of Learning to Read (Module 1)*. Sopris West.

National Institute for Educational Policy Research. (n.d.). *TALIS 2018: Teaching and learning international survey in Japan*. NIER. https://www.nier.go.jp/kenkyukikaku/talis/pdf/talis2018_points.pdf

National Institutes of Health (NIH). (2018, March). *Autism spectrum disorder.* https://www.nimh.nih.gov/health/topics/autism-spectrum-disorders-asd/index.shtml

National Institutes of Health (NIH). (2020, March 5). *What are reading disorders?* https://www.nichd.nih.gov/health/topics/reading/conditioninfo/disorders

Olympus Co. (1999, September 14). *ScanTalk, scan talk reader, & sound print kobo*. https://www.olympus.co.jp/jp/news/1999b/nr990823r300j.jsp

Organisation for Economic Cooperation and Development. (2018). *PISA: Programme for international student assessment*. OECD. http://www.oecd.org/pisa/

Reveal, H. P. (n.d.). *From print to a whole new reality*. https://www8.hp.com/us/en/printers/reveal.html

Schleicher, A. (2020). *Teaching and Learning International survey: TALIS 2018 Insights and Interpretations.* OECD. http://www.oecd.org/education/talis/TALIS2018_insights_and_interpretations.pdf

Stade, L. (2018, June 2). *The importance of student-teacher relationships: Build the bond.* Linda Stade Education. https://lindastade.com/the-importance-of-student-teacher-relationships/

The DAISY Consortium. (n.d.). *Creating the best way to read and publish.* https://daisy.org

W3C. (n.d.). *HTML & CSS*. https://www.w3.org/standards/webdesign/htmlcss.html

KEY TERMS AND DEFINITIONS

Dot-Code Reader: Multimedia files, such as movies, can be linked to the dot codes overlaid with the GM Authoring Tool and "Post-it-like" sticker icons. The linked multimedia is replayed on the screen of iOS or Windows OS devices by touching the dot codes with a dot-code reader.

EPUB 3 eBook: An electronic book (or e-book) is a book publication made available in digital form, consisting of text, images, or both, readable on the flat panel display of computer or other electronic devices like tablets and smartphones. The users need two types of software: software to create eBooks and software to read the created eBooks.

File Linker: This software can create a standalone application on Windows OS. A maximum of four multimedia-like movies can be linked to each "Post-it-like" sticker icon or dot code overlaid using a GM Authoring Tool. Touching the dot codes with a speaking-pen (G-Speak) or a dot-code reader (G-Pen BT) replays the corresponding multimedia on a screen of a PC.

GCV (Gridmark Content Viewer): This software can create content available on iOS devices, such as the iPad and the iPhone. Using G-Pen Blue with Bluetooth functionality, users can replay the multimedia, such as movies, photos, audio files, and web pages, on the screen of an iOS device.

GM Authoring Tool: This software can overlap the dot codes on users' designed document at any place, in any size, and in any number. A maximum of four audio and multimedia files can be linked to each unique set of dot codes.

Multimedia-Enabled Dot Code: Invisible dot codes developed by Gridmark Inc. are a novel two-dimensional code technology consisting of extremely small dots. Each "Post-it-like" sticker icon or dot code overlaid with a GM Authoring Tool program can be linked to up to four multimedia, such as movies, in addition to four audios/sounds. A simple touch by a speaking-pen and dot-code reader on the dot codes enables a link between the paper and the digital content.

Partnership: Every year, the university students have volunteered to help younger students with various difficulties learn in the class, and help schoolteachers develop self-made teaching materials, by using the recently developed information communication technologies such as the multimedia-enabled dot codes, EPUB 3 eBooks with media overlays, and Augmented Reality.

School Activity: School activities can be improved through the use of original and individual self-made teaching materials and aids developed with newly developed technologies, tailored to each individual student. Collaboration with the university students is crucial to realizing 21st century technological classes in schools.

Sound Linker: A software that can create content for a speaking-pen. A maximum of four audio files can be linked to each "Post-it-like" sticker icon or dot code overlaid using the GM Authoring Tool program; a recently updated one enables users to link a maximum of ten audios/sounds to each dot code. Created content is copied onto an MicroSD card in a speaking-pen. By touching the icon with a speaking-pen, such as G-Speak and G-Talk, a user replays the corresponding audio.

Speaking-Pen: The G-Speak and G-Talk speaking-pens reproduce original voices and sounds simply by touching the "Post-it-like" sticker icons or overlaid dot codes printed directly on a paper. These speaking-pens connected to Windows PCs can also replay multimedia, such as movies, using a standalone application created using the File Linker program.

Chapter 9

Connecting Pre-Service Teachers to Astronauts and the International Space Station Through an Educational Downlink:

An Out-of-This-World Experience

Franklin S. Allaire
University of Houston-Downtown, USA

Becky Kamas
NASA Office of STEM Engagement, USA

ABSTRACT

Pre-service teachers, particularly those focusing on early childhood and elementary education, consistently view science as one of the disciplines they felt the least comfortable teaching. Connecting students, pre-service teachers, and in-service teachers with STEM experts, such as astronauts, through downlinks or other virtual visits, creates invaluable learning opportunities in and out of the classroom. In February 2017, pre-service teachers at UHD connected with Astronaut Joseph Acaba while he was on the International Space Station through an Educational Downlink. It was an opportunity for pre-service teachers to learn first-hand about living and working in space, what life is like for an astronaut, and how to translate the research happening in space into their everyday lives and classrooms. The experience of interacting with Acaba had a profound short- and long-term impacts for the pre-service teachers directly involved in the downlink and shaped how they approach teaching, science education, and NASA-related opportunities for their students.

DOI: 10.4018/978-1-7998-4966-7.ch009

INTRODUCTION

"Station, this is the University of Houston-Downtown. How do you hear me?" A hush fell over the crowd as astronaut Joseph (Joe) Acaba responded, confirming the connection. Suddenly, the crowd erupted with thunderous applause as teacher candidates in Houston, Texas, spoke live with the International Space Station, the culmination of the past semester's activities.

Sometimes, the stars do align to connect students with scientists. Starting in Fall 2017, teacher candidates enrolled at the University of Houston-Downtown's (UHD) teacher preparation program was linked with Education Specialists from NASA's STEM on Station activity. This connection kicked off a semester full of interactions, including hands-on workshops featuring NASA content, the development of experiments for astronauts to perform on the space station, and the ultimate science connection - a downlink with an astronaut.

An in-flight education downlink, or downlink, is an exciting way to connect students of all ages with astronauts aboard the International Space Station. A downlink has the power to change and shape the lives of everyone involved. This chapter explores the steps we, the leads at our respective institutions, took to prepare and engage the teacher candidates in a downlink hosted at the university. We explain the rationale that guided the processes and the challenges faced along the way while sharing the short- and long-term benefits directly from the student participants themselves. It is our hope this chapter will inspire and guide others to pursue similar activities at their institutions.

"Houston, We've Got a Problem"

Teacher candidates enrolled in UHD's teacher preparation program, particularly those focusing on early childhood and elementary education, consistently view science as one of the disciplines they felt the least comfortable teaching. This trend aligns with the *National Survey of Science and Mathematics Education* (2018), in which only 39% of elementary teachers in self-contained classrooms felt well prepared to teach science. Elementary Science Methods courses support teacher candidates with in-class hands-on activities, experiments that engage them in the inquiry process, and opportunities to teach science content to their peers. However, despite these experiences, and heavy investment in science and STEM (science, technology, engineering, and mathematics) at the PK-12 levels, many teacher candidates still do not feel comfortable with science content or confident in their ability to effectively teach as they enter student teaching (Brigido, Borrachero, Bermejo, & Mellado, 2013).

Studies involving elementary teacher candidates and beginning teachers show that a lack of confidence in their ability to teach science negatively impacts both the quantity and quality of science instruction (Brigido et al., 2013; Gunning & Mensah, 2011). Further, Holzberger, Philipp, & Kunter (2013) and Palmer (2006) note that teachers with low confidence will give up after minimal effort or even avoid a content area altogether. Conversely, Huinker and Madison (1997) explain that "the more positive the impact on pre-service elementary teachers' [confidence] during their teacher preparation program, the more likely it is that these individuals will engage in effective teaching behaviors in the future" (p.109). There is scant research examining the root causes of this lack of confidence. However, journal entries and class discussions have shed light on how TCs' negative science-related experiences have impacted them personally and professionally. Additionally, studies such as Frenzel et al. (2015) and Allaire, et al. (In Press) are beginning to examine how strong discipline-specific negative emotions, particularly anxiety, impact the quality and quantity of science teaching.

Several factors likely contribute to gaps in teacher candidates' science knowledge and skills. First, the majority of students in UHD's teacher preparation program seek early childhood or elementary certification. Unfortunately, an all-too-common misconception by students seeking these certification areas is that they will not have to "do science." In fact, in written reflections, teacher candidates have admitted that one of the reasons they sought early childhood or elementary certifications is their lack of science, mathematics, or social studies knowledge. Secondly, UHD has a large, non-traditional population, and although undergraduates must take several science courses to graduate, courses such as chemistry or physics are not required. As a result, students may not have had an in-depth physical science course since high school. Finally, during their field experiences, teacher candidates seeking early childhood or elementary certification see and hear from in-service teachers and administrators that science is not emphasized or prioritized in lower grades due to testing. Therefore, teacher candidates have limited opportunities to engage in science teaching. In follow-up discussions and surveys, teacher candidates admitted that their limited science content knowledge likely impacted the types of experiments they ran along with their ability to collect and analyze data.

Bandura (1997) suggested that enactive mastery experiences - authentic successes at dealing with a particular situation - are among the most influential sources of self-efficacy as they provide authentic evidence of whether an individual can be successful at a specific task. This has led to discussions on how best to enhance science methods courses to develop and support the teaching self-efficacy of teacher candidates. Connecting students, teacher candidates, and in-service teachers with STEM experts, such as astronauts, through classroom visits, downlinks or other virtual means, creates invaluable learning opportunities in and out of the classroom (Adedokun, Parker, Loizzo, Burgess, & Robinson, 2011; Richmond, 2017). In general, STEM-related subject matter experts can share exciting developments in various fields of study and humanize both scientists and scientific research (Houseal, Abd-El-Khalick, & Destefano, 2014; Peker & Dolan, 2012). More specifically, connecting teacher candidates at UHD with Acaba while he was on the International Space Station, was an opportunity to learn first-hand about living and working in space, what life is like for an astronaut, and how to translate the research happening in space into their everyday lives and classrooms.

The Year of Education on Station

In 2017, NASA announced the back-to-back missions of astronauts - and former classroom teachers - Joe Acaba and Ricky Arnold. Their missions would align almost perfectly with the 2017 - 2018 school year, giving birth to the Year of Education on Station (YES). Led by NASA's STEM on Station, a team charged with connecting students and educators to the International Space Station, a plan was developed to capitalize on this unique opportunity. But to understand just how special YES was, you have to start back at the beginning, when NASA first sought to put a teacher into space.

NASA's Johnson Space Center has been the home of human spaceflight for over half a century. The Space Shuttle captured the nation's attention as it frequently lifted off from Florida. However, a great tragedy occurred on January 28, 1986, when the Space Shuttle Challenger exploded shortly after liftoff, killing all on board, including Christa McAuliffe. McAuliffe was NASA's first teacher in space and planned to teach science lessons during her mission.

In the 1990s, NASA created the Educator Astronaut Project to carry on the goals of the Teacher in Space Project, including elevating the teaching profession and inspiring students. Barbara Morgan, who was selected as McAuliffe's back-up, remained involved with NASA following the Challenger accident and continued to work with (at that time) NASA's Office of Education. Morgan was selected as a Mission Specialist in 1998 and, following two years of astronaut training, began official astronaut duties in 2000. She traveled to space on STS-118 in August 2007 as a full-fledged astronaut.

Following Morgan's selection to the Astronaut Corps in 1998, an additional call for educator-astronauts went out. The 2004 Astronaut Class included Joe Acaba, Ricky Arnold, and Dottie Metcalf-Lindenberger, all classroom teachers. Acaba and Arnold flew their first mission on STS-119 in March 2009, and Metcalf-Lindenberger followed on STS-131 in April 2010. Acaba also spent six months on the International Space Station in 2012.

Before arriving at NASA, Acaba taught one year of science at Melbourne High School and for years of math and science at Dunnellon Middle School, both in Florida. Arnold taught middle school mathematics in Maryland, college preparatory biology and marine environmental science in Morocco, middle and high school science in Saudi Arabia, and middle school mathematics in Indonesia and Romania. These two former teachers were poised to help bring the education community onboard the International Space Station during their mission. Acaba was onboard from September 2017 to March 2018. Arnold launched to the space station in March 2018 following Acaba's return to Earth and remained on board until October 2018.

During YES, astronauts on the ISS, including Acaba and Arnold, conducted 64 downlinks with educational organizations in 29 different states plus Washington, DC, Puerto Rico, and Canada, reaching over 300,000 students and educators. Acaba and Arnold also filmed 12 "STEMonstration" videos highlighting various STEM concepts, such as Newton's Laws, in the space station's microgravity environment. Partnerships with Microsoft and the National Science Teaching (nee Teachers) Association helped spread YES resources and opportunities across the country. A partnership with Challenger Learning Centers allowed Arnold to fulfill the mission of the original Teacher in Space, Christa McAuliffe, by filming the lessons she planned to teach from space. And, as you'll learn about throughout this chapter, STEM on Station worked with UHD and their teacher candidates to equip them to inspire and engage the next generation of spaceflight explorers.

The ultimate, lofty objectives of YES were to inspire students in STEM, help teachers stimulate their students' interests in these subjects, and through their students' pursuit of their dreams, advance American achievements in discovery, innovation, and exploration. The coordinated efforts of YES enabled STEM on Station to reach education audiences across the country, engage with influencers and media, and increase the awareness of the International Space Station and its mission.

What is a Downlink?

A downlink is an opportunity for students to speak with astronauts aboard the International Space Station. For twenty minutes, astronauts are live, answering questions about what it is like to live and work in space. To the host organization, the connection is almost as simple as any video conference. The end connection just happens to be about 220 miles up, going 17,500 miles per hour!

Although the time spent talking with the astronauts is only 20 minutes, a downlink encompasses much more than just that conversation. Organizations must develop a comprehensive education plan to help prepare the audience. This plan typically includes a suite of activities designed to provide students enough background to formulate in-depth, thoughtful questions for the astronauts. This can also include field trips or bringing in subject matter experts to talk with students. Moreover, organizations often use the momentum to continue activities beyond the downlink, such as the creation of after-school STEM clubs. Often, organizations choose to situate their downlink around a theme, such as plants in space or how the space station is preparing us to go back to the Moon. Selecting a theme allows organizations to focus their activities, and therefore questions, making for a well-focused experience.

Organizations must also put together an outreach plan to help share the downlink and surrounding activities with their community. A downlink is a time to shine a spotlight on an organization and showcase the unique opportunities afforded to the students. As part of the outreach plan, organizations may reach out to local, state, or federal officials or other VIPs. Organizations also work closely with NASA to put together joint traditional and social media plans to spread the word.

Of course, a connection to space can't happen with the appropriate technology. Fortunately, connecting to the space station is as simple as connecting to the Johnson Space Center via video conference. Television producers and audio engineers at NASA work with Mission Control to connect to the astronauts and pushes the downlink out live on NASA Television for anyone to tune in. STEM on Station works closely with organizations to ensure all technical aspects are in place to ensure a successful connection for the downlink.

Any interested U.S. educational organization can apply to host a downlink through a proposal process covering each of the aspects noted above. Proposals are typically accepted twice a year. A downlink with the International Space Station can help inspire and engage students in STEM.

UHD and Pre-Service Teacher Participants

The University of Houston-Downtown (UHD) was founded in 1974 and is the second-largest university in Houston, serving the educational needs of the nation's fourth-largest city. UHD is a comprehensive four-year university with over 14,000 students. The university offers 44 undergraduate and eight master's level degree programs. The Interdisciplinary Studies of Education B.A. is the second most popular undergraduate degree program, accounting for 10% of the undergraduate population. UHD is a federally-designated Hispanic and Minority-Serving Institution, with approximately 40% of all undergraduates identifying as Hispanic or Latinx. Over 65% of Hispanic undergraduates identify as female. Hispanic females are also the student demographic with the highest graduation rate at 23% (University of Houston-Downtown, 2019).

The Department of Urban Education is part of the College of Public Service. Located in downtown Houston, the department and faculty embrace their roles in preparing undergraduates to be competent elementary and secondary teachers in urban and multicultural settings and leaders in Houston metropolitan school districts. The department offers preparation for both undergraduate and post-baccalaureate students seeking early childhood, elementary, and secondary core certification with supplemental English as a Second Language (ESL) or Bilingual certification. The teacher preparation program includes theory, methods, and field-based courses. Additionally, students work with in-service teachers within selected public schools in several Houston metropolitan districts.

The data presented in this chapter was collected at two different times. The statistical data was collected at the end of the Fall 2017 semester. Undergraduate students who were enrolled in the Science Methods courses were invited to participate in a voluntary online survey about the inclusion of NASA activities (i.e., NASA Educational Specialists coming to class, developing and running an experiment, and engaging in NASA-generated hands-on activities) in the Science Methods course. The survey was based on similar instruments used by NASA Specialists for gauging in-school activities. Undergraduates were asked to rate a series of statements on a Likert-like scale (one = strongly disagree; five = strongly agree), including "I learned something new about NASA and/or the International Space Station," "I will be more effective in teaching STEM concepts introduced in this NASA experience," and "participation in the STEM on Station program provided a focus for learning about the inquiry process in my Science Methods course." This provided Science Methods instructors and NASA Specialists with a window into how the UHD undergraduates received the inclusion of NASA activities.

The qualitative data provided throughout this chapter in the form of block quotes were collected via email surveys from some of the participants approximately two years following the downlink. These surveys enabled the authors to share the educational downlink's personal and professional long-term impacts on the then-undergraduates. Downlink participants were sent a short survey with open-ended questions and invited to write as much or as little as they desired for each of the questions. Some examples of questions are, "What did you like most/least about the NASA-related activities in our science methods class?" "How did it feel to be selected to represent UHD as a participant in the Educational Downlink?" and "How has the Educational Downlink influenced how you teach, what you teach, and the opportunities you try to create for your students?" The quotes used in this book are used with permission with names changed to protect students' anonymity.

Origins of the Partnership

Like many opportunities, the partnership between NASA and UHD was a combination of enthusiastic individuals from both organizations and specific circumstances bringing them together. Partnerships begin for different reasons. They generally follow a pattern that includes initiation, clarification, implementation, and completion (Sargent & Waters, 2004; Buys & Bursnall, 2007). Dr. Juan Sánchez Muñoz, then President of UHD, had been lifelong friends with astronaut Joe Acaba. Acaba served as a Flight Engineer aboard the International Space Station for Expedition 53/54, beginning in September 2017 and lasting until March 2018. Acaba is a former secondary math and science teacher, and he expressed interest in working with teacher candidates at UHD during YES. Working with UHD was an opportunity for Acaba to share his content and teaching knowledge, experience, and expertise during his mission. The President approached the Interim Dean of the College, who, in turn, reached out to Dr. Franklin Allaire, the science lead in the Department of Urban Education at UHD (and an author of this chapter).

During Summer 2017, we met at NASA Johnson Space Center along with the department chair and Interim Dean. The purpose was to meet NASA Education Specialists from STEM on Station and learn about YES and other NASA educational resources and opportunities. Science methods instructors and NASA Education Specialists developed a structure for the partnership and created a tentative schedule for the 2017-2018 academic year. There were three overarching goals throughout the planning process. First was strengthening teacher candidates' science content knowledge and science teacher self-efficacy. The second was increasing the quality and quantity of hands-on opportunities for teacher candidates to engage in the inquiry process. The third was introducing pre-service to and engaging them in NASA's educational resources. NASA and the University agreed to four central aspects of the partnership:

NASA Educational Specialists would visit Science Methods classes and work with instructors to introduce and engage teacher candidates in NASA-related educational resources. In particular, Education Specialists would showcase lessons and laboratory plans available online at the STEM on Station website.

Students enrolled in Science Methods courses would design and conduct scientific experiments using only similar resources and materials available to Acaba aboard the International Space Station. NASA provided instructors with a list of materials available to astronauts to demonstrate various scientific principles. Materials included different sized magnets, gyroscopes, metal and plastic Slinkys, and masses of different shapes, sizes, and densities. Student-generated experiments and results were to be shared with Acaba, who would then perform some of the experiments in space. Through recorded videos, he would share feedback and results with the students to help them refine their experiments and compare results on Earth to those in low-Earth orbit.

The university would host a NASA Open House and invite area elementary and secondary schools to engage in hands-on STEM activities related to NASA's mission and the International Space Station. Due to tight school budgets, few schools in the Houston-metro area visit Johnson Space Center's official visitor center, Space Center Houston, despite their proximity. The philosophy guiding this part of the partnership was that if students could not go to NASA and experience it for themselves, then we would bring NASA to them. We worked with the Department of Urban Education to contact and invite eight urban schools with existing ties to the university. NASA Education Specialists and the International Space Station Program Office members would support various activities and hands-on exhibits to engage the students.

UHD would host a downlink between teacher candidates and Acaba. During the downlink, teacher candidates would have the opportunity to talk live with Acaba and ask questions about science content, teaching, and life aboard the space station.

PREPPING FOR THE DOWNLINK

Just as it takes a dedicated team of professionals to support astronauts during their mission to the International Space Station, a downlink is not and should not be a single-person operation. In addition to working with personnel across multiple NASA offices, preparation for the downlink at UHD required faculty and staff members' coordination across multiple departments and administrative levels. Individuals from the Event Planning and Technology departments reserved space for the event and coordinated with their NASA counterparts to ensure UHD met all of the space and technology requirements. UHD's Office of University Relations published online articles, sent press releases to local news outlets, and coordinated social media messaging to build support and excitement for the downlink.

The bulk of in-class coordination - integrating NASA themes and activities into the science methods courses - fell to course instructors from the Department of Urban Education.

COORDINATING SCIENCE METHODS COURSES, INSTRUCTORS, AND STUDENTS

Maintaining a cohesive schedule and calendar for this project, especially concerning the submission of student work to NASA and scheduling NASA Education Specialists for in-class visits, was challenging for all of the instructors involved. The situation was further complicated because four different instructors taught multiple course sections on three separate campuses.

Nonetheless, Science Methods instructors and NASA personnel endeavored to craft a schedule for the Fall and Spring semesters, including student experiments, classroom visits, and video messages from Acaba. Schedules were approved. Courses were modified. During the first week of classes in August 2017, instructors shared with PSTs that they would be working with NASA Education Specialists and astronaut Joe Acaba as part of their science methods courses and culminating with a downlink in February 2018. Students and instructors alike were excited about this opportunity!

Hurricane Harvey made landfall about one week after classes started for the Fall 2017 semester on August 25, 2017. Rain and subsequent flooding necessitated the closing of all campuses and the cancellation of a week's worth of classes. The cancellation of classes, in turn, required drastic modifications to courses and assignment schedules. The carefully crafted plans agreed upon by UHD and NASA had to be modified to fit the post-Harvey reality of the fall semester. These changes, while necessary, impacted almost all other aspects of the partnership. The Lone Star College-Kingwood UHD campus located in Atascocita, TX, was so severely flooded that it remained unusable for the rest of the semester. As a result, the instructor and students based at that campus were unable to participate in the NASA-based experiments.

Additionally, not all course instructors had the same enthusiasm for the partnership between UHD and NASA. As a result, the amount of class time given to teacher candidates to develop, perform, and refine experiments varied significantly depending on the instructor. While frustrating, the overall experience was positively received by the teacher candidates and has since led to better communication and coordination between Science Methods instructors.

Educational Specialists' In-Class Engagement

Each of the instructors who participated in the partnership arranged to have NASA Education Specialists from the STEM on Station team visit Science Methods course sections under their instruction. The only exception was a single section of Science Methods offered at a satellite campus, severely impacted by flooding from Hurricane Harvey. The in-class engagement was an opportunity for NASA Education Specialists to teach teacher candidates where and how to access online resources. NASA Education Specialists engaged students in hands-on activities. They introduced them to various online resources such as searchable databases of lesson and activity plans and the plethora of resources on the STEM on Station website. The specialists also took teacher candidates on a virtual tour of the International Space Station through Google Expeditions. The tour of the space station served several purposes. First, the virtual tour was a way to focus attention on the International Space Station and Joe Acaba while allowing teacher candidates to have an immersive learning experience. Secondly, it showed teacher candidates how to integrate this type of learning experience and technology into their future classrooms. In this respect, the specialists took time to show the teacher candidates how to use Google Expeditions beyond providing just the tour. Teacher candidates learned how to guide students through a virtual tour of the space station, point out particular features in the modules, and use the available "cheat sheets" containing information and age-appropriate questions to guide student learning.

Science Experiments

Besides engaging in NASA's classroom-ready labs and activities, teacher candidates experienced the inquiry process first-hand by designing and running their experiments. One of the International Space Station's primary purposes is to conduct scientific investigations to benefit life on Earth and prepare us for future spaceflight beyond low-Earth orbit. Therefore, teacher candidates were challenged to create, run, refine, and rerun experiments using similar school-based resources and materials available onboard the space station. Materials included magnets, different-sized containers, gyroscopes, parachutes, elastic bands, paper, metal, plastic Slinkys, and masses of different shapes, sizes, and densities. Like the astronauts - and their future students - teacher candidates, followed safety rules and wore proper safety equipment throughout their experiments.

Teacher candidates were originally going to send their experiment procedures to the International Space Station, where Acaba would choose some to replicate and provide feedback in a recorded video. Without a doubt, Acaba was enthusiastic about working with the teacher candidates. However, several changes to his schedule, including spacewalks, crew changes, resupply ships, and scientific experiments, impacted the amount of interaction between him and the teacher candidates. He was only able to record one video for the students, doing a quick demonstration of how things work in the microgravity environment. Still, it did not feature any of their experiments. This was detrimental to the teacher candidates' enthusiasm, particularly as many followed him on social media. Although they were very excited about working with an astronaut, the realities of dynamic human spaceflight saw their interest in the project waning.

The teacher candidates' science experiments were generally successful, with each group running and refining their experiment through two iterations. Success, however, was not measured in whether an experiment worked or whether their hypotheses were supported. Instead, success came from the teacher candidates' participation and the experience with the inquiry process. The experiment designs included:

measuring changes in the swing of a pendulum or the oscillation of a spring using objects of different masses and densities,

investigating the surface tension of liquids,

testing the range of magnetic fields, and

experimenting with Newton's Second ($F = ma$) and Third (action/reaction) Laws.

Even though the teacher candidates had resources to develop and run their experiments, many found the inquiry process to be quite challenging. Some teacher candidates had difficulty in identifying specific, measurable independent, and dependent variables. Additionally, some used vague or fuzzy dependent variables like "hot" or "cold" without specifying a temperature. Therefore, it was unclear what they were measuring.

Following their experiments and their in-class interactions with NASA Education Specialists from STEM on Station, teacher candidates were asked to develop and submit potential questions for the downlink. Questions were submitted digitally. Any pre-service teacher who had participated in YES through their science methods courses in Fall 2017 could submit a question for the downlink. More than 50 teacher candidates submitted questions and volunteered to participate. From that list, 15 questions from 11 undergraduates and one graduate student were chosen. All who were chosen expressed a mix of excitement and honor. Julia, one of the undergraduates chosen to participate in the downlink, noted that "it was an honor and a pleasure to be selected to represent my university on such an important platform." Karen, another downlink participant, made the connection between the downlink and the science methods coursework: "Being [chosen] to interact with astronaut Acaba gave practical life to our theoretical project ideas."

Both students and questions were purposefully selected to ensure a broad range of topics were covered during the 20-minute downlink. The TCs selected to participate represented as many course sections, campuses, and demographics at UHD as possible.

NASA-UHD Open House

To support the build-up to the downlink, UHD hosted a NASA Open House on Thursday and Friday leading up to the downlink. The Open House gave UHD students access to NASA displays, including a spacesuit and astronaut food, a virtual reality tour of the International Space Station through Google Expeditions, NASA and space-themed artwork, a scale model of the International Space Station, and a variety of educational and technical experts from NASA's Johnson Space Center. Eight elementary and secondary schools with existing relationships with the Department of Urban Education were invited to participate. Over 125 elementary and secondary students, teachers, and administrators visited the campus and participated in the Open House. The event was also open to any UHD students.

In addition to the displays, students from the invited schools participated in hands-on, STEM-focused activities led by NASA Education Specialists. Students investigated Newton's Laws and explored radiation and the electromagnetic spectrum. Students and teachers also received a "briefing" from STEM on Station's Activity Manager (and chapter author) about the mission and life aboard the International Space Station. While there was no formal evaluation of the Open House, conversations with student and teacher participants were overwhelmingly positive, with one student querying if "we [NASA and UHD] could do it again next week."

DAY OF THE DOWNLINK

The day of the downlink finally arrived on February 13, 2018. As students, staff, faculty, and administrators gathered, they learned about NASA and the International Space Station through the same interactive displays and artwork used during the Open House. UHD's downlink was attended by over 300 students, faculty, staff, and administrators, including then-President Dr. Juan Sánchez Muñoz and the Provost. UHD's proximity to NASA's Johnson Space Center also enabled the participation of several officials, including the Deputy Director of Johnson Space Center's External Relations Office and NASA's Director of University Partnerships and Collaboration Office. The downlink was also simultaneously broadcast on both NASA TV and UHD's website. Additionally, the event was published by various organizations, including the University of Houston System and Houston's NBC-affiliate news station, KPRC.

After a few speeches and last-minute preparations, the audience and participants made their way into the auditorium. The excitement of the moment was palpable as over six months of time, effort, and logistics came to fruition. A hush fell over the room as Mission Control checked-in with Acaba aboard the International Space Station to ensure he was ready for the downlink. "University of Houston-Downtown, this is Mission Control Houston. Please call Station for a voice check." Then-President Muñoz initiated the downlink by asking, "STation, this is the University of Houston-Downtown. How do you hear me?" Aboard the International Space station, about 220 miles above the Earth, orbiting at 17,500 miles per hour, floating in microgravity in the Japanese module, Kibo, Joe Acaba replied, "University of Houston-Downtown, welcome to the International Space Station. I hear you loud and clear." Students, faculty, staff, and administrators alike erupted into excited cheers! The downlink had begun.

For the next 20 minutes, teacher candidates asked Acaba pointed questions about his teaching, teaching career, science concepts and content, and life as an astronaut aboard the International Space Station. Some of the questions posed to astronaut Acaba included:

How did your family react when you told them you were accepted into the astronaut program?

Figure 1. A NASA Official uses a scale model to discuss parts of the International Space Station with a teacher during the NASA-UHD Open House for K-12 schools
Courtesy of the University of Houston-Downtown" or "Copyright 2018, University of Houston

Who or what inspired you to become a math teacher?

Pendulum-related experiments were pretty popular last semester. Students looked at factors such as the length or weight of the pendulum and their effects on the speed or the length of time a pendulum would swing. Pendulums rely on gravity, which is not present on the space station. Are there ways to simulate the effects of gravity on the experiments you're performing?

Many of us are preparing to be elementary school teachers. What advice would you give to teacher candidates to make science exciting and engaging for younger students?

What do you and the other astronauts do for special occasions, such as birthdays, American and other countries' holidays, aboard the space station?

Everyone who participated in the downlink and interacted directly with Acaba experienced a range of emotions. The teacher candidates, who generated and asked their questions, summed up the experience succinctly:

Bethany: "The atmosphere was exhilarating! It didn't hit me how big of a moment this was until the room filled with reporters along with staff and students. So many people were there that it made my nerves skyrocket (hehe), but by the end, I was smiling ear to ear with excitement. I felt like I was able to represent and participate in an event that reflected so positively onto my school community."

Figure 2. Then-Dean of the College of Public Service, Dr. Leigh Van Horn (left), takes a virtual tour of the International Space Station with some UHD students
Courtesy of the University of Houston-Downtown" or "Copyright 2018, University of Houston

Julia: "My actual participation in this event was incredibly amazing. I was given the opportunity to self-select a question I was interested in, which made my participation feel personal and relevant. Mr. Acaba answered all of our questions thoroughly and with the passion of an educator."

Karen: "Participating in the Educational Downlink with Astronaut Acaba was the highlight of my academic year at UHD. Here was an actual middle school science teacher living and working on the International Space Station!"

Irene: "It was an unforgettable moment that I'm able to share with my students and kids."

As any teacher will tell you, 20-minutes can go by very quickly. It seemed that it was even more true with the downlink. As soon as it started, it was already time to sign-off so Joe could return to working on scientific research. In his closing remarks, Acaba encouraged teacher candidates to "stick with it" and exclaimed that "there's nothing more important than being an educator." With a final microgravity-assisted somersault, Acaba signed off, and Mission Control concluded the downlink.

POST-DOWNLINK

The downlink was an amazing experience. One pre-service teacher who participated said it was "a dream come true - to talk to an astronaut." In this final section, we explore some of the lessons learned from the downlink and overall partnership, the impact the downlink had on the students who participated, and how UHD and NASA have continued to work together to support and inspire teacher candidates.

Figure 3. Elementary student artwork from the NASA-UHD Open House
Courtesy of the University of Houston-Downtown" or "Copyright 2018, University of Houston

Lessons Learned

While the partnership and the downlink were successful, not everything went according to plan. There were several natural and human-made factors which forced us to adapt. We strongly encourage educational organizations to explore the feasibility of hosting a downlink. There are three significant lessons individuals from both UHD and NASA in this university-community partnership learned throughout the process.

Figure 4. Mr. Matthew Wallace, a NASA Educational Specialist, poses with a spacesuit from the NASA-UHD Open House
Courtesy of the University of Houston-Downtown" or "Copyright 2018, University of Houston

First, support is critical. Buys and Bursnall (2007) explained that when universities and community organizations enter into partnerships, they approach it with different goals, resources, and institutional factors. The downlink and associated activities, such as the in-class projects and Open House, took time, energy, and resources. Arranging the downlink required careful and frequent communications between organizations. We were very fortunate that there was mutual recognition by NASA and UHD that a downlink between an astronaut who was a previous classroom teacher and teacher candidates would significantly raise the university's profile, NASA's STEM on Station activity, and the Year of Education on Station. As a result, both organizations made the partnership a priority and provided the support each

Figure 5. Then-UHD President, Juan Muñoz, welcomes his good friend Astronaut Joseph Acaba to UHD at the start of the Educational Downlink
Courtesy of the University of Houston-Downtown" or "Copyright 2018, University of Houston

needed to ensure success. The level of support did not go unnoticed by the teacher candidates. Karen noted that she "appreciated what a big deal UHD made of the downlink event. To celebrate teachers in space and the educational opportunities it can create is one of the reasons I chose UHD for my graduate education. I hope UHD continues to host similar events." The support for the downlink and partnership was enhanced by the fact that we, the authors, developed a great collegial relationship, providing quick and open communication.

A second lesson learned is that you, and each member of your team, must pay attention to all of the details. There were a lot of moving parts to both the downlink and the partnership as a whole. Naturally, it is up to the host organization as to how big of a deal they want to make of an opportunity. Of course, we recommend turning a downlink into a large-scale event that reaches as much of the student population and the surrounding community as possible. The downlink alone required the coordination of faculty, staff, and administrators across multiple university departments. The additional facets of the partnership - the in-class experiments and work with NASA Education Specialists as well as the Open House - required careful coordination across multiple course sections and campuses as well as outreach to community partners. We strongly suggest taking thorough and detailed notes, asking questions as soon as something is happening that you don't understand, communicating with stakeholders regularly within and between organizations, and delegating specific tasks to individuals. An undertaking of this magnitude requires its own "Mission Control" at the host organization.

The third and final lesson learned is to be flexible - or at least as flexible as possible. Again, there are a lot of moving parts when it comes to hosting a downlink. Some aspects are under your control, and other elements that are totally out of your control. There may also be factors - like a hurricane or a government shutdown - that force you to change your original plans. What complicates matters, at least in our experience, is that there is a vision for what the downlink will look and feel like. While this vision

Figure 6. An audience of NASA officials and UHD administrators, faculty, and students watch the live downlink as participating UHD students ask Astronaut Acaba questions about teaching, being an astronaut, and life aboard the International Space Station
Courtesy of the University of Houston-Downtown" or "Copyright 2018, University of Houston

is essential, be prepared for reality to not precisely match what you have in mind. Human spaceflight, among many of the other factors mentioned, is dynamic. The only constant changes and that can be challenging in specific environments and for certain institutions.

Student Impact

When the dust settled, Science Methods instructors surveyed teacher candidates in-class NASA experiences and their participation in the downlink. Teacher candidates who responded to the voluntary survey were overwhelmingly positive about their experience. Over 86% of respondents agreed or strongly agreed that participating in the Year of Education on Station was a good investment of time in their Science Methods courses. Similarly, 87% of respondents agreed or strongly agreed that participation in the Year of Education on Station provided a focus for learning about the inquiry process, and 92% enjoyed learning about the inquiry process through the Year of Education on Station. Seventy-seven percent of respondents noted their experiences with NASA inspired them to bring NASA-related content into their classrooms. Finally, 90% of respondents felt more confident, and 64% thought they would be more effective at teaching STEM concepts introduced during their NASA experience.

It is unclear why some teacher candidates did not feel they would be more effective at teaching STEM because of the NASA experience. Unfortunately, we were unable to clarify TCs' survey responses due to the anonymity of the surveys used and their deployment over multiple classes. Several factors could have influenced their responses. As explained earlier in this chapter, many TCs were not comfortable with teaching science. If we view the 39% reported in the NSSME+ as a baseline, an increase to 64% is quite an improvement. As the partnership between UHD and NASA continues, we will work to improve upon

Figure 7. A UHD pre-service teacher excitedly asks a question to Astronaut Acaba during the Educational Downlink
Courtesy of the University of Houston-Downtown" or "Copyright 2018, University of Houston

that percentage. Secondly, as was noted previously, not all instructors approached this opportunity with the same enthusiasm level. Therefore, lower ratings could have been a direct result of an instructor who did not put in the time or effort to make it exciting or memorable. In this way, the experience becomes an important lesson in how influential teachers' attitudes towards topics can be for their students. Finally, while it may be hard to imagine, there are students (K12 and undergraduate) who are not excited by NASA and the ISS. As the partnership continues, we will strive to make better connections to branches of science beyond physics and earth/space science.

Beyond the survey, teacher candidates described the short- and long-term impact of the downlink, including students in their current classrooms:

Bethany: "The entire experience had a positive impact on me and my outlook as a teacher and a student. From the NASA activities to the Educational Downlink, I was reminded of the excitement that can come from learning. This opportunity allowed me to see the significance of making learning fun and how impactful we can be with students. As an educator, I feel we tend to focus heavily on what students are expected to know versus what they would like to know, so it's important to allow students to learn opportunities based on their interests, which can help us find and make connections through doing so. I hope to impact and support my students' desires to learn the way that was for me!"

Julia: "The Educational Downlink made me feel extremely connected to my profession and a renewed sense of wonder and excitement that followed me into my first year of teaching. Although I teach Language Arts in elementary school, I have still been able to bring this experience into my classroom and share it with my students in meaningful ways."

Karen: "It was an experience I was able to use in teaching my own 7th-grade science students. I was able to breathe more life into the Space and Technology unit for my students. A passionate teacher can influence more students than a teacher bored with the required curriculum. I have been finding ways to identify and expand the interest and knowledge my students have for microgravity. It has been a fun adventure."

Irene: "Participating in the Educational Downlink has had a positive impact on my life. It has opened my eyes to see how little science is taught in schools, especially for the younger grades. It has shown me how important it is to teach science in every grade."

The Mission Continues

Even though years have passed since the beginning of this partnership, the mission continues. Building a sustainable partnership requires both parties to seek opportunities to work together. We have built a lasting, positive relationship that continues to bring NASA and UHD pre-service students together each semester (Allaire & Kamas, 2020). NASA Education Specialists from the STEM on Station team join Science Methods instructors in class to showcase NASA resources such as Google Expeditions, Commercial Crew Virtual Reality experiences, NASA's Eyes, the Office of STEM Engagement website containing over 1,300 classroom-ready lessons and labs, and ways to connect with the NASA Educator Professional Development Collaborative. In-class and virtual engagements align with suggestions from Adedokun et al. (2011) and Richmond (2017) and introduce teacher candidates to some of the lessons and activities developed by NASA and its partners. Activities include "Mass vs. Weight: Air-Powered Mass," "Eggstronaut Parachute Challenge," and "Sound on a String." These and many activities are available online, free of charge, and align with the Next Generation Science Standards. These on-going interactions allow UHD and NASA to continue the partnership, making each other natural "go-to" organizations when an opportunity presents itself.

As the teacher candidates graduated and ventured into classrooms of their own, they continued to be inspired by astronaut Acaba and their downlink experience. Bethany, who is now an elementary teacher, explained, "The entire experience had a positive impact on me and my outlook as a teacher. From the NASA activities to the downlink...this opportunity allowed me to really see the significance of making learning fun and how impactful we can be with students." Similarly, Karen noted, "The educational downlink experience inspired me to reconnect with my interest in space exploration and education." Former teacher candidates have connected their students to NASA resources, thus inspiring and engaging the next generation of spaceflight explorers. Julia shared that "participating in the Educational Downlink with astronaut Acaba provided me with the experience and excitement needed to promote and stimulate a love of science in future generations. It gives me hope to think that one day one of my students might choose to be a scientist or an astronaut because of what I was able to share with them from my time at UHD."

REFERENCES

Adedokun, O., Parker, L., Loizzo, J., Burgess, W., & Robinson, J. (2011). A field trip without buses: Connecting your students to scientists through a virtual visit. *Science Scope*, *34*(9), 52–57.

Allaire, F. S., & Kamas, B. (2020). Science That's Out of This World: Connecting Elementary Pre-Service Teacher Science Methods Courses to NASA and the International Space Station. *Connected Science Learning, 2*(2).

Allaire, F.S., King, J.P., & Frenzel, A.C. (in press). *Development and Validation of the Science Teaching Emotion Scales (Sci-TES)*. Academic Press.

Bandura, A. (1997). Self-efficacy: The exercise of control. *The Freeman*.

Brígido, M., Borrachero, A., Bermejo, M., & Mellado, V. (2013). Prospective primary teachers' self-efficacy and emotions in science teaching. *European Journal of Teacher Education*, *36*(2), 200–217. doi:10.1080/02619768.2012.686993

Buys, N., & Bursnall, S. (2007). Establishing university-community partnerships: Processes and benefits. *Journal of Higher Education Policy and Management*, *29*(1), 73–86. doi:10.1080/13600800601175797

Frenzel, A., Becker-Kurz, B., Pekrun, R., & Goetz, T. (2015). Teaching this class drives me nuts!-Examining the person and context specificity of teacher emotions. *PLoS One*, *10*(6), e0129630. Advance online publication. doi:10.1371/journal.pone.0129630 PMID:26053623

Gunning, A., & Mensah, F. (2011). Pre-service elementary teachers' development of self-efficacy and confidence to teach science: A case study. *Journal of Science Teacher Education*, *22*(2), 171–185. doi:10.100710972-010-9198-8

Holzberger, D., Philipp, A., & Kunter, M. (2013). How teachers' self-efficacy is related to instructional quality: A longitudinal analysis. *Journal of Educational Psychology*, *105*(3), 774–786. doi:10.1037/a0032198

Houseal, A. K., Abd-El-Khalick, F., & Destefano, L. (2014). Impact of a student-teacher–scientist partnership on students' and teachers' content knowledge, attitudes toward science, and pedagogical practices. *Journal of Research in Science Teaching*, *51*(1), 84–115. doi:10.1002/tea.21126

Huinker, D., & Madison, S. (1997). Preparing efficacious elementary teachers in science and mathematics: The influence of methods courses. *Journal of Science Teacher Education*, *8*(2), 107–126. doi:10.1023/A:1009466323843

Palmer, D. (2006). Sources of self-efficacy in a science methods course for primary teacher education students. *Research in Science Education*, *36*(4), 337–353. doi:10.100711165-005-9007-0

Peker, D., & Dolan, E. (2012). Helping students make meaning of authentic investigations: Findings from a student-teacher–scientist partnership. *Cultural Studies of Science Education*, *7*(1), 223–244. doi:10.100711422-012-9385-3 PMID:23828722

Richmond, G. (2017). The power of community partnership in the preparation of teachers. *Journal of Teacher Education*, *68*(1), 6–8. doi:10.1177/0022487116679959

Sargent, L. D., & Waters, L. E. (2004). Careers and academic research collaborations: An inductive process framework for understanding successful collaborations. *Journal of Vocational Behavior*, *64*(2), 308–319. doi:10.1016/j.jvb.2002.11.001

KEY TERMS AND DEFINITIONS

Educational Downlink: In-flight education downlinks, or downlinks, are opportunities for students and educators to interact. with astronauts aboard the International Space Station during a live, 20-minute question-and-answer session.

NASA: The National Aeronautics and Space Administration.

Science: The intellectual and practical activity encompassing the systematic study of the structure and behavior of the physical and natural world through observation and experiment.

Pre-Service Teachers: A pre-service teacher is a college student who is pursuing a teacher certification.

Scientist: An individual who engages in the scientific process.

Space Station: The International Space Station is a modular space station in low Earth orbit. It is a multinational collaborative project involving five participating space agencies: NASA (United States), Roscosmos (Russia), JAXA (Japan), ESA (Europe), and CSA (Canada).

STEM: Science, Technology, Engineering, and Mathematics.

Teacher Preparation: A teacher preparation program is designed to prepare both undergraduate and graduate students to become licensed teachers.

Section 4

Chapter 10
Training Students to Conduct Outreach in an Informal Science Learning Environment

Laura Wagner
The Ohio State University, USA

Leslie C. Moore
The Ohio State University, USA

Kathryn Campbell-Kibler
The Ohio State University, USA

ABSTRACT

This chapter reviews a science outreach training course designed for the Language Sciences Research Lab, a working research facility embedded inside of the Center of Science and Industry (COSI). The course integrates training in informal science education theory, background in a specific science area (language science) and extensive hands-on work at COSI where students engage museum visitors in interactive language-focused activities. Key components of the course are described and a formal evaluation of the course outcomes is reported. The authors discuss how the course enhances an existing partnership between COSI and Ohio State University, as well as ways in which its core elements have been adapted to other fields.

INTRODUCTION

Academic research has long had a reputation for being done inside of an ivory tower, with little regard for how it connects with the concerns of society or is perceived by the general public. In recent years, however, there has been a growing interest in making such academic work more accessible (Baron, 2010; Meredith, 2010; Wu, 2017) and training scientists to communicate effectively with the broader society (Besley and Tanner, 2011; Neeley et al, 2014). Fostering public understanding of science has the

DOI: 10.4018/978-1-7998-4966-7.ch010

potential to foster a more pro-science societal climate at a time when trust in science is a matter of some concern (Pew Research Center, 2019). Moreover, there is evidence that the process of engaging with the public may help researchers clarify their goals and provide a source of inspiration for new research (Lidz & Kronrod, 2014). However, effective public engagement, like any skill, is one that benefits from training in core techniques and opportunities for practice. Informal science institutions (ISI), such as science museums, are well attended by the public and serve as important venues for learning across the lifespan (Falk & Dierking, 2010). As such, they can play an important role in the advancement of science communication by scientists: they are key sites for accessible science communication to happen, and ISI educators can support the training of scientists and future scientists in the theory and practice of science communication.

We describe in this paper an Ohio State University (OSU) training course that teaches undergraduate and graduate students to engage in accessible public communication about language science topics within a science museum, the Center of Science and Industry (COSI). The course has a strong experiential learning component (Moore, 2010; Perrin, 2014) and reflects aligned goals of a longstanding partnership between OSU and COSI, devoted to promoting both the practice of science (through the collection of scientific data) and public understanding of science. We begin by contextualizing the course within this larger partnership and then discuss the course more specifically.

The PARTNERSHIP

The training course is taught in conjunction with the Language Sciences Research Lab (https://u.osu.edu/thebln/language-pod/), an OSU working research lab embedded within COSI (Wagner et al., 2015). Affectionately known as the "Language Pod," the lab is dedicated to the study of language and the dissemination of knowledge about language sciences. It is one of three research pods, called the Labs in Life, located in COSI's Life exhibition on the human body and mind. These glass-enclosed research spaces allow museum visitors to observe actual scientific research as it is occurring in a research-in-real-time experience (see Figure 1). Many science centers invite scientists to carry out experiments at the museum so that visitors may have firsthand exposure to the research process (King et al 2018), but COSI is the rare science center that has permanent working research laboratories integrated into the museum's public floor. The Labs in Life were created jointly by COSI and OSU to showcase Ohio State research and foster interaction between researchers and the public, with the goals of creating excitement about science and encouraging children and youth to pursue careers in science. Researchers from the College of Optometry and the School of Education's Exericise Science program were previous residents of the exhibit. Currently, the exhibit hosts researchers from the Communications department and the College of Pharmacy, in addition to the Language Pod.

The research projects conducted within the Language Pod are run by OSU faculty from multiple departments, with the majority coming from the departments of Linguistics, Psychology, and Speech & Hearing Sciences. The faculty work in a collective fashion, with each faculty member taking responsibility for their own research projects. The specific topics studied by the Language Pod have included the development of children's perception of regional dialect differences, the online processing of intonation and stress, the conceptual-semantic processing of plural marking, children's understanding of verb endings, and the ways that adults' expectations about individuals influence how their speech is perceived. Visitors to COSI are able not only to see this real research being conducted, but also have the

Figure 1. The Language Sciences Research Lab testing area, as seen from the floor of the COSI museum.

opportunity to participate in these studies themselves. Such participation is a popular activity at COSI, and approximately 3000 visitors volunteer to serve as research participants each year.

Beyond its research mission, the Language Pod is committed to promoting public understanding of language research. This outreach mission is accomplished in part through explanations of the ongoing research projects. Each study is accompanied by a debriefing explanation (typically provided by a student research assistant) which addresses the research process on several levels, including the research content (What problem is this researcher studying?), the research methods (How is this researcher doing her study?), the larger research goal (Why is this researcher studying this part of language?), as well as the scientific method more generally (How do we ask and answer a question scientifically?). In addition, the lab has developed a suite of brief, hands-on demonstration activities that showcase fundamentals from the field of language science. These activities are conducted by students on the floor of the museum with visitors and are central to the students' experience in the outreach training course.

The Labs in Life exhibit which houses the Language Pod is part of a longstanding relationship between COSI and OSU that encompasses a wide range of programs and interactions. The OSU-run public radio station, WOSU, has a permanent exhibit at COSI and does occasional programming with the museum. In addition, as part of COSI's Interactive Video Conferencing Program, a surgery team member from OSU's Wexner Medical Center narrates live surgeries to thousands of high school students and the general public throughout the United States. The Labs in Life partnership takes place within and strengthens this larger relationship between the institutions. The Language Pod, established in 2012, benefits both COSI and OSU's missions in direct and indirect ways. On the one hand, COSI guests have opportunities to engage with working scientists – and the actual practice of science – in robust and meaningful ways. They learn about how science is practiced and see a diverse and dedicated team engaged in the day-to-day details of constructing scientific knowledge. On the other hand, OSU researchers gain access to a broader range of potential participants than the traditional undergraduate student, especially including children and older adults. In addition, researchers are able to share their research and their passion for the language sciences with a broader public and to develop their science communication skills.

Museum-university partnerships can be mutually beneficial and enriching for a wider public, and that has been our experience in the Language Pod. However, to be part of science center-university partnership is to work across fields, which presents challenges along with benefits. Callanan (2012), a developmental psychologist who has collaborated with a children's museum for several years, gives good counsel on how to approach such partnerships: find overlapping goals, get to know the organization, recognize museum staff as professionals, recognize that your outsider status has both benefits and drawbacks, give something back, and keep an open mind. In pursuing the core mission of the Language Pod – to advance the study of language and disseminate knowledge about the language sciences to a broader public – we are continuously learning how to achieve our goals in ways that COSI recognizes as supporting their goals and recognizing their institutional needs and constraints. As researchers, we have learned to adapt our data collection practices to the science center's priorities with regards to visitors' experiences, which meant that study tasks need to be fairly brief and, if possible, visually interesting for the viewing public. In our outreach work to promote interest in and excitement about language as an object of study, we have learned not only about the hows and whys of informal science education, but also about the whens and the wheres that were preferred by the science center. Cultivating and communicating respect for our COSI colleagues, the institution, and visitors is essential to the success of our work. As Douglas (2016: 66) notes, "we must think hard about how we approach these relationships, how we value the knowledge and expertise that others bring, and how we inhabit the complex interrelated roles of academic researcher, university teacher, and public ambassador for our subject".

One aspect of this partnership is ensuring that the lab offers opportunities for interaction as much as possible during the museum's open hours. Even the most productive of research labs does not need to be open as frequently as a museum needs to be. But while signs and slideshows on a TV offer some engagement during quiet moments, the key benefit of the lab as an exhibit lies in its people. One of the central motivations for the training course was to substantially increase the number of hours that the lab is staffed with OSU personnel ready to engage with the public. Beyond the additional hours, the course also enables a variety of opportunities for interaction between OSU and COSI. COSI staff provide regular orientation and occasional training for the OSU students. In addition, students from the OSU course are invited to help staff special COSI functions (such as Deaf Awareness Day). The course also expands the pool of scientists and future scientists who participate in the OSU-COSI partnership.

An additional function of the course was that it provided a mechanism through which at least some of the lab's faculty researchers could get OSU recognition for their time investment in the lab. While the COSI museum does provide the researchers with a valuable resource (research participants), the additional time and effort needed to create museum-friendly research studies and to train students to serve as public ambassadors of research in addition to data collectors is substantial, and indeed far more than what is required to simply run a typical lab on campus. Moreover, while OSU financially compensates COSI for the use of the space (through annual rent payments), the university does not provide the faculty with any additional compensation for their time. The course, therefore, allows the faculty member teaching it to acquire teaching credit for training students to staff the exhibit and for generally insuring the day-to-day maintenance of the space during the semester.

Of course, beyond the course's utility in maintaining the OSU-COSI partnership, it also serves a primary function of providing students with a rich educational experience that fosters their understanding and interest in research as well as in accessible science communication with the public – a mission common to both OSU and COSI.

THE TRAINING COURSE

Overview

The training course has been a feature of the Language Pod from the first semester it was fully functional as a research lab (Fall semester, 2012). The course was conceptualized from the outset as an experiential learning course (Moore, 2010; Perrin, 2014) in which students would connect their traditional classroom knowledge with real-world experiences engaging with the public. The course has three main goals: 1) increase students' knowledge about language and their confidence in engaging in scientific discussions about language, 2) train the students to present a successful language science demonstration, 3) improve COSI visitors' experience as a result of interacting with a student. Students learn about ongoing research projects within the Language Pod and about critical background concepts from linguistics. Students also learn about informal science learning and are trained in how to communicate language sciences information to the COSI public, including both child and adult visitors. Students spend a significant portion of their time at COSI actively engaged in explaining the language research to visitors, conducting demonstration activities about language science, and assisting in recruitment for the research studies. Three OSU faculty (the first, second, third authors) teach the course in rotation, and each semester the instructor revises the syllabus, building on the work of, and lessons learned by, the other instructors. All three of the instructors conduct research at COSI and have developed strong commitments to and relationships with the institution, including federally-funded research and training programs in which COSI is a partner.

At its heart, the course is about preparing students to be effective language science communicators and educators in informal contexts. Students learn about all the language science demonstrations that are used at the Language Pod, and each student becomes adept in a subset of these demonstration activities over the course of the semester. Students are supported as they develop their presentation skills through structured and informal observations and feedback by the instructor, as well as through structured peer observations. In addition to doing outreach at COSI, students complete several small assignments that are designed to develop their knowledge of language sciences and informal science learning and their ability to communicate about language using the tools of informal science education. There are three substantial written assignments: a discussion of a language myth and how one might dispel it in conversation with a COSI visitor; the comparison of two demonstration activities with respect to their content, goals, and structure; and a write-up of the guidelines for a new demonstration activity or an expansion on an existing one.

In the design and implementation of this course, we have been inspired by COSI and informal science institutions in general. Free choice is a key feature of informal science learning that we strive to honor in the course by encouraging students to put their energies into the readings, language myths, and demonstrations that interest them most. At the same time, we challenge students to push themselves beyond their comfort zone. One of the benefits of teaching at COSI is that the science center is a very welcoming and supportive space for novices to explore and learn by doing. We emphasize this to students and provide them with multiple opportunities in class and at COSI to practice their outreach skills and receive extensive formative feedback before they are formally assessed at the end of the semester.

The Informal Science Education Component

As engaging with the general public in an informal learning environment is the fundamental focus of the course, it is critical for students to learn about core concepts in informal science learning and key strategies of success in the informal science setting. We center our informal science discussions on the Strands Approach to understanding and evaluating the goals and outcomes of informal science education (Bell, et al. 2009). The six strands are listed below.

- **Strand 1:** Spark interest and excitement.
- **Strand 2:** Help people understand scientific content and knowledge.
- **Strand 3:** Support people as they (learn to) engage in scientific reasoning.
- **Strand 4:** Encourage people to reflect on science.
- **Strand 5:** Empower people to engage in scientific practice.
- **Strand 6:** Inspire people to identify with the scientific enterprise.

Each strand reflects a goal for an informal learning encounter, and we treat them as ordered by importance. Thus, students are encouraged to satisfy Strand One (spark interest and excitement) first and foremost: If the COSI visitors do not think that our language science activities are interesting and fun, then they are far less likely to learn from them, either at the museum or after they go home. Often for the youngest visitors (e.g., pre-school aged children) Strand One is the only goal that is accomplished with our activities. For most visitors, however, students are expected to aim for Strands 2 and 3 (helping people understand content and supporting them to engage in scientific reasoning). The written guidelines that accompany our demonstration activities are organized with these strands in mind and clearly articulate core content points to include (especially the "critical take-home message") and suggest interactions that would elicit and support scientific hypothesis testing from the visitors. The additional strands are considered aspirational for the students in the course, although they are more achievable by the research assistants who work with the visitors as participants within the studies.

In addition to this general framework, we also spend time helping students with some very particular practical skills, such as how to avoid using jargon terms when explaining the language science. Technical terminology can foster an air of expert credibility, but it also serves as a barrier for understanding and interest (Bullock et al, 2019; McDonnell et al., 2016). As interest and understanding reflect our two most important learning goals (see Strands One and Two, above), we emphasize the importance of demonstrating phenomenon through the use of videos, pictures and games, and practice re-phrasing complex ideas using only simple terms. The use of jargon terms does help us address the later strands (e.g. Strands Four, Five, and Six) by introducing visitors to scientific terminology, but students are taught to introduce such terms only if they feel they have already accomplished the first two strands.

Our informal science training component also includes specific discussion about how to appropriately navigate the diversity found within an informal learning environment (Fenichel & Schweingruber, 2010). Because our research focus is on language science, an easy entry point to this discussion is the importance of respecting linguistic diversity among the museum guests. Many of the language myths that students learn about and learn how to dispel are concerned with language variation. Students learn how to engage critically and productively with widely held false beliefs, for example that some languages and dialects are better (more pure, logical, complex, beautiful) than others. Our demonstration activities lend themselves to discussions with the visitors about their own linguistic experiences, and students are encouraged to foster such discussions in ways that lead to new understandings about how language functions not only as a means of communicating content but also as a marker of social identity. More generally, we discuss the range of individuals who attend museums such as COSI (cf. Falk & Storksdieck, 2010) and talk about ways in which our own biases may lead us to approach some visitors over others.

Training in Language Science

While our principles can be applied across a wide range of disciplines (see below for adaptations to other fields), Linguistics and Language Science are a natural place to develop new science outreach techniques (Honda & O'Neil, 1993). Because language is such an integral part of people's day to day lives, they already have beliefs, opinions, questions and observations on the topic. Museum visitors are also able to themselves produce linguistic data and experience linguistic phenomena immediately and directly.

The training course is cross-listed in the departments of Linguistics, Psychology and Teaching and Learning and attracts students from other departments as well. Many students come from the department of Speech & Hearing Sciences, since the practice of explaining complex speech-related concepts to a lay audience is highly useful training for future speech-language pathologists. But the course has attracted students from across the university with an interest in science outreach, many of whom have had no prior interest or training in linguistics. Accordingly, one important learning objective is to provide all students with the subject matter understanding necessary to conduct the demos and discuss the relevant linguistics concepts with visitors.

Our training in language science has three major goals. First, we want to ensure that all students have a baseline understanding of the core questions and principles of linguistics and related language science fields. This includes issues like the prescriptive/descriptive divide and the different elements of language processing (e.g. phonetics, phonology, morphology, etc). Second, we need to train students in the concepts underlying the specific demonstration activities in our collection, to ensure that they can conduct and discuss the activities accurately. Finally, we educate students about commonly held language myths and train them on providing fact-based responses to them.

For this training, we assign students reading from linguistics books aimed at a popular audience, such as David Crystal's *How Language Works: How Babies Babble, Words Change Meaning, and Languages Live or Die*. In addition to providing some of the needed information, this text offers another view into how linguistics can be made accessible and interesting for informal learning. We also provide students with access to introductory linguistics materials, such as the OSU Linguistics Department's textbook *Language Files*, and encourage them to use these and other resources to deepen their understanding of specific topics or demonstration activities that catch their interest.

Each activity is accompanied by written guidelines which provide an overview of the language science being demonstrated. The guidelines also contain links and citations to additional reference materials, and students are encouraged to pursue additional reading. Students are allowed to choose which demonstration activities they wish to use at COSI, and they are encouraged to select a handful of specific demonstration activities most relevant to their own interests. Example guidelines can be found on our website, along with a few videos of the activities in action: https://u.osu.edu/thebln/language-outreach/

Finally, students are also introduced to current linguistics and language science research through their exposure to the research conducted in the Language Pod. This occurs through formal and informal means. Formally, all students are trained in human subjects protection and added to the lab IRB protocol to allow them to recruit participants for studies being conducted in the lab. As part of this training, students are provided with information sheets about each study and are given presentations from the faculty members conducting the studies. Informally, students interact regularly with the research assistants and researchers conducting the studies. Many of the research assistants are former students from the course, and students often plan to look for research assistant positions in the lab, where their skills in interacting with museum patrons are highly valued. As a result, their day-to-day activities in the Language Pod give them a direct view into the work of linguistics and language science research, supplementing their more basics-focused classroom training.

The Practicum

The informal science and language science material is covered in a classroom setting. Ultimately, however, learning how to engage with the public is a skill that can only be mastered by practice, and thus the centerpiece of the course is hands-on time at the museum. Figure 2 and Figure 3 show students at the museum engaging with child visitors in some of our activities. Students spend approximately 30 – 40 hours over the semester at the museum working with the visitors.

The partnership with COSI has been critical for managing the logistics of having the students on the floor. From the perspective of the museum visitors, the students are perceived as being part of the "COSI team" and their special status as OSU students in a learning experience does not exempt them from representing COSI as an institution. To make sure that the students understand their responsibilities as part of the public face of COSI, they receive an "onboarding" session from COSI staff which covers practical information about the museum, expectations about good team behavior, and a discussion about best practices for working with the museum visitors in a variety of situations. From COSI's perspective, the students are similar in function to volunteers, individuals ranging from high school students to retired science professionals who work on the floor of the museum with visitors.

The students engage in the practicum portion of the class in a very independent manner which promotes autonomy and self-reliance. They schedule their museum shifts at their own convenience (with the help of a scheduling program) which allows them to flexibly accommodate their other courses and their work schedules. Moreover, while there is frequently direct supervision at the lab by the instructor or another lab faculty member, most students work at least some of their museum shifts wholly on their own. More generally, the sheer number of interactions that students engage in with the museum public means that they are required to operate in a very self-sufficient manner.

The course includes several assignments designed to foster a sense of self-efficacy when conducting the educational activities. For example, one assignment centers on providing peer feedback to another student. By watching a fellow classmate engaging with the public, students are encouraged to reflect on their own skills and the interaction itself is intended to create a mutually supportive peer support group for the course. Another assignment designed to foster confidence is the "Demo Without a Net" assignment. Once students have gained a minimal level of comfort and skill on the floor, they are required to do one of the activities (a "demo") that they have never tried before – ideally, one which they feel is outside of their personal comfort zone. This assignment encourages students to pay extra attention to the written guidelines provided for each activity, and pushes them to learn new content material. When writing about

Figure 2. A course student demonstrating to children their ability to determine the location where a sound is coming from.

their experience with the new activity (the write-up is required for the grade), students often write about how they overcame their fear of the new material and ended up finding the new activity a surprisingly effective way to engage with the public; occasionally they write about their failures and their efforts to provide a positive visitor experience even when the demonstration activity takes an unanticipated and problematic turn. Regardless of how the specific experience goes for an individual student, the goal of the assignment is to demonstrate that fear – or even failure – can be a valuable learning experience.

THE FORMAL EVALUATION

In considering the effectiveness of the course, our museum partners were primarily interested in the extent of our reach with museum guests– that is, how frequently were our students on the floor working with the public and how many visitors did our students engage with. As university educators we wanted to investigate more thoroughly what the impact of the class was on our students. As we noted above, COSI is a supportive space for novices to explore and learn by doing. COSI rightfully prides itself on providing supportive spaces not just for visitors to learn about science, but also for COSI team members (staff and volunteers of all ages) to learn and grow as science educators. One of our goals in conducting an evaluation study of our students' perspectives was to provide evidence of the kinds of growth that are afforded by experiences as a science educator at COSI. Moreover, as scientists, we adopted a structured experimental approach. We undertook our formal evaluation study after the course had been taught for three years.

Figure 3. A course student demonstrating to a child how his vocal cords vibrate when he makes speech sounds.

The Student Perspective

Participants

Students enrolled in the course during two different semesters were invited to participate and all 32 students agreed to complete at least some portion of the study (pre-course survey, post-course survey, or field observation). These students were mostly women (N = 25) and white (N = 27). The majority were juniors in college (N = 18), although 3 graduate-level students were enrolled. The three most common majors of the students were Speech & Hearing Sciences (N = 13), Psychology/Neuroscience (N = 11) and Linguistics (N = 4). An additional 23 students from previous semesters also completed the post-course survey. These students were also mostly women (N = 20) and white (N = 20) and came from a similar set of majors: Speech & Hearing Sciences (N = 12), Psychology/Neuroscience (N = 7), and Linguistics (N = 4).

Method

Because the research investigators on the evaluation included the current instructor of the course, special care was taken to avoid undue coercion to participate. All study interactions (informational, consent, assessment collection) were conducted by an individual unconnected with the course who emphasized the fact that all responses would be de-identified and the instructor would have no access to any research information, including the number of participating students, until after all grades for the course were submitted.

There were three points of assessment for the students: a pre-course survey (administered during the first week of class), a field observation (undertaken during the last 3 weeks of the semester), and a post-course survey (administered during the last week of class). The students currently enrolled in the course were given all three assessments while the students from previous classes only completed the post-course survey.

The pre-course survey asked students to reflect (in an open-ended fashion) about their expectations for the course. It also contained a 16 item ratings task where students were asked to rate their level of belief of statements concerning language science, including several common myths (e.g. *It is illogical to have two negatives in a sentence; Sign languages are essentially pantomimes).*

The post-course survey asked for more open-ended reflections about students' experience in the course, specifically requesting information about what they had liked, disliked, and learned over the semester. It also contained the same ratings task about language science content that was given in the pre-course survey.

In addition, the post-test contained a 12 item self-assessment in which students rated their interest in informal science as well as in language research, their ability levels in core interaction skills (e.g. interacting with the public, recruitment for research), and their interest in various professional opportunities. Students rated themselves both with respect to their current attitudes (the post-test) and retrospectively about their attitudes at the beginning of the semester (see Sneider & Noam, 2019).

Finally, students who were currently enrolled in the course were observed in action while doing a demonstration activity on which they had been trained. The observations were conducted by two individuals not directly connected with the class, including a post-doctoral fellow in the Education school and a staff member at the museum who worked regularly on the floor with visitors. The observers used a 3-point rubric (Satisfactory, Below Satisfactory, Above Satisfactory) and graded each student along three dimensions: Accuracy, Engagement, and Appropriateness for the visitor group. The scores were averaged across observers.

Results and Discussion

We first asked whether the students had learned the practical outreach skills that are the core focus of the course. Based on the independent observations of the students on the floor conducted towards the end of the semester, they did. The performance scale ranged from 1 to 3, and students averaged scores that placed them in the Satisfactory to Above Satisfactory range along all three dimensions: Accuracy ($M = 2.2$), Engagement ($M = 2.3$), and Appropriateness ($M = 2.2$). Performance on the three dimensions was positively correlated (Pearson R's ranged from .28 to .48; all p's $< .02$) suggesting that students' skill with outreach was being mastered in a holistic fashion. More subjectively, a majority of students listed the creation of outreach materials and their use with the visitors as their favorite class activities (17/27 students just completing the class; 15/23 students from previous semesters). Students also rated themselves as having significantly improved at public speaking and interacting with the public over the course of the class (For students just completing the class: $t(27) = 10.4$, $p < .001$; For students from previous semesters: $t(23) = 10.8$, $p < .001$). See Figure 4 for the mean scores. In addition, the majority of students (18/27 students just completing the class; 16/2 students from previous semesters) indicated that they viewed their improved public speaking skills as something they would likely use in the future.

Looking more generally at students' attitudes about informal science education, students rated their own appreciation of its importance as increasing after taking the class (For students just completing the class: t (27) = 7.14, $p < .001$; For students from previous semesters: t (23) = 6.9, $p < .001$). See Figure 4 for the mean scores. However, students' enthusiasm for informal science education and outreach appears to be somewhat short-lived. While most of the students (17/27) just completing the course expressed strong positive interest in continuing to do more informal science outreach work, only a few students from previous semesters (3/23) had done any additional outreach work or expressed a positive interest in doing more in the future.

By contrast, the course was more successful at solidifying students' interests in conducting research. Students not only rated their interest in research as higher after taking the class (For students just completing the class: t (27) = 5.1, $p < .001$; For students from previous semesters: t (23) = 3.7, $p < .001$), but they actively sought out research positions. See Figure 4 for the mean scores. There were 6 students who found RA positions while taking the class; among students from previous semesters, 12/23 had gotten involved in a specific research project and 16/23 had plans to get an advanced degree in a language sciences research field.

As noted, Figure 4 shows how students (combining the data from both the current and former students) assessed themselves after taking the course. The 12 questions were grouped into four classes: core interaction skills (public speaking and engaging with the public in various forms) interest in language and linguistics, interest in informal science education, interest in pursuing course-related career opportunities. Recall that the assessment included both a post-test (reflecting students' assessments after taking the course) and a retrospective pre-test (reflecting students' assessments of how they used to feel prior to taking the course).

In terms of students' knowledge about language science and language-related myths, the class did significantly improve their knowledge: t (27) = 3.96, $p < .001$. However, this increase should be seen in the context of students having strong scores at the beginning of the course. Students successfully rejected most of the myths at the pre-test, and the increases generally reflected an increase in their certainty (e.g., a shift from "probably not true" to "definitely true"). The scores of the students from previous semesters were comparable to those who had just completed the course.

The Visitor Perspective

In parallel to the student evaluation, we also investigated the impact that the students had on the museum visitors. The visitor evaluation was conducted during the same two semesters that the student evaluation was conducted, and thus the student demonstrators that the visitors rated were the same students discussed in the previous section.

Participants

One hundred visitor groups were surveyed immediately after they had engaged with one of the student demonstrators from the class. The average size of each visitor group was 8 people, each containing at least one adult and one child. The average age of the adults was 39.7 years and the average age of the

Figure 4. Mean ratings by all students (current and former) on four classes of self-assessment questions. The scale ran from 1 – 10, with higher numbers indicated increased interest or perceived ability. The assessment was given after taking the class and the "pre" scores reflect a retrospective pre-test to their interest/abilities prior to taking the course.

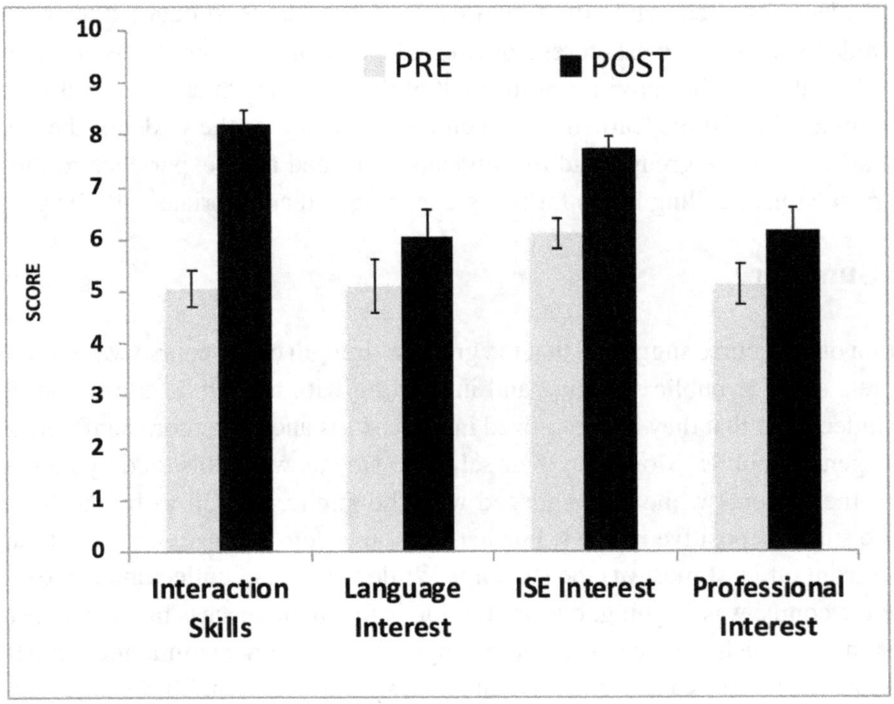

children was 8.2 years. The majority of the groups (74) contained individuals who had previously visited the museum before.

Method

A researcher monitored the students in the course who were engaging with the public. Once a family had completed the activity, they were approached and asked if they would be willing to answer a few questions about their experience. Each group was encouraged to choose a designated responder, although that person was encouraged to take input from the whole group. Each group was asked to tell the research about the activity they had just completed and was explicitly asked what they had learned from it. Groups were also asked to provide ratings about the activity and the student who conducted it, as well as about their interest in language science, research, outreach activities and the museum more generally.

Results and Discussion

The visitor groups saw six different language activities conducted by 18 different students across the whole set of interviews. Overall, the visitors did learn the expected content from the activity they saw. When asked what they had learned, 67% of visitor groups accurately described various aspects of the activity and 42% correctly reported the intended scientific message from the activity. There were differences in how well visitors learned from the different activities, with some activities leading visitors to provide more vague descriptions while others led visitors to discuss specific scientific content. However, these differences were confounded with the skill of the students conducting the activities: the activities that led to the most detailed learning from visitors were conducted by students with the highest scores from the independent observers while the activities that led to the least detailed visitor learning were conducted by students with the lowest scores from the independent observers. Thus, it is impossible to tell whether it was the nature of the activities or the skill of the students which promoted visitor outcomes.

Beyond how much the visitors learned from their interactions with the students, they gave them high marks in general. 97% of the groups said the students enhanced their experience to the museum and students were rated at near ceiling levels for being engaging and appropriate with the group.

Evaluation Summary

Our investigation of the course suggested that the greatest strength of the course was the way it improved students' concrete skills at public speaking and interacting with the public. Based on their own self-assessments, students felt that they had improved in these skills and were more confident in their ability to talk with the general public. Moreover, their self-assessments were bolstered by independent expert observations of the students while they engaged with the public, as well as by the fact that museum visitors gave the students positive reviews. Further, most museum visitors were able to learn language science content from (at least most of) the students. While we did not collect interim data on students' progress while the course was ongoing, our impression is that their success in the course depended just the kinds of elements which have been regularly implicated as elements in a good experiential learning class. Our course promotes a sense of student empowerment (Perrin, 2014): students are trusted to work very independently throughout the term, and are encouraged (and guided) to develop their own style with the activities. Moreover, are course fosters a strong sense of "craft" (Moore, 2010) through repeated practice and regular feedback. We treat public engagement as a skill that can be learned and constantly improved. Not all the students master the skill equally well, but they all improve their abilities through practice.

The course is somewhat less successful in helping students identify and maintain career goals, although it does promote positive attitudes about informal science education, at least in the short term. Long term follow-up with former students will be necessary to see what the lasting effects of the class are on students' life trajectories.

ADAPTING THE MODEL

We appreciate that many aspects of our model are difficult to replicate: We are grateful for the physical infrastructure of the Language Sciences Research Lab itself, which is made possible by an extensive, ongoing partnership between COSI and OSU, supported at the highest levels in both institutions. However, our training course does not depend on having the full infrastructure and can be effective in a wide variety of situations. Indeed, we have already facilitated the extension and adaptation of our training course in two quite different ways.

First, we have worked with faculty from the OSU department of Evolution, Ecology and Organismal Biology (EEOB) to develop a parallel course that brings students into the COSI museum to engage with the public around evolutionary biology topics. COSI recently partnered with the American Museum of Natural History to acquire an extensive exhibit on dinosaurs, which provides a natural home-base for activities featuring evolutionary theory. Faculty and students in EEOB are highly motivated to engage with the public about evolution, and developed a set of hands-on activities that would be suitable for the museum context. For example, one of their activities presents visitors with sets of bone fossils and encourages the visitors to group them in ways that highlight functional and species similarities.

While the science content of this course is quite different, the informal science training for the EEOB students is virtually identical to what is done in our course. In fact, the EEOB course was explicitly modeled on our syllabus: students read the same core papers on informal science theory and practice in engaging with the public, and use many of the same assignments, such as one geared towards reducing the use of jargon terms. Moreover, the EEOB course also includes a hands-on practicum at COSI and the students are required to spend a significant amount of time on the museum floor interacting with the public. Their use of museum resources is restricted to their outreach activities, and their presence on the floor (both in terms of specific location and in terms of amount of time) is negotiated with COSI on a semester-by-semester basis.

The EEOB version for the course has created a new link in the COSI-OSU partnership that is valuable for both sides. The EEOB department has expanded their access to public engagement opportunities, contributes to a pro-engagement climate within their department, and improved their curricular training to facilitate student skills in working with the public. From the perspective of COSI, the EEOB students enhance a popular exhibit and also help cultivate a stable of experts with knowledge that is highly relevant to a new exhibit, as well as an interest in using that knowledge to engage with the public. While some aspects of the development of the EEOB course were supported by the larger COSI/OSU partnership, it does not depend on the physical lab space of Labs in Life and could be easily replicated in any relevant museum setting.

In a completely different sort of adaptation, we have also been working with a national outreach coalition to bring our language science activities into national venues, such as the USA Science & Engineering Festival and the American Association for the Advancement of Sciences Family Science Days. The national coalition we are part of is called "Language Science for Everyone" and as the name suggests, its goal is to promote public engagement with language science. The core members of the coalition are faculty working on language science from several different universities; moreover, the coalition has a mandate to extend its opportunities and training broadly, including faculty and students who engage the public in different ways or to those who have never done it before at all.

The venues where the engagement happens are also quite similar to the situation the course was designed for. The national science fairs which the coalition favors provide the same kind of engagement environment as the COSI floor: brief, free-choice interactions with individuals of a variety of ages and backgrounds. Moreover, as the scientific focus of the coalition is the same as the course (language science), the activities designed for the students in the course are well suited for this group at these events.

Where this adaptation is different is in the extent of the training. The national events last for only a few days and each individual typically works less than 8 hours with the public over the full course of an event. Thus, the individuals working at these events get only a tiny fraction of the practicum time compared to students in the course, and preparation training is also limited to a matter of hours. Nevertheless, these events still serve as a valuable training opportunity.

Typically, we try to insure that at least half (or more) of the students working an event at any one time have had some prior training or experience doing public engagement, typically at the Language

Pod. For these students, the events serve as a kind of capstone experience, allowing them to see how their skills translate into a slightly different context and are valued by a new range of people. More importantly, these students serve as strong peer mentors, modeling good practices and acting as a resource for individuals who are new to the process.

At each event, we also seek out students and faculty from other universities, often including those near the site of the event, to introduce them to our techniques and activities. The training goal for these students (and faculty!) who are new to public engagement is largely inspirational. The relatively brief preparatory training provides a glimpse of the informal science training provided in the course, and serves to demonstrate that there are evidence-based best practices for public engagement. Individuals are also given some time to learn a few of our demonstration activities, including reading the written guidelines and practicing with one another. And of course, the event itself serves the role of the practicum portion of the course.

The outcome of this abbreviated adaptation of the course is not the creation of students (or faculty) who are as accomplished as students who took the full version. Instead, the goal is to show these newcomers what is involved with public engagement and set them on the path of learning more.

Finally, while we have not ourselves taken steps (yet!) to adapt this course into a K-12 learning environment, we see our approach as being a natural extension to other efforts currently ongoing to introduce language science to students in this age group. Language science is a topic that is naturally engaging for many people – a fact which we put to good use when interacting with the general public – and can serve as a basis for promoting an understanding of scientific thinking and an understanding of social diversity. Currently, there are several scientist groups who have developed language science and linguistics curricula that are appropriate for working with high school, middle school, and even elementary school students (see the range of papers in Denham & Lobeck, 2010). Moreover, we are further inspired by projects such as the School Kids Investigationg Language in Life and Society (SKILLS) program (Buckholtz, et al., 2014) which engage students in the actual practice of sociolinguistic data collection within their own peer groups and local communities. These related efforts demonstrate that K-12 students are capable of not only learning the relevant scientific concepts we use, but of productively using those concepts in an experiential way with others.

CONCLUSION

Engaging with the public is not a traditional part of scientific training, but is nevertheless an important skill to develop, for both scientists and their students. Informal science learning institutions such as science centers offer an environment that is naturally suited for bringing together scientists and the public, but it takes care and thoughtfulness to build a working partnership that works for all sides. Our training course provides one example of how university researchers can work with an informal learning institution in a mutually beneficial way.

ACKNOWLEDGMENT

We thank Megan Johanson for her help in developing and executing the formal evaluation of the course. This research received no specific grant from any funding agency in the public, commercial, or not-for-profit sectors.

REFERENCES

Baron, N. (2010). *Escape From the Ivory Tower: A Guide to Making Your Science Matter*. Island Press.

Bell, P., Lewenstein, B., Shouse, A. W., & Feder, M. F. (Eds.). (2009). *Learning Science in Informal Environments: People, places, and pursuits*. The National Academies Press.

Besley, J., & Tanner, A. (2011). What science communication scholars thing about training scientists to communicate. *Science Communication*, *33*(2), 239–263. doi:10.1177/1075547010386972

Bucholtz, M., Lopez, A., Mojarro, A., Skapoulli, E., VanderStouwe, C., & Warner-Garcia, S. (2014). Sociolinguistic justice in the schools: Student researchers as linguistic experts. *Language and Linguistics Compass*, *8*(4), 144–157. doi:10.1111/lnc3.12070

Bullock, O. M., Colón Amill, D., Shulman, H. C., & Dixon, G. N. (2019). Jargon as a barrier to effective science communication: Evidence from metacognition. *Public Understanding of Science (Bristol, England)*, *28*(7), 845–853. doi:10.1177/0963662519865687 PMID:31354058

Callanan, M. A. (2012). Conducting cognitive developmental research in museums: Theoretical issues and practical considerations. *Journal of Cognition and Development*, *13*(2), 137–151. doi:10.1080/15248372.2012.666730

Denham, K., & Lobeck, A. (Eds.). (2010). *Linguistics at school: Language awareness in primary and secondary education*. Cambridge University Press. doi:10.1017/CBO9780511770791

Douglas, F. (2016). Sociolinguistics in the Museum: Enrichment, Engagement and Education. In R. Lawson & D. Sayers (Eds.), *Sociolinguistic Research: Application and Impact* (pp. 66–86). Routledge.

Falk, J. H., & Dierking, L. D. (2010). The 95 percent solution. *American Scientist*, *98*(6), 486–493. doi:10.1511/2010.87.486

Falk, J. H., & Storksdieck, M. (2010). Science learning in a leisure setting. *Journal of Research in Science Teaching*, *47*(2), 194–212.

Fenichel, M., & Schweingruber, H. A. (2010). *Surrounded by Science: Learning science in informal environments*. The National Academies Press.

Honda, M., & O'Neil, W. (1993). Triggering science-forming capacity through linguistic inquiry. In The View from Building 20: Essays in Linguistics in Honor of Sylvain Bromberger. Cambridge, MA: MIT Press.

King, D., Ma, J., Armendariz, A., & Yu, K. (2018). Developing interactive exhibits with scientists: Three example collaborations from the life sciences collection at the Exploratorium. *Integrative and Comparative Biology*, *58*(1), 94–102. doi:10.1093/icb/icy010 PMID:29697800

Lidz, J., & Kronrod, Y. (2014). Expanding our reach and theirs: When linguists go to high school. *Language and Linguistics Compass*, *8*(10), 449–463. doi:10.1111/lnc3.12094

McDonnell, L., Barker, M. K., & Wieman, C. (2016). Concepts first, jargon second improves student articulation of understanding. *Biochemistry and Molecular Biology Education*, *44*(1), 12–19. doi:10.1002/bmb.20922 PMID:26537537

Meredith, D. (2010). *Explaining Research.* Oxford University Press.

Moore, D. T. (2010). Forms and issues in experiential learning. *New Directions for Teaching and Learning*, *2010*(124), 3–13. doi:10.1002/tl.415

Neeley, L., Goldman, E., Smith, B., Baron, N., & Sunu, S. (2014). GRADSCICOMM Report and Recommendations: Mapping the Pathways to Integrate Science Communication Training into STEM Graduate Education. In *Conference proceedings.* COMPASS.

Perrin, J. (2014). Features of engaging and empowering experiential learning programs for college students. *Journal of University Teaching & Learning Practice*, *11*(2), 2.

Pew Research Center. (2019). *Trust and mistrust in Americans' views of scientific experts*. Full report available at https://www.pewresearch.org

Sneider, C., & Noam, G. G. (2019). The common instrument suite: a means for assessing student attitudes in STEM classrooms and out-of-school environments. *Connected Science Learning, 11*.

Wagner, L., Speer, S. R., Moore, L. C., McCullough, E. A., Ito, K., Clopper, C. G., & Campbell-Kibler, K. (2015). Linguistics in a science museum: Integrating research, teaching, and outreach in a language sciences research lab. *Language and Linguistics Compass*, *9*(10), 420–431. doi:10.1111/lnc3.12164

Wu, K. (2017). Why can't scientists talk like regular humans. *Scientific American*. Retrieved from https://blogs.scientificamerican.com/guest-blog/effective-communication-better-science/

ADDITIONAL READING

Allen, S., & Gutwill, J. P. (2009). Creating a program to deepen family inquiry at interactive science exhibits. *Curator (New York, N.Y.)*, *52*(3), 289–306. doi:10.1111/j.2151-6952.2009.tb00352.x

Bauer, L., & Trudgill, P. (Eds.). (1998). *Language myths.*

Crystal, D. (2007). *How language works.*

Culicover, P. W., & Hume, E. (2013). Teaching about how language works to diverse populations. *Language and Linguistics Compass*, *7*(10), 523–533. doi:10.1111/lnc3.12042

Dawson, H., & Phelan, M. (2016). *Language files: Materials for an introduction to language and linguistics*. The Ohio State University Press. doi:10.26818/9780814252703

Denham, K., & Lobeck, A. (Eds.). (2014). *Linguistics at School: Language Awareness in Primary and Secondary Education*. Cambridge University Press.

Fitzgerald, C. M. (2010). Developing a service-learning curriculum for linguistics. *Language and Linguistics Compass*, *4*(4), 204–218. doi:10.1111/j.1749-818X.2010.00196.x

McKee, C., Zimmer, E., Fountain, A., Huang, H. Y., & Vento, M. (2015). Public outreach in linguistics: Engaging broader audiences. *Language and Linguistics Compass*, *9*(9), 349–357. doi:10.1111/lnc3.12153

KEY TERMS AND DEFINITIONS

Free-Choice Learning Environment: A location, such as a science center or a museum, where individuals choose for themselves what learning activities they will engage with and for how long they will engage in them.

Informal Science Learning: Learning that individuals do on scientific topics that is done without formal classroom instruction or classroom assessment practices.

Language Myth: A commonly held belief about language – how it works, how it should be used, what it tells us about individuals – that has been discredited by Language Science.

Language Science: An alternative term for Linguistics, but one that is often considered to be more inclusive of scientists working in departments such as Psychology, Speech & Hearing Sciences, and Computer Science.

Linguistics: The formal study of language, including its history, structure, representation, and use.

Public Engagement: The practice of educating the general public in a way that fosters mutual interaction allows for positive contributions from both sides.

Strands of Science: An approach to informal science education set forth by the National Research Council which identifies six key goals, or strands, for an educator to bring to their interactions.

Chapter 11
Real-World Connections With Conservation Science Through Zoos and Aquariums

Joy Kubarek
Inform Evaluation and Research, USA

Brian Johnson
Inform Evaluation and Research, USA

ABSTRACT

Zoos and aquariums have a long history of providing conservation education to students and teachers. As the conservation work of zoos and aquariums has grown, so have the opportunities to connect students and teachers to the work of these scientists. This chapter highlights a number of exemplary zoo and aquarium programs that connect students and teachers to scientific resources, career opportunities, and firsthand experience doing conservation science. These examples are not exhaustive of all the connections to scientists that zoos and aquariums provide, but rather they are representative of the types of experiences and subsequent outcomes that are possible. The chapter concludes with a discussion of opportunities for the future, including recommendation for robust evaluation and research of these programs.

INTRODUCTION

Zoos and aquariums have a long history of serving students and teachers. However, *how* they serve them has evolved in recent years. Zoos and aquariums are moving beyond the basic field trip and one-off programs toward more immersive and authentic science learning experiences. Today, more than 220 accredited zoos and aquariums reach tens of millions of students and teachers every year (AZA, 2020). The strengths of zoos and aquariums as sites for informal science learning are their live animals readily available to spark curiosity and observation; authentic representation of global habitats otherwise inaccessible to many learners; presentation of real-world issues such as wildlife conservation and climate change; and ongoing scientific research which is becoming more and more visible and integrated in

DOI: 10.4018/978-1-7998-4966-7.ch011

their education efforts. The combination of all of these elements makes zoos and aquariums uniquely situated to provide learning partnership opportunities that connect teachers and students to real-world science and conservation.

EDUCATION AT ZOOS AND AQUARIUMS

The premise for zoos and aquariums dates back to 3,000 B.C. as civilizations in Egypt, China, and India attempted to house live animal collections, primarily of exotic species not commonly encountered in these native lands. Around the time of the European Renaissance, these animal collections evolved into menageries where a diverse range of species was arranged taxonomically for public viewing. The primary purpose at the time was recreation and entertainment with very little educational or scientific intent present (Kisling Jr., 2001). It would not be until the mid-19th century that these menageries transformed into zoological gardens more reminiscent of the settings found today. The zoological garden was seen as a more advanced menagerie, one with naturalistic holding areas and arranged ecologically or geographically. With this change also came improved education, conservation, and research. In 1929, the New York Zoological Society's Bronx Zoo was the first zoo to recruit docents specifically for the task of working with teachers and their classrooms (Kisling Jr., 2001). Soon after, zoos and other informal science institutions began to hire teachers as staff to conduct educational programs at their facilities. Today more than ever, zoos and aquariums are being asked to demonstrate their scientific and educational value to the community at large (Carr & Cohen, 2011).

During the 1970s, the number of zoos and aquariums was flourishing in the United States, and the Association for Zoos and Aquariums (AZA) was established to validate the field and provide a unifying organization with standard policies and procedures (Kisling Jr., 2001). Today, AZA defines a zoo or aquarium as:

a permanent institution, which owns and maintains wildlife, under the direction of a professional staff, provides its animals with appropriate care and exhibits them in an aesthetic manner to the public on a regular basis. The institution, division, or section shall further be defined as having as their primary missions the exhibition, conservation, and preservation of the earth's fauna in an educational and scientific manner (2013, p.3).

These settings have progressed from arenas for entertainment to conservation-oriented organizations playing a key role in the survival of threatened and endangered species. In addition to conservation and research conducted by zoos and aquariums, their animal collections also inspire and educate visitors. The live animals housed in naturalistic exhibit spaces provide a unique opportunity for visitors to observe and experience wildlife firsthand. Zimmerman (2009) states:

...live animal exhibits are a zoo's heart and its organizing principle. They are fundamental in helping people to understand and care. The best exhibits are not only cognitive tools, but also powerful esthetic experiences almost magically connecting their municipal visitors with the beauty of distant nature (p.15).

As the authoritative organization for most zoos and aquariums in the United States, AZA details not only standards for animal care and welfare, conservation, and research, but also for education. AZA-accredited zoos and aquariums must demonstrate that they meet the minimum requirements for such standards. Requirements include having education be a key component of the institution's mission, having a written education plan, employing a paid staff person who is trained or has experience in educational programming as the leader of the education department, participating in partnerships with organizations and individuals who may expand the scope of their education work (e.g., schools or community groups), and assessing the impact of their education programs (AZA, 2013).

In 2017, the AZA Conservation Education Committee presented a position paper outlining the ways that zoos and aquariums provide STEM education for K-12 students and teachers. The paper says zoos and aquariums "provide STEM programming in real world life science applications using unique and engaging topics about care and conservation of wildlife and wild places." The document also provides guiding questions for zoos and aquariums to consider when developing or reflecting upon the STEM programs they provide. Included in these questions is whether the program provides an opportunity for students and teachers to conduct research and formulate solutions. The paper recognizes the unique role that zoos and aquariums can play in providing these opportunities to K-12 students and teachers. It also marks a shift from the more traditional modes of zoo or aquarium education using direct instruction and biofacts to engage students toward a more student-centered learning approach anchored in scientific processes.

It is clear zoos and aquariums have evolved through the years and are now seen as a centerpieces of communities and assets to formal K-12 education systems. Through scientific resources, authentic learning experiences, and direct access to real-world scientific inquiry, zoos and aquariums fill a unique niche for science education.

SCIENCE AND RESEARCH AT ZOOS AND AQUARIUMS

Just as education has evolved in zoos and aquariums, so have their scientific research and supporting activities. Compared with other informal learning settings, zoos and aquariums are unique in that they engage people with living organisms and ecosystems (rather than phenomena) and have responsibility to work toward protecting and conserving them. This has translated into a burgeoning set of ongoing research activities. AZA tracks member organizations' mission-based research activities on an annual basis. The *Annual Report on Conservation and Science* synthesizes and communicates research work done each year, including conducting and supporting research studies, allocating staff and funds toward research, and publishing in peer-reviewed publications. In 2018, AZA member organizations reported carrying out more than 1,300 research projects and spending more than $25 million on research activities (AZA, 2020).

From field conservation efforts to sustainable practices, zoos and aquariums engage in a diverse range of scientific activities. Field conservation research includes population monitoring, animal reintroduction, habitat assessment and protection, human-wildlife conflict mediation, conservation biology, and more. Within their own facilities, accredited zoos and aquariums are required to set goals for their sustainability practices, including using renewable resources, sourcing food locally, recycling and composting, and conserving energy. Both within zoo and aquarium facilities and in the natural environment, numerous studies look at animal welfare, care, and health. In 2018, zoos and aquariums conducted research in more than 50 different countries. In 2015, AZA launched the Saving Animals From Extinction (SAFE) program to help conserve some of the world's most vulnerable animal populations. SAFE prioritizes collaboration among AZA member facilities and builds on existing recovery plans to implement strategic conservation and public engagement activities. As this brief summary of science and research activities at zoos and aquariums makes clear, these institutions have ample opportunity to connect people - including students and teachers - to the scientific work they do.

CONNECTING STUDENTS AND TEACHERS TO CONSERVATION SCIENCE

A closer look at the work of zoos and aquariums reveals a rich palette of programs that partner students and teachers with zoo and aquarium scientists and those who support scientific activities. It is important to note that each zoo and aquarium highlighted in this chapter may have its own definition of partnership. Some define partnership as a mutually beneficial relationship between two parties with a shared goal. For others, partnerships are less formal connections to resources and expertise. In the broadest sense, as mission-based organizations, zoos and aquariums aim to advance their conservation goals through increased interest, awareness, and active participation in conservation efforts. Students and teachers, on the other hand, need access to science and scientists to bolster their science instruction. The intersection of these ambitions - zoos and aquariums advancing their mission and students and teachers accessing science - is where the partnerships emerge.

The remainder of this chapter showcases exemplary student-teacher-scientist partnerships at zoos and aquariums. Instead of being an exhaustive review, the chapter identifies and explains three strategies that are common across these partnerships: (1) connecting to resources; (2) connecting to careers; and (3) connecting to field work and research. In addition, evaluation or research findings associated with these partnerships are shared when possible. However, substantial research of these programs is lacking for various reasons. Most zoos and aquariums lack the resources and expertise to execute a full social science research study of their programs, nor are they ultimately attempting to generate new generalizable knowledge about their partnership approaches. Additionally, many of these programs serve small populations and do not have sample sizes sufficient to conduct a robust research study. Zoos and aquariums have increased their evaluation efforts, however, to examine program effectiveness as it pertains to their internal goals and intended outcomes for programs and audiences.

Connecting to Careers

Today, career-focused programs for youth and children are commonplace at zoos and aquariums. These programs connect students with traditional science-based careers such as zookeepers, veterinarians, horticulturists, and conservation scientists, but also with non-science-based careers that use scientific information as part of their role, including policy advocates, communications professionals, and development officials. The benefits of exposing youth to STEM careers are becoming well documented. For example, Habig, Gupta, Levine, and Adams (2020) found that exposure to and connecting with STEM professionals through out-of-school time experiences at informal science education institutions had a positive impact on STEM persistence. Similarly, Shin, Levy, and London (2016) found that STEM role model exposure contributed to increased interest in STEM, a sense of belonging in STEM, and self-efficacy. At some zoos and aquariums, career-focused programs include more intensive experiences such as structured mentoring with STEM professionals. Studies have shown that STEM mentoring can increase knowledge of science processes and interest in STEM careers, as well as break down preconceived ideas of what constitutes "science" or who is a "scientist" (NRC, 2009). Additionally, mentoring programs can be an important strategy for supporting students from underrepresented groups in STEM (Estrada, Hernandez, & Schultz, 2018).

Houston Zoo, Zoo Crew

Houston Zoo's Zoo Crew program helps teens ages 13-17 learn more about wildlife, conservation, and associated careers through an innovative summer and year-round program. Teens are selected to participate in the program through a competitive application and interview process. Approximately 150 teens each year take part in either a five-week summer program or a program that runs the duration of the school year. During the program, teens have the opportunity to lead wildlife-focused education activities for Zoo visitors, serve as mentors for young children during the Zoo's annual summer camp, learn about animal care as Junior Zookeepers, and help manage the Zoo Crew program as part of a Teen Leadership Council. Exploring STEM careers is the dedicated focus of one week of activities during each program cycle. During this Career Week, the teens meet with zoo staff from a range of departments, visit field-based conservation researchers at their research sites in the Houston area, and reflect on their own career interests and goals.

Zoo Crew aims to support the Zoo's strategic priorities by "cultivating the next generation of conservation heroes." The decades-old program has worked with hundreds of teens since its inception. Starting in 2016, however, the program changed from a traditional teen volunteer experience into a more structured educational program for participants. Zoo Crew is designed to align with outcomes from the Houston Zoo Conservation Education Framework, including understanding how personal actions affect wildlife, applying critical thinking skills to understand wildlife conservation issues, and developing teamwork and communication skills.

Despite the program's long history, no formal evaluation of the Zoo Crew program had ever taken place until 2018. Working with an external evaluator, the Zoo undertook a yearlong evaluation to be better able to communicate about the program's outcomes and impacts, and to identify areas for programmatic improvements (PEER Associates, Johnson, B., & Kubarek, J., 2019). Traditional evaluation methods (e.g., interviews and surveys) were combined with innovative approaches (e.g., photovoice and embedded assessments) to create a rich and varied dataset for analysis. Current and past Zoo Crew participants were invited to take part in the evaluation. Key findings from the evaluation included:

- Teens increased their understanding of how personal actions can positively and negatively affect wildlife.
- Teens increased their understanding of the complexity of wildlife conservation issues.
- Teens made progress toward developing and applying critical thinking skills.
- Teens gained a range of professional and life skills, including communication and collaboration.
- Teens were exposed to a range of conservation-related careers, and many further focused their career pathways.
- Teens carried the Zoo Crew experience into their personal and professional life years after they finished the program.

Lincoln Park Zoo, Malott Family Research Apprenticeship Program

The Lincoln Park Zoo's Malott Family Research Apprenticeship Program (RAP) provides Chicago-area high school students the opportunity to work alongside zoo researchers and engage in authentic scientific research for eight weeks each summer. Engaging with professionals from a variety of scientific careers is an explicit goal of the program. Participating teens learn about science career paths while gaining hands-on experience in conservation research. Participants are mentored by zoo scientists and other staff who provide support to fully integrate the experience into teens' academic goals, career plans, personal strengths, and professional development.

Participating teens work closely with zoo scientists in one area (e.g., urban wildlife or endocrinology). A supervising scientist teaches and coaches them on content, methods, field and laboratory skills, and data analysis. Teens then practice these skills throughout the summer as they collect data to answer a research question that they identify. Research projects have focused on butterfly and dragonfly counts, bird counts, giving up density studies (GUDS), hormone analysis, and animal behavior analysis (ethograms). At the end of the summer, teens analyze their data, discuss interpretations, and present their findings through presentations and scientific posters. Throughout the apprenticeship, the teens also work alongside laboratory technicians, field staff, post docs and college interns.

The RAP program has been frequently evaluated throughout its seven-year history using methods such as mind maps, surveys, and interviews. In 2019, the evaluation included analysis of a pre/post discussion prompt focusing on science identity; a letter writing activity to describe their overall experience; and a word cloud to sum up the RAP program. The evaluation found that after the RAP experience, participants described science as more of a process and highlighted the outcomes of scientific activities. The evaluation also found that the program increased participants' interest in science and research.

Wildlife Conservation Society, Project TRUE (Teens Researching Urban Ecology)

The Wildlife Conservation Society (WCS) operates four zoos and an aquarium in New York City. In 2014, WCS launched Project TRUE (Teens Researching Urban Ecology) with the goal of increasing the number of underserved and underrepresented youth pursuing STEM majors and STEM careers. During the annual summer program, high school students are mentored by WCS educators and Fordham University undergraduate and graduate students. With support from their mentors, the teens conduct authentic urban ecology field research throughout New York City. Research projects have investigated the effects of runoff on water quality in an urban river; the relationship between bat activity and urban environmental factors such as artificial light, noise level, and insect richness; and the impacts of urbanization on American eel populations. At the end of each summer's research activities, participants share their findings at the annual Bronx Science Consortium Poster Symposium

To gauge the effectiveness of Project TRUE in increasing the number of underserved and underrepresented youth pursuing STEM degrees and careers, WCS and Lifelong Learning Group carried out a four-year longitudinal research study with funding support from the National Science Foundation. The study investigated the impact of hands-on STEM experience and mentoring within Project TRUE on interest and participation in STEM. The findings suggested that Project TRUE participants maintained a strong STEM interest into college. For example, immediately after participating in Project TRUE, the vast majority (84%) of students said they planned to pursue a STEM major in college. In the fall semester of their freshman year, 87.5% of Project TRUE past participants said they were majoring in a STEM discipline. That percentage decreased to 77% in their second year of college and to 57% by their third year, but these rates are similar to observed rates of STEM persistence for Latinx and Black students (Bauer-Wolf, 2019). Overall, the data suggested that the research experience provided by Project TRUE had positive effects on interest in STEM careers that were sustained even two years after the program (Beauchamp et al., in Review).

Connecting to Field Work and Research

Conservation field work and research are mainstays of zoos and aquariums and provide an opportunity to immerse students and teachers in scientific research firsthand. Early immersive experiences conducting scientific research have been shown to positively influence an individual's interest and subsequent pursuit of careers in science. For example, Roberts and Wassersug (2009) did a substantial retrospective study of former high school students who had engaged in a summer scientific research program compared with university students. They found that the high school students were significantly more likely ($p<.005$) to enter and maintain a career in science compared with students whose first research experience did not occur until university. The Next Generation Science Standards shifted K-12 science education from being narrowly focused on science content knowledge to now engaging and subsequently developing students' understanding and proficiency in scientific practices (National Academies, 2013). Kuhn et al. (2017) found that high school students who engaged in activities of scientific inquiry demonstrated more advanced investigation, analysis, and argumentation skills compared with their peers who had not engaged in such activities. However, the majority of K-12 schools in the United States today lack resources, access to expertise, and research context to fully engage students in scientific practices. Zoos and aquariums can fill a niche role in this regard, providing students and teachers with authentic, real-world experiences alongside researchers and scientists.

Shedd Aquarium, Teen Science Expedition: Bahamas

One example of this partnership between students and field researchers comes from the John G. Shedd Aquarium in Chicago, Illinois. The Shedd Aquarium has offered an immersive field research experience--Teen Science Expedition: Bahamas (formerly known as High School Marine Biology)--for high school students since 1974. The program introduces them to marine biology by exploring the ecology and biodiversity of the Bahamas. The students live and sleep aboard Shedd's research vessel, the R/V Coral Reef II. Each day, the students explore coral reefs, seagrass beds, and mangrove forests to learn about the lives of marine animals and the impacts of humans on these ecosystems. In addition to exploration of the Bahamian marine environment, students meet scientists conducting research in the area and even carry out their own research projects during the week on the research vessel. Students carry out research activities and practice various field research techniques, including underwater transects, beach seines, plankton tows, and reef surveys.

Students taking part in the Teen Science Expedition visit the Bimini Biological Field Station to meet with researchers, tour the lab, and learn about Bahamian sharks and rays. They see firsthand the work required to conduct field research and study these animals. Students also meet with researchers from the Dolphin Communication Project. They learn about the two species of dolphins this research team studies in the Bahamas, how they track individual animals, and what equipment they use to study the animals underwater. Both connections to scientists shed light on the context of field-based research and demystify the reality of field research.

The targeted outcomes for the program include applying scientific knowledge in an authentic environment, recognizing that there is more than one "correct" answer for science or conservation issues, using evidence and explanations to design solutions for science or conservation issues, applying systems thinking when studying and solving complex problems, understanding various stakeholder perspectives in ecological issues, and exhibiting 21st-century skills.

Evaluation of the program has used pre-post surveys, rubrics to assess student-generated projects and presentations, social media analysis, and content analysis of student scientific journal entries. The evaluation findings suggest that students shifted their perspectives on conservation issues and solutions, as well as improved their ecological knowledge and ability to conduct scientific research. Students demonstrated a statistically significant improvement in their understanding of the interaction of environmental and human factors that contribute to conservation issues from pre-program ($M=4.25$, $n=20$) to post-program ($M=4.75$, $n=20$; $p<.05$). Students also had a statistically significant improvement in understanding how different solutions affect different stakeholders from pre-program ($M=3.85$, $n=20$) to post-program ($M=4.70$, $n=20$; $p<.05$). Students also self-reported gains in 21st-century skills such as working with others, applying relevant data and reasoning to problems, communicating new ideas to others, and adapting ideas to different situations.

Houston Zoo, Zoo Crew: Galapagos Island Ecology

The Houston Zoo also partners high school students with field researchers through their Zoo Crew travel opportunities locally and abroad. In 2018 and 2019, a small group of teens traveled with zoo staff to the Galapagos Islands to connect with the zoo's research partner, Ecology Project International (EPI). Together, the zoo and EPI support field research focused on the Galapagos giant tortoise. During their visit, the Zoo Crew teens first learned appropriate techniques for data collection with the giant tortoise. They then worked alongside field researchers to collect data including weighing tortoises, tagging tortoises in the field, and accessing GPS tracker data for monitoring tortoises. The teens also met with researchers at the Charles Darwin Research Station and conducted marine surveys to support local citizen science efforts. Houston Zoo and EPI also support the Mola Mola Eco Club in the Galapagos Islands. The club was started by a local biologist who was passionate about engaging the next generation of conservation advocates in conservation action on the islands. The club engages teens in conservation projects such as beach clean ups, ecological monitoring, awareness campaigns, and other citizen science activities. The Houston teens engaged in some of these same activities with them while visiting the Galapagos.

EPI conducted an evaluation of the experience for the teens. The evaluation focused on measures of environmental literacy to understand the influence of the experience on teens' ecological knowledge, disposition towards taking conservation action, and skills and abilities to analyze environmental or scientific problems. Ecological knowledge included measures of teens' understanding of ecological systems, environmental issues, and conservation action strategies. Dispositions toward taking action included measures of environmental sensitivity, attitudes and concerns, personal responsibility, locus of control, and motivation or intention to act. Lastly, skills and abilities included identifying and asking scientific research questions, designing and collecting data, analyzing and interpreting data, constructing explanations, articulating and presenting findings, and engaging in scientific discourse or arguments. The evaluation methods consisted of a pre-post survey for teen participants and accompanying adults.

Results showed improvement across all three areas of environmental literacy (ecological knowledge, dispositions toward conservation action, skills and abilities to analyze problems), with the greatest gains in ecological knowledge. The survey also assessed scientific method competencies, with positive gains across all six components of understanding and application. The greatest gains were observed in designing a study and collecting data. While limited by a small sample size, the findings suggest that the teens' exposure to scientists doing field research and then actually participating in field research had positive effects on science skills, competencies, and attitudes.

Mote Marine Laboratory, High School Internships

Mote brings teens into their labs for multiple experiential learning opportunities focusing on marine science research. The partnership program includes three tiers that expose high school students to real-world marine science research being conducted at the lab. The first tier (High School Internship program) is a competitive afterschool program for students interested in marine science. Participating students are required to attend regular workshops on marine science topics and complete volunteer hours including working with education staff. In the second tier (High School Advanced Program), students work in teams with a science advisor to support a research project related to the scientists' research. Students engage in project set up, data collection, and analysis. The experience culminates in an end-of-year presentation to help strengthen their communication skills. Lastly, in the third tier of the program (High School Research Experience), students who previously completed the other two tiers take part in an intensive research experience working with Mote scientists. Students run their own project independently with the guidance of the scientists.

These programs provide benefits for both the students and the scientists. The scientists gain experience mentoring the next generation of marine science researchers, raise awareness of Mote's research and conservation efforts, and practice communicating science to a younger audience. The students participate in real-world research gaining practical knowledge and skills to conduct research of their own, develop teamwork and communication skills, and apply critical thinking and problem solving skills.

Houston Zoo, Saving Wildlife School Partnership

Lastly, even elementary and middle school students can gain exposure to the work of zoo and aquarium field researchers. The Houston Zoo's Saving Wildlife School Partnerships program connects students with the zoo's global wildlife research partners as well as their on-site staff who support scientific activities at the zoo. In the partnership, multiple grades from a Houston area school participate in partnership activities that support saving wildlife. For example, elementary schools participate in mascot-related activities in which they learn about their school's mascot (e.g., rhinos, lions, tigers) from zoo staff and connect with scientists doing conservation research. Elementary students from a school with a lion mascot connected with researchers from the Niassa Lion Project in Mozambique. The students used web-based video conferencing to meet with researchers in Mozambique and learn more about lions and issues threatening their existence. Students also met with the zoo's animal care staff who work with lions to learn about - and see firsthand - lion behavior. Students then engaged in multiple fundraising activities to support the conservation efforts in Mozambique. Other schools choose to learn about pollinators and build pollinator habitats on school grounds and in their communities. A key feature of the pollinator activities is to work with the zoo's horticulture staff and correctly choose pollinator-friendly plants for their gardens.

While students may not have been immersed in research activities in the same way as the previous examples in this section, the Saving Wildlife School Partnerships program does expose them to the work of scientists and can potentially raise awareness and interest in science and conservation. In 2019, an external evaluation of the program was carried out to better understand the impacts of the Partnership program. The methods consisted of teacher interviews, parent surveys, and student written assessments. Findings demonstrated that students changed their perception of their connection to animals and the environment, were better able to articulate why it was important to save animals in the wild, and were more motivated to help save wildlife.

Connecting to Resources

One of the most traditional connections zoos and aquariums make with students and teachers is providing access to scientific information and resources. Elementary teachers, in particular, are often tasked with teaching subjects (e.g., science) in which they have very little to no formal education (National Academies of Science, 2010). Even in middle school and high school, science teachers have little time and money to procure current, relevant scientific resources to use in their instruction. Braund and Reiss (2007) contend that science which is taught strictly with the resources and within the confines of the school is not authentic. They argue that students need to be connected to real-world science, science that occurs outside of school. Places such as zoos and aquariums can provide this important connection. Previously, zoos and aquariums attempted to address this need by creating lesson plans to share with teachers. Over the years, however, zoos and aquariums realized that rather than lesson plans, teachers needed access to rich resources and experts who could help bring science to life for students. Subsequently, professional development and teacher resources evolved, as shown in the following examples.

North Carolina Aquarium at Pine Knoll Shores, SciREN

The North Carolina Aquarium at Pine Knoll Shores first implemented the Scientific Research and Education Networking Event (SciREN) in 2013. Started by two doctoral students from the University of North Carolina Chapel Hill and an educator from the aquarium, the event aims to connect K-12 educators with local scientists in a trusted scientific venue to catalyze collaboration, awareness, and sharing of resources. Scientists at the event represent a variety of disciplines, with a focus on marine science. Participating scientists come from the aquarium, University of North Carolina, Duke University's Marine Lab, North Carolina State University, and local non-profit environmental organizations.

Each year, SciREN brings hundreds of educators together with scientists to learn about their current research and opportunities to integrate this research into their classrooms or other teaching environments. As a result of the event, teachers include the research in their lesson plans, bring scientists into their classrooms to meet with students, engage students in scientific experiments related to the research, and have their students analyze the scientists' data. Examples of research topics covered in classroom activities include wave patterns, oyster research, bird feeding behavior, seagrass ecology, animal adaptations, land erosion, sonar of whales, and the use and creation of remotely operated vehicles. The SciREN event has been so successful that a new non-profit organization was formed, and the networking events have expanded to new locations.

Evaluation of the event has focused on attendance and satisfaction metrics. However, the evaluation findings suggest that SciREN is helping educators forge new connections with scientists and impacting instruction. For example, over the past three years, on a five-point Likert-type scale, 96% of participants agreed that they "made good connections to scientists" and that they would "use at least one lesson" in their classroom.

Mote Marine Laboratory, Teacher Professional Development

The Mote Marine Laboratory in Sarasota, Florida, offers professional development workshops for teachers that aim to create connections to scientists similar to that of the SciREN events. During the workshops, teachers learn about current research and scientific processes; gain access to data sources/research publications; tour the lab and experimental facilities; and learn about the implications for the research. The thematic workshops first began in 2016 when a local foundation provided funds to improve science education and use of technology in the local community. Themes include a variety of marine science topics such as coral reef ecology, local coastal environments, and marine invertebrates.

The intended outcomes for the workshops include increasing teachers' knowledge of regional marine science issues and research; increasing confidence in using current science research in lesson plans; increasing awareness of resources available; and ongoing and sustained use of scientific research in classroom instruction. Evaluations of the workshops suggest that these outcomes are being met. Every year, 95% or more of teachers overwhelmingly agreed that the workshops were high quality and relevant, enhanced their understanding of the given topic, provided useful material for use in their classrooms, and helped them feel more comfortable teaching the content to their students. Teacher open ended responses indicated intent to integrate the content and lessons into their classroom instruction, including interdisciplinary connections. Teachers applauded the focus on hands-on activities for student learning.

Wildlife Conservation Society, Field Sight

WCS has a long history of providing professional development opportunities for teachers. In recent years, WCS's teacher professional development has expanded to online formats. In 2017, with input from the organization's Professional Development Teacher Advisory Council and funding support from the Laura J. Niles Foundation, WCS launched Field Sight, an online resource to provide students and teachers with examples of field-based wildlife conservation. Field Sight uses video stories to frame conservation issues and then accompanies each video with a curriculum aligned with the Next Generation Science Standards. Conservation projects highlighted on Field Sight are selected from WCS's conservation work in more than 60 countries, with an emphasis on researchers from groups underrepresented in STEM. In 2020, there were seven stories available on Field Sight, including units on the Grauer's gorillas, bison genetics, and spatial analysis in conservation. The resources are available in English, Spanish, and the dominant in-country language of the spotlighted conservation projects.

The resources are designed to allow students to apply the knowledge that they learn in the videos in practical ways. For example, following a video story about how WCS promotes conservation leadership in individuals and communities, students are then asked to reflect on what matters to them in order to identify a conservation project that they would like to take action on in their local communities.

To date, evaluation of Field Sight has focused on web analytics. As of Spring 2020, Field Sight had 10,000 unique users, 52,000 page views, and 2,400 video views. The NGSS-aligned curriculum materials have been downloaded 5,100 times. Future evaluation is being planned to investigate how teachers are using the resources in their classrooms and what aspects of the program are most effective in engaging students.

DISCUSSION AND CONCLUSION

This chapter has described a range of efforts taking place at zoos and aquariums to connect students and teachers to real-world science. Over the past several decades, zoos and aquariums have strengthened their roles as conservation and research organizations. This emphasis, combined with the fact that accredited zoos and aquariums reach more than 50 million students each year (AZA, 2020), suggests that these institutions are well placed to strengthen teacher-student-scientist connections and engagement. This chapter has identified three primary strategies zoos and aquariums use to build these partnerships: providing access to scientific resources, creating greater understanding of science careers through training and mentoring, and providing opportunities to learn about and experience conservation in action locally and around the world.

The common thread in all of these strategies is a focus on authentic, real-world learning opportunities. Connecting students and teachers with science at zoos and aquariums provides access to individuals and institutions who are investigating real-world questions and striving to use that research to drive conservation solutions. In the examples highlighted earlier in this chapter, students are not just reading about science in a textbook. Instead, they are seeing, hearing, and interacting with scientists locally and around the world to learn more about their work and in some cases even contribute to it. This focus on "authentic science" is an important, yet often underutilized or ignored, approach in STEM education (Burrows et al., 2016). Braund and Reiss (2006) even suggest that many students are being "turned off" of science through traditional classroom teaching, but that these same students become engaged and excited about science when the learning opportunities shift to informal learning settings such as zoos and aquariums. They argue that school-based science is often disconnected from the real world, and that informal learning institutions create a bridge to real-life understanding and application of STEM learning.

Zoos and aquariums are also well positioned to engage students at a pivotal time in their development, especially related to STEM learning and career interests. As noted earlier, tens of millions of students engage with zoos and aquariums every year. Research has shown that students make decisions about whether or not to pursue STEM careers early on in their high school years, and perhaps as early as middle school (Tai, Liu, Maltese, & Fan, 2006). However, students at this age may lack exposure to STEM career possibilities or may perceive these careers as outside the options available to them. Wyss, Heulskamp, and Siebert (2012) note that "the overall stereotype of the older white male chemist persists" (p. 503). Therefore, showcasing a range of STEM professionals in non-traditional roles may help students identify with real-life professionals working in these fields. As one of the sources for informal science learning in elementary and secondary students' lives, zoos and aquariums are uniquely positioned to reach children and young adults at this critical point in their STEM learning journey.

However, the limited data to support the outcomes these partnership programs are trying to achieve suggest a need for more robust evaluation and research about these efforts taking place at zoos and aquariums. While zoos and aquariums are using established best practices in STEM education to develop innovative experiences for students and teachers, the degree to which those experiences are meeting intended outcomes is less evident due to a lack of consistent evaluation. We recommend that zoos and aquariums providing these opportunities for students or teachers to connect with scientists also carry out systematic evaluation, particularly summative evaluation, to identify program outcomes and impacts. These evaluations should use mixed methods, recognizing the value of both quantitative and qualitative data to understand the depth and breadth of the program's effects. When possible, we recommend these evaluations use validated instruments to substantiate the findings in the broader field of research in science education. Lastly, beyond evaluation there is a need for more research across zoos and aquariums doing similar programs in order to contribute to the already burgeoning knowledge base on the power of connecting students and teachers to scientists in other settings. Research of programs of this nature are lacking in zoos and aquariums, yet as the examples in this chapter suggest, their contribution to the learning and engagement of students and teachers with scientists is substantial.

ACKNOWLEDGMENT

We would like to thank the following individuals for providing detailed information about the highlighted programs and for carrying out this work: Stephanie Bohr (Lincoln Park Zoo), Meredith Bruhn (North Carolina Aquarium), Aly Busse (Mote), Halley Gardner (Shedd Aquarium), Dave Johnston (Wildlife Conservation Society), Samantha Norton (Shedd Aquarium), Su-Jen Roberts (Wildlife Conservation Society), Melanie Sorensen (Houston Zoo), and Karen Tingley (Wildlife Conservation Society).

REFERENCES

Association of Zoos and Aquariums (AZA). (2013). *2013 accreditation standards and related policies.* Retrieved from: https://www.aza.org/accreditation/

Association of Zoos and Aquariums (AZA). (2020). *Research and science*. Retrieved from: https://www.aza.org/research-and-science

Bauer-Wolf, J. (2019). Early departures. *Inside Higher Ed*. Retrieved from: https://www.insidehighered.com/news/2019/02/26/latinx-black-college-students-leave-stem-majors-more-white-students

Beauchamp, A.L., Roberts, S.J., Aloisio, J.M., Wasserman, D., Heimlich, J.E., Lewis, J.D., Munshi-South, J., Clark, J.A., & Tingley, K. (in review). *The Impact of Research and Mentoring Experiences on STEM Persistence during the Transition to College.*

Braund, M., & Reiss, M. (2006). Towards a more authentic science curriculum: The contribution of out-of-school learning. *International Journal of Science Education*, *28*(12), 1373–1388. doi:10.1080/09500690500498419

Burrows, A., Dipompeo, M., Myers, A., Hickox, R., Borowczak, M., French, D., & Schwortz, A. (2016). Authentic science experiences: Pre-collegiate science educators' successes and challenges during professional development. *Problems of Education in the 21st Century, 70*, 59-73.

Carr, N., & Cohen, S. (2011). The public face of zoos: Images of entertainment, education and conservation. *Anthrozoos: A Multidisciplinary Journal of the Interactions of People & Animals, 24*(2), 175-189.

Estrada, M., Hernandez, P. R., & Schultz, P. W. (2018). A Longitudinal Study of How Quality Mentorship and Research Experience Integrate Underrepresented Minorities into STEM Careers. *CBE Life Sciences Education, 17*(1), ar9. doi:10.1187/cbe.17-04-0066 PMID:29351912

Habig, B., Gupta, P., Levine, B., & Adams, J. (2020). An Informal Science Education Program's Impact on STEM Major and STEM Career Outcomes. *Research in Science Education, 50*(3), 1051–1074. doi:10.100711165-018-9722-y

Johnson, B., & Kubarek, J. (2019). *Summary of Evaluation Findings: Zoo Crew*. Retrieved from: https://www.PEERAssociates.net

Kisling, V. N. Jr. (2001). *Zoo and aquarium history: Ancient animal collections to zoological gardens*. CRC Press.

Kuhn, D., Arvidsson, T. S., Lesperance, R., & Corprew, R. (2017). Can engaging in science practices promote deep understanding of them? *Science Education, 101*(2), 232–250. doi:10.1002ce.21263

National Academies of Science. (2010). *Preparing Teachers: Building Evidence for Sound Policy*. The National Academies Press.

National Research Council. (2009). *Learning Science in Informal Environments: People, Places, and Pursuits. Committee on Learning Science in Informal Environments. Board on Science Education, Center for Education. Division of Behavioral and Social Sciences and Education*. The National Academies Press.

Roberts, L. F., & Wassersug, R. J. (2009). Does doing scientific research in high school correlate with students staying in science? A half-century retrospective study. *Research in Science Education, 39*(2), 251–256. doi:10.100711165-008-9083-z

Shin, J. E. L., Levy, S. R., & London, B. (2016). Effects of role model exposure on STEM and non-STEM student engagement. *Journal of Applied Social Psychology, 46*(7), 410–427. doi:10.1111/jasp.12371

Tai, R., Liu, C., Maltese, A., & Fan, X. (2006). Planning early for careers in science. *Science, 312*(5777), 1143–1144. doi:10.1126cience.1128690 PMID:16728620

Wyss, V. L., Huelskamp, D., & Siebert, C. J. (2012). Increasing middle school student interest in STEM careers with videos of scientists. *International Journal of Environmental and Science Education, 7*(4), 501–522.

Zimmerman, A., Hatchwell, M., Dickie, L., & West, C. (Eds.). (2009). Zoos in the 21st century: Catalysts for conservation? Cambridge University Press.

KEY TERMS AND DEFINITIONS

Aquariums: A visitor-based institution caring for and engaging people with aquatic animals and environments.

Conservation: Working toward the protection and sustaining populations of animals and environments.

Education: The teaching and learning of concepts, facts, and issues.

Science: Study of the physical and natural world and engaging in processes to gain understanding of surroundings and phenomena.

Zoos: A visitor-based institution caring for and engaging people with both terrestrial and aquatic animals and environments.

Chapter 12
The Potential for Science Centers in Student–Teacher–Scientist Partnerships:
Explorations From One Museum

Joe E. Heimlich
https://orcid.org/0000-0003-2586-5732
Center of Science and Industry (COSI), USA

Andy Aichele
Center of Science and Industry (COSI), USA

Frederic Bertley
Center of Science and Industry (COSI), USA

ABSTRACT

Science centers and museums have been heavily invested in helping scientists engage in reaching broader publics. Starting with understanding the types of learning that happen in these institutions, the chapter explores the roles of science centers in society as defined by the Association of Science and Technology Centers. The authors explore how those roles play out in COSI, a large science center in Columbus, Ohio. Each of the roles identified is tied to programs related to the interface among school-aged youth, teachers (formal and informal educators), and scientists. The authors explore these roles by describing each program and then laying out the theoretical foundations and the desired outcomes of the engagement.

DOI: 10.4018/978-1-7998-4966-7.ch012

INTRODUCTION

Science centers, in the US and elsewhere, have been heavily invested in helping engage scientists in reaching broader publics. Major efforts such as Portals to the Public (e.g. Pacific Science Center, 2018; Selvakumar & Storksdieck, 2013; Stylinksi, Storksdieck, Canzoneri et.al., 2018) demonstrated and continues to reveal some of the ways that scientists can better directly reach and communicate with wider publics. But connecting students and scientists and teachers is a challenge for these informal institutions where school groups are often more a "mass engagement experience" or a one-off in-school experience, and less about deep learning. Yet many science centers and museums are increasingly trying to position themselves as resources for connecting learners of all ages and contexts to science and scientists.

The potential in this role is vast, although it puts the museum in a dramatically different role than being the 'educator', the more traditional role of the museum albeit one that has been poked, teased, and niggled at for decades by many institutions and individuals who see potential beyond the historic role. Potential roles include serving as a platform for discussions and debates that enable citizens to inform and participate in the democratization of science (Bandelli & Konijn, 2011), explaining the social and political roles of science and technology in contemporary life (Grinell, 1988), addressing current science issues such as climate (Hamilton, 2012; Swim,), and focusing on social exchange and participation (Schwan, Grajal & Lewalter, 2014) and learning for fun (Packer, 2006). As the role of convener emerges more fully as an important role of a science center, the potential to become a way of connecting youth and teachers (and other audiences) with scientists in meaningful ways are increasingly visible.

This chapter explores some of the potential roles for science centers as a conduit for connecting students, teachers and scientists. We look at programs conducted at one science center and critically consider the evaluation findings for indicators of the appropriateness and uniqueness of the science center as the convener for the connection. In this chapter we will include brief descriptions of (and intended outcomes for) several programs conducted at COSI, a large science center located in Columbus, Ohio. Along with evidence of degree of success from the evaluations we will share some of the expected findings along with insights that may be a bit more surprising. The programs include a growing effort called "Color of Science," a distance learning program where students watch surgeries and talk with the surgeons, dedicated "content days" where scientists from a particular area of focus (e.g. space science; heart health; brain science) are present to share their research, and working laboratories where research scientists from a local university have full scale research labs conducting studies using the guests, including school groups, as subjects (often the control population).

We conclude the chapter with imagining this slight movement toward greater connection among these three groups as a major shift in the role of science centers with schools—becoming places beyond field trips and more as safe, engaging spaces where science is humanized, the process of STEM is the focus rather than the textbook knowledge usually shared, and where civic and consumer science literacy subsume the focus on scientific literacy and free the learning to be what facilitates excitement and interest in STEM in youth and teachers' daily lives.

THE ROLE OF SCIENCE CENTERS IN LEARNING

Within the construct of the science ecosystem of learning, there are four major "actors" upon the child who is the focus of the ecosystem: the home, school, after-school and summer programs, and STEM-focused institutions (Traphagen & Traill, 2014). Broadening the science learning ecosystem to include adults, there are also elements such as hobby clubs, conferences, and community organizations/activities (Corin, Jones, Andre, Childers & Stevens, 2017). In the particular learning ecosystem related to science, science museums, depending on their size and relationship to the larger community, play a more or less important role in being an anchor in the STEM-focused institutions of a community, especially in communities where components of the science-learning ecosystem for children is seen through collaborative lenses.

Though it has been challenged in the past, there is strong evidence science learning is indeed an outcome of visiting science centers, natural history museums, zoos, aquariums, and other venues for science-based leisure experiences (Ballantyne, Packer, Hughes & Dierking, 2007; Falk & Storksdieck, 2010; Mujtaba, Lawrence, Oliver & Reiss, 2018; National Research Council, 2009). Further evidence shows that visits to these types of places is both social and interactive (Pekarik, Doering & Kams, 1999) specific to the type of context. And the idea that pleasure and learning cannot be combined in a visit is well challenged in the literature (Hein, 1995; Packer, 2006) and in studies at COSI (Heimlich & Deedrick, 2011).

As an industry, science centers and museums are vastly different while very much alike.[1]

The Association of Science and Technology Centers (ASTC), the industry membership association, identifies four roles for science museums (see Table 1). For each of these roles, we want to examine programs at COSI that attempt to use engagement of students/youth, teachers/caregivers, and scientists in order to better serve that role. The roles and the programs aligned are:

Table 1. Roles of a science center as defined by ASTC and COSI programs explored

ASTC role	COSI program
Support lifelong learning	Science Days
Connect science and society	COSI Big Science Celebration (COSI SciFest) Interactive Video Conference
Engage diverse communities	Color of Science
Partner to tackle global and local challenges	Labs in Life

Science centers and museums, and most informal science learning organizations and contexts, seek to draw people across ages into deeper, more relevant science learning. With each institution playing a different role in the community science learning ecosystem (Broffenbrenner, 1977; Uden et al., 2007). For some, science centers are a vital player in this ecosystem serving as a destination, and an anchor for the informal science learning through cross ecosystem partnerships and relationships (Falk & Dierking, 2018).

We also subscribe to the understanding that the organizations in the ecosystem act upon the learner, or the child, in these learning ecosystems and push back that individuals have agency in their lifecourses and the resulting learningscapes they create for themselves as they move not among those organizations/institutions within the informal science learning ecosystem, but all the other learning ecosystems and structures including arts and culture, humanities, sports and leisure, literature, and daily life. These learningscapes (Heimlich & Reid, 2017; Heimlich, 2019) are the structures through which individuals make meaning across all the aspects of their lives.

Connecting Scientists with Students and Teachers at COSI

As with all elements in the informal and larger science learning ecosystems, each of the following programs can provide outcomes that differ between individuals, especially as related to social role and, in this case, scientist, teacher, and student. Although in the evaluation studies of these programs, the outcome focus has been primarily on the learner, the evidence from the studies does indicate important learning outcomes for all participants. We acknowledge that though the following are examples of bringing together scientists with adults and children including teachers and students, not all the programs reach the strength of partnership that could be desired. The roles of teachers, scientists, and even the youth are different in each program, especially when we hold them up against the four components of partnerships found in Wojton's (2009) study of museum-school partnerships: mutual goals, communication plan, key leader support, and four interpersonal elements of personal responsibility, honesty, communication at the intimate level, and trust. We also acknowledge that in many of these programs, COSI has partnerships with the scientists and the schools or the teachers. But for many of the programs, the primary role for the museum is connecting the teachers and students with the scientists.

Any of the data reported in the following descriptions are from internal evaluations conducted on the programs. While some evaluations were very robust, some program evaluations on older programs were sometimes less and in other cases, the evaluations were focusing on specific outcome questions that did not relate to the bringing together of the teacher, scientist, and students.

Support Lifelong Learning: Science Days

Science Days type of activities have a lot of different names across institutions and organizations. For COSI, Science Days are content 'take-over' days when the atrium and hallway spaces, lectures, and programs become dedicated to a particular science topic. These days are usually initiated by community organizations, societies, university departments, etc., and then organized collaboratively with COSI. The museum functions as on a normal day, but there is additional content on the particular topic presented through hands-on experiences, individual dialogues with scientists and professionals, passive displays of the science topic, and presentations of differing types. Science Days are variable in terms of breadth, depth, and quality of the experience. Yet the goal for these programs clearly aligns with the action of connecting science with the larger community. In this case, the role of the museum is mostly passive, and is as a forum for connecting youth, scientists, and teachers and/or caregivers.

Much has been written about informal science institutions and their roles in lifelong, life-wide, and life-deep learning (National Research Council, 2010). Science days provide an integration of informal and nonformal learning (Mocker & Spear, 1982) which, for the visitors, creates a learning structure that is engaging, variable, and can meet the needs of many different groups. A challenge for any institution is creating an experience that can appeal to a wide array of visitors with wildly varying levels of interest and knowledge in any subject by intentionally scaffolding the experience for the individual through the one-on-one interaction with the scientist, researcher, or representative (Yoon et al, 2013). Science days types of experiences help bridge the need for more information for some visitors while still providing an interesting experience for someone with less interest or knowledge. These days also provide a forum to allow local organizations, researchers, universities, and agencies to collaborate and share their expertise while deepening the relationship with the institution.

The target audience for COSI's Science Days is the membership visitor and the daily guest. Studies have shown that these Science Days provide 'surprises' for members as the experiences and content are varied from the more routine experience at the museum. For these days, the museum usually does not target an audience (with exceptions for days such as focused on social service organizations or individuals with specific conditions), but the experiences are meant to bring the scientists into the community to engage with the intergenerational groups.

There are annual science days that are more than one day, and in which the science museum takes on a greater role of engagement: Big Machines, and Farm Days. These week-long events feature large-scale equipment, hands-on activities, and engagement activities with professionals in the field are conducted on the grounds of the science center. Most attendees to these events do visit the museum though engagement in the event itself can occur without admission. Although visitors appreciate these experiences and programs, there is not a strong, consistent understanding of construction and agriculture as STEM and the individuals with whom the youth and adults connect are not necessarily seen as STEM professionals.

The ongoing questions for this type of program are two-fold: one is purely operational—does having this event continue to bring in larger audiences to the science museum? The second is more programmatic. Evaluations of the various events reveal no surprises: members highly value the days when something more or something different is happening as it creates a different visit experience; the big events do draw audiences that might not normally visit the science museum; and learning is very much tied to entry interest. Those interested in the topic are most likely drawn into the experiences with a learning outcome over an experience outcome. In any of these evaluations, teachers with students were not distinguished from adults with youth as the evaluands.

Connect Science and Society: COSI Big Science Celebration

In his essay "Salvaging science literacy, Feinstein (2010) offers a shift from thinking about science literacy and instead thinking about science usefulness in daily life. He builds on Shen's (1975) distinction among practical science literacy, or science that helps people make daily decision, from civic science literacy which ties to science-related policy issues and cultural science literacy which is how society views or may appreciate science as a human achievement. Some researchers have offered that it is trust in the science findings (Ryoko, 2018) or perhaps it is asking what makes a scientist credible and how the scientists themselves might think (Van Montagu, 2018).

In its inaugural year, *COSI Science Festival and Big Science Celebration* had over 140 community partners across the festival, and 125 science exhibitors at the celebration, with 760scientists and STEM professionals involved and 40,000 visitors across the events!

Due to the closure of museums related to the pandemic in 2020, the second year quickly was shifted to a virtual celebration and included eighteen communities as partners, forty virtual events, a statewide bioblitz citizen science project; delivered through and with local, regional and national partners. The overall impact included 119,000 content views and 790,000 media impressions.

One of the stated purposes for COSI's Science Festival and Big Science Celebration is to connect with people where they live, lounge, and learn. To that end, the events around the 12 partner communities in year 1 plus the BSC were designed to engage people in science when and where they usually would not. Although there was an explicit school component to the festival, and teachers were an intended audience for the BSC, for the inaugural BSC the evaluation focused on the overall event and did not seek out teachers. And the virtual experiences in year 2 attempted to create unique experiences and opportunities that would likewise not be possible for individuals to engage in.

In the measures of the event, generally science learning was clear, although the experience of the Big Science Celebration overwhelmed the individual learning events. Importantly for the institution, it was discovered that the zip codes of those attending the BSC *did not* mirror by proportion the usual visiting population from central Ohio. Indeed, there were proportionally more visitors from some of COSI's target partner urban neighborhoods than there were from the suburbs where the larger proportion of visitors and members come from, suggesting the BSC and the science festival community events were important for expanding the reach of the science center. One indicator of the importance of the BSC to COSI's goal of connecting visitors with real science and STEM professionals across disciplines was the extrapolation of 760 STEM professionals not including COSI staff who directly interacted with attendees to the celebration. One quality check was that after defining for themselves what a quality interaction entailed (and for most it was spending enough time to complete a hands-on activity), the exhibitors overwhelmingly felt they had quality interactions over the course of the day meaning there were constant, meaning full connections of scientists with visitors. And their intention to participate more in future events was also high. As one exhibitor said: "Informal STEM education. This is not just a game. Content is important. What we do as scientists is important."

Connect Science and Society: Interactive Video Conferencing (IVC)

IVC has been a part of informal education institutions for four decades. From the satellite programming in the 1980s to the multi-user conference rooms of the 2020s, with video, interactive television along the way, internet-supported interactive video conferencing is the centerpiece of what is referred to as the fourth generation of distance learning (Passerini & Granger, 2000). Formal and informal institutions have relied on distance learning using live hosts to reach distant populations (Markham et al, 2016), distributed populations (Rao & Giuli, 2010), rural communities (Feil et al., 2008), and to expand their reach and visibility (Dennison, 2008; Motamedi, 2001).

For more than 20 years, COSI's IVC programs have been an important tool for connecting students and their teachers with scientists, doctors, and COSI educators for a live, interactive two-way experience. COSI's Wireless Workshops stream directly into classrooms, community centers and homes to provide students experiences of hands-on learning. STEMM Connections (STEM plus medicine) provide students with career exploration opportunities. In the programs with the live scientists or doctors, the program is about connecting the teacher, the student, and the scientist, even though it is not in a partnership.

Responding to the new challenges and opportunities of the pandemic in 2020, COSI partnered with Game Arena to present virtual camp experiences wherein STEM education was merged with eSports gaming. COSI used this partnership to provide opportunities for campers to learn, receive coaching, and play alongside experts. The pilot program provided COSI with essential insights that will guide COSI's online programs and to deepen partnership models moving forward.

Engage Diverse Communities: The Color of Science

There is ample evidence that women and minorities are tremendously underrepresented in STEM careers and in STEM pipeline. An ongoing challenge is how to address the tremendous differences in cultures and acculturation among and between minorities and gendered 'other' populations. It is also well understood that assuming homogeneousness among any group of women or of a minority can obscure important racial and ethnic differences suggesting the importance of critical consideration of intersectionalities in why/not individuals enter or remain in STEM careers (Johnson, 2011). So how can a program possibly serve to help move youth into seeing themselves in STEM careers?

There is a tremendous amount of research on [the lack of] women and minorities in STEM professions. For children, one of the big challenges is that the do not see people like them as STEM professionals. For example, being a scientist is constructed by some students as a highly gendered and racialized profession (Wong, 2015). Another is that children through high school see scientists as stereotypically male and white (Farland-Smith, 2017) and this type of stereotype can affect positively or negatively a child's choice of future careers (Ōzgelen, 2017). Another challenge is that a lack of science "identity" can hinder minority students' sense of belonging and performance in college science courses (Chen, Binning, Manke, Brady, McGreevy Betancur, Limeri & Kaufmann, 2020). And such identity relies on family support (Nugent, Barker, Welch, Grandgenett, Wu & Nelson, 2015) and seeing a STEM career as family compatible (Weisgram & Diekman, 2017) as well as interest, role models, mentors, and peer support (e.g. Bruce & Bridgland 2014; Inzer & Crawford, 2005; Zimmerman, M.A., Bingenheimer & Notaro, 2002).

As one effort to help address some of these concerns, COSI implemented *The Color of Science*, a comprehensive science interest and literacy program that invites the public to engage with the stories of remarkable individuals who have impacted our society through their scientific work and achievements. These individuals include women, persons of African American, Latinx/Hispanic, Asian Pacific, and Native American heritages, members of the LGBTQ+ community, and persons with disabilities. The program has two key components: a panel discussion event featuring local scientists, and a student "passport" experience at COSI, which is a hands-on STEM career exploration program geared towards middle school students. As conducted, the passport program is again creating an intersection of teacher, student and scientist rather than a true partnership; the potential for deepening the connections is strong. The Color of Science celebrates diversity and showcases the science, technology, engineering, and math (STEM) fields as an inclusive space already filled with important contributions from diverse scientists, showing underrepresented populations that STEM is for everyone.

In the pilot year of the program at COSI, attendees' written comments about their The Color of Science experience indicate that it influenced some to think differently about STEM and how they think of themselves in relation to science. Some attendees felt that the event helped to affirm their own value as well as other persons of color in the STEM field. For one attendee, the event "made me fall in love again with science." Some of those attending noted that "science is very diverse" while others felt "inclusion is necessary to STEM". Evaluation data for all components of The Color of Science reported very high satisfaction, interest, and outcome scores.

Partner to Tackle Local and Global Issues: Labs in Life

A challenge for teaching science and very much present in science centers and museums, is the dual pressures of teaching the basic body of knowledge that underlies a scientific discipline or that is necessary for a person to be having civic, social, and/or consumer science literacy with the reality that science is not about what is known, but is about knowing and new knowledge. This balance suggests that understanding the scientific processes is not simply knowing "the steps" of whichever model of process is being taught, but that they actually engage in or experience the research enterprise. For museums, visitors must be involved in the research process and not simply see it from a distance as there is a sizeable inferential gap between observing and interpreting (Rennie & Johnston, 2004).

Public participation in scientific research was a label coined to cover the range of approaches in citizen science, participatory action research, and community-based participatory research which serve as "intentional collaborations in which members of the public engage in the process of research to generate new science-based knowledge" (Shirk, Ballard, Wildermann, *et al.*, p. 29).

Within the PPSR literature, there is support that science knowledge and understanding can be enhanced through such interactive learning experiences for the nonscientists in the projects (Bell, Lewenstein, Shouse, & Feder, 2009; Bonney et al., 2009).

Prainsack (2014) contrasted data gathering versus being a subject in a study and offers that in both the objective should be to question whether or not citizen science changes how we assess expertise and authority in creating scientific knowledge, and how it does or does not support participation in science. Woolley et al. (2016) examined the multiple and potentially conflicting meanings of public engagement when the people involved are the subjects of the research. They conclude that explication of the varied meanings and uses of "citizen" would help reveal the participant's role in participation and engagement, making the meaning in use of the label citizen-scientist clearer.

On the other hand, many studies have looked at recruitment into research studies. Studies have explored many reasons why individuals would participate including the relationship of the study participants to the research personnel (Penckofer, Bynr, Mumby & Ferrans, 2011); risks in participation (Bentley & Thacker, 2004); financial inducements (Almieda, Azevedo, Nune et al., 2007) and such incentives both good (Resnik, 2001) and questionable (Tishler & Bartholomae, 2003); and for student learning (Shi, 2006). But few studies have looked at voluntary participation as subjects in research during leisure pursuits, such as visiting a museum.

COSI features three, fully-working research laboratories within the Life Exhibition area that were created jointly by COSI and The Ohio State University (OSU). The research labs are designed in such a way to put research in the midst of the guest experience and allow the public to interact with the researchers by observing, asking questions, and participating where appropriate. At the time of writing this chapter, researchers are from the areas of Pharmacology, Media Communications, and Linguistics. Working in partnership with the OSU Vice President of Research, COSI jointly selects researchers and their projects that are interesting and accessible to the public. The partnership has been extremely beneficial to both COSI and OSU. COSI guests are able to engage with working scientists in a robust and meaningful way, while OSU researchers have found that they can greatly expand the amount of research they can conduct by working on-site at COSI. The pods conduct programs with classes and have special field-trip experiences that engage the students in the labs and then in activities around the museum that demonstrate the science concepts explored in the program. Although the programs were designed with guidance from teachers, they are again more focused on connecting the players than partnering for something deeper.

While evaluations rate the exhibition highly, a more interesting measure of the "pods" in the Labs in Life is the academic productivity of the labs. One of the three labs had, in its first four years, 7 peer-reviewed journal articles about the pod and the experience of having the lab in the museum and about teaching undergraduate and graduate students to engage in sharing scientific information and experiences in an informal setting. The lab also created a for-credit course that is mandatory for students working with the lab, and optional for any other students across the university. In terms of articles related to the studies being conducted, during the same time period they had 12 peer-reviewed articles published in academic journals and 34 peer-reviewed conference presentations and posters.

CONCLUSION

Although we would love to imagine that these are unique programs and experiences, we know they are not. Every science center or museum has programs and experiences designed to connect youth and teachers with scientists, and they all do them slightly different, but in similar ways. Many museums have, like COSI, conducted evaluations of their programs, and find that their intended outcomes are met. Yet the actual evidence of these experiences is either 1) fleeting and temporal; or 2) anecdotal in the long-term. Knowing that these experiences are a part of the learning of the youth it is important to understand the impact of these types of programs as part of the suite of experiences in each youth's life. Further, all of the programs have inherent potential for building partnerships among teachers, students, and the scientists. As with all organizations, the time required for building and maintaining the relationships is challenging to a museum, but certainly a target to more critically consider.

RESEARCH NEEDS

Evaluations are important sources of data for understanding if and in what ways a program is/not working. Evaluations are primarily a tool for program improvement. And because an evaluation is looking at the whole of a program, some important questions about individuals are missed. An example: the evaluation of Passport program for *The Color of Science* showed two respondents who, in their responses to the questionnaire they completed at the end of the experience, revealed they chose to not engage with *any* of the STEM professional women throughout the building; these two youth also chose to not engage with the museum experiences and rated the whole day poorly. While overall the experience was a positive one for the young women involved, it was a wasted day for those two who may or may not have even known each other. The research needs for these programs are great and exceed the capacity of most institutions to do the studies as time and access is constrained in most museums. The false ideal of the 'gold standard' control group study and the challenges in doing true longitudinal studies force institutions to rely on the evaluations and trust that engaging with STEM professionals through the science center or science museum will have more than a positive, fleeting reaction but do make a difference in the lives of the youth. The anecdotal testimonies from adults years later suggest these are more than "one-off" stops on at least some of the youths' learningscapes; research is needed to understand *who* benefits in *what* ways and *how* can the institution better support more youth in their understanding of science and scientists/STEM professionals. Further, our collective understanding of the nonformal and informal structures of learning are based primarily on isolating the experiences in nonformal/informal learning. Yet meaning-making about life is not made from individual experiences, but an integration of the constant experiences of living providing all sorts of input. Better understanding of how an experience in one of these types of programs fits with the life of the youth or adult engaging in the teaching/learning exchange is needed.

PROGRAMMATIC AND INSTITUTION NEEDS

The bottom line across the approaches mentioned above and others activities in science museums trying for similar outcomes is twofold: one is the goal of engaging people of all ages and situations in science – meeting mission *and* generating funds for operational supports. The second is the more esoteric and difficult bottom line to both realize and measure for success – creating opportunities to humanize science and scientists, to make technology accessible by engaging with those who develop it, to bring math and engineering and those who do them into the realm of accessible and relevant STEM. The institution's goal in connecting scientists to students and teachers does have an element of pipeline intention, but far more important is the breaking of boundaries between textbook science and scientific endeavors. A piece of that is, of course, seeing scientists as diverse and real people solving real world problems. But another important aspect is the intention of discovery of the importance of STEM in society and in every aspect of an individual's life.

An important potential role for science museums in this work of creating connections among these audiences of students/teachers/scientists, is in creating a breadth of opportunities and potential structures (and affiliated outcomes) for doing this convening. Not every child should or will become a scientist, or an engineer, mathematician, or go into technology. But every child (and adult) should see STEM as a living, dynamic field full of diverse individuals with a wide range of backgrounds and experiences who collectively are contributing to how we understand our world, live in it, and move forward into the future.

This research received no specific grant from any funding agency in the public, commercial, or not-for-profit sectors.

REFERENCES

Almieda, L., Azevedo, B., Nunes, T., Vaz-da-Silva, M., & Soares-da-Silva, P. (2007). Why healthy subjects volunteer for phase I studies and how they perceive their participation? *European Journal of Clinical Pharmacology, 63*(11), 1085–1094. doi:10.100700228-007-0368-3 PMID:17891536

ASTC.org. (2020). *Our strategic direction: The work of our members.* www.astc.org/about/strategy/

Ballantyne, R., Packer, J., Hughes, K., & Dierking, L. (2007). Conservation learning in wildlife tourism settings: Lessons from research in zoos and aquariums. *Environmental Education Research, 13*(3), 367–383. doi:10.1080/13504620701430604

Bandelli, A., & Konijn, E. (2011). An experimental approach to strengthen the role of science centers in the governance of science. In J. C. Marstine (Ed.), *The Routledge Companion to Museum Ethics* (pp. 164–173). Routledge.

Bentley, P., & Thacker, P. G. (2004). The influence of risk and monetary payment on the research participation decision making process. *Journal of Medical Ethics, 30*(3), 293–298. doi:10.1136/jme.2002.001594 PMID:15173366

Bonney, R., Ballard, H., Jordan, R., McCallie, E., Phillips, T., Shirk, J., & Wilderman, C. (2009a). Public participation in scientific research: defining the field and assessing its potential for informal science education. A CAISE Inquiry Group Report. Washington, DC: Center for Advancement of Informal Science Education (CAISE).

Bronfenbrenner, U. (1977). Toward an experimental ecology of human development. *The American Psychologist*, *32*(7), 513–531. doi:10.1037/0003-066X.32.7.513

Bruce, M., & Bridgeland, J. (2014). The mentoring effect: Young people's perspectives on the outcomes and availability of mentoring. Washington, DC: Civic Enterprises with Hart Research Associates for MENTOR: The National Mentoring Network.

Chen, S., Binning, K. R., Manke, K. J., Brady, S. T., McGreevy, E. M., Betancur, L., Limeri, L. B., & Kaufmann, N. (2020). Am I a science person? A strong science identity bolsters minority students' sense of belonging and performance in college. *Personality and Social Psychology Bulletin*. Advance online publication. doi:10.1177/0146167220936480 PMID:32659167

Corin, E. N., Jones, G. M., Andre, T., Hilders, G. M., & Stevens, V. (2017). Science hobbyists: Active users of the science-learning ecosystem. *International Journal of Science Education. Part B*, *7*(2), 161–180.

Dennison, J. (2008, March). Understanding the utilization of interactive videoconferencing in the K12 setting. In *Society for Information Technology & Teacher Education International Conference* (pp. 336-339). Association for the Advancement of Computing in Education (AACE).

Falk, J. H., & Dierking, L. D. (2018). Viewing science learning through an ecosystem lens: A story in two parts. In D. Corrigan, C. Buntting, A. Jones, & J. Loughran (Eds.), *Navigating the changing landscape of formal and informal science learning opportunities* (pp. 9–29). Springer. doi:10.1007/978-3-319-89761-5_2

Falk, J. H., & Storksdieck, M. (2010). Science learning in a leisure setting. *Journal of Research in Science Teaching*, *47*(2), 194–212.

Farland-Smith, D. (2017). The evolution of the analysis of the draw-a-scientist test. In P. Katz (Ed.), *Drawing for Science Education* (pp. 169–178). Brill Sense. doi:10.1007/978-94-6300-875-4_15

Feil, E. G., Baggett, K. M., Davis, B., Sheeber, L., Landry, S., Carta, J. J., & Buzhardt, J. (2008). Expanding the reach of preventive interventions: Development of an internet-based training for parents of infants. *Child Maltreatment*, *13*(4), 334–346. doi:10.1177/1077559508322446 PMID:18843143

Feinstein, N. (2011). Salvaging science literacy. *Science Education*, *9*(1), 168–185. doi:10.1002ce.20414

Grinell, S. (1988). Science centers come of age. *Issues in Science and Technology*, *4*(3), 7–75.

Heimlich, J. E. (2019). Queering STEM learningscapes. In W. Letts & S. Fifield (Eds.), *STEM of desire: Queer theories in science education* (pp. 161–176). Brill.

Heimlich, J. E., Deedrick, R., & Yocco, V. (2012). *COSI membership study 2011: Final report.* Columbus, OH: COSI Center for Research and Evaluation.

Heimlich, J. E., & Reid, A. (2017). Environmental learning. In M. A. Peters (Ed.), *Encyclopedia of educational philosophy and theory*. Online. doi:10.1007/978-981-287-588-4_447

Inzer, L. D., & Crawford, C. B. (2005). A review of formal and informal mentoring: Processes, problems, and design. *Journal of Leadership Education*, *4*(1), 31–50. doi:10.12806/V4/I1/TF2

Johnson, D. R. (2011). Women of color in science, technology, engineering, and mathematics (STEM). *New Directions for Institutional Research*, *152*(Winter), 75–85. doi:10.1002/ir.410

Markham, C. M., Rushing, S. C., Jessen, C., Gorman, G., Torres, J., Lambert, W. E., ... Peskin, M. F. (2016). Internet-based delivery of evidence-based health promotion programs among American Indian and Alaska native youth: A case study. *JMIR Research Protocols*, *5*(4).

Mocker, D. W., & Spear, G. E. (1982). *Lifelong learning: Formal nonformal informal and self-directed. Information Series No. 241*. Eric Clearinghouse for Adult and Vocational Education.

Motamedi, V. (2001). A critical look at the use of videoconferencing in United States distance education. *Education*, *122*(2), 386–394.

Mujtaba, T., Lawrence, M., Oliver, M., & Reiss, M. J. (2018). Learning and engagement through natural history museums. *Studies in Science Education*, *54*(1), 41–67. doi:10.1080/03057267.2018.1442820 PMID:29951162

National Research Council. (2009). *Learning science in informal environments: People, places, and pursuits*. The National Academies Press.

Nugent, G., Barker, B., Welch, G., Grandgenett, N., Wu, C. R., & Nelson, C. (2015). A model of factors contributing to STEM learning and career orientation. *International Journal of Science Education*, *37*(7), 1067–1088. doi:10.1080/09500693.2015.1017863

Ōzgelen, S. (2017). Primary school student5s¢ views on science and scientists. In P. Katz (Ed.), *Drawing for Science Education* (pp. 191–203). Brill Sense. doi:10.1007/978-94-6300-875-4_17

Pacific Science Center & Institute for Learning Innovation. (2018). *Portal to the public: Implementation manual & catalog of professional development elements*. Institute for Learning Innovation.

Packer, J. (2006). Learning for fun: The unique contribution of educational leisure experiences. *Curator (New York, N.Y.)*, *49*(3), 329–344. doi:10.1111/j.2151-6952.2006.tb00227.x

Passerini, K., & Granger, M. J. (2000). A developmental model for distance learning using the Internet. *Computers & Education*, *34*(1), 1–15. doi:10.1016/S0360-1315(99)00024-X

Pekarik, A. J., Doering, Z. D., & Karns, D. A. (1999). Exploring satisfying experiences in museums. *Curator (New York, N.Y.)*, *42*(2), 152–173. doi:10.1111/j.2151-6952.1999.tb01137.x

Penckofer, S., Byrn, M., Mumby, P., & Ferrans, C. E. (2011). Improving Subject recruitment, retention, and participation in research through Peplau's Theory of Interpersonal Relations. *Nursing Science Quarterly*, *24*(2), 146–151. doi:10.1177/0894318411399454 PMID:21471039

Prainsack, B. (2014). Understanding participation: the 'citizen science' of genetics. In B. Prainsack, S. Schicktanz, & G. Werner-Felmayer (Eds.), *Genetics as social practice* (pp. 147–164). Ashgate.

Rao, K., & Giuli, C. (2010). Reaching remote learners: Successes and challenges for students in an online graduate degree program in the Pacific Islands. *The International Review of Research in Open and Distributed Learning*, *11*(1), 141–160. doi:10.19173/irrodl.v11i1.785

Rennie, L. J., & Johnston, D. J. (2004). The nature of learning and its implications for research on learning from museums. Science Education, 88(1), S4-S16.

Resnik, D. B. (2001). Research participation and financial inducements. *The American Journal of Bioethics*, *1*(2), 54–56. doi:10.1162/152651601300169112 PMID:11951894

Ryoko, A. (2018). Trust—What connects science to daily life. *Health Physics*, *115*(5), 581–589. doi:10.1097/HP.0000000000000945 PMID:30260848

Schwan, S., Grajal, A., & Lewalter, D. (2014). Understanding and engagement in places of science experience: Science museums, science centers, zoos, and aquariums. *Educational Psychologist*, *49*(2), 70–85. doi:10.1080/00461520.2014.917588

Selvakumar, M., & Storksdieck, M. (2013). Portal to the public: Museum educators collaborating with scientists to engage museum visitors with current science. *Curator (New York, N.Y.)*, *56*(1), 69–78. doi:10.1111/cura.12007

Shen, B. S. P. (1975). Science literacy and the public understanding of science. In S. B. Day (Ed.), *Communication of scientific information*. S. Karger and A.G. Basel. doi:10.1159/000398072

Shi, L. (2006). Students as research participants or as learners? *Journal of Academic Ethics*, *4*(4), 205–220. doi:10.100710805-006-9028-y

Shirk, J. L., Ballard, H. L., Wilderman, H. L., Phillips, C. C., Wiggins, T., Jordan, A., Mc, R., Callie, E., Minarchek, M., Lewenstein, B. V., Krasny, M. E., & Bonney, R. (2012). Public participation in scientific research: A framework for deliberate design. *Ecology and Society*, *17*(2), 29. doi:10.5751/ES-04705-170229

Stylinski, C., Storksdieck, M., Canzoneri, N., Klein, E., & Johnson, A. (2019). Impacts of a comprehensive public engagement training and support program on scientists' outreach attitudes and practices. *International Journal of Science Education, Part B.*, *8*(4), 340–354. doi:10.1080/21548455.2018.1506188

Tishler, C. L., & Bartholomae, S. (2003). Repeat participation among normal healthy research volunteers: Professional guinea pigs in clinical trials? *Perspectives in Biology and Medicine*, *46*(4), 508–520. doi:10.1353/pbm.2003.0094 PMID:14593220

Traphagen, K., & Traill, S. (2014). *How cross-sector collaborations are advancing STEM learning:* Working paper.

Uden, L., Wangsa, I. T., & Damiani, E. (2007, February). The future of E-learning: E-learning ecosystem. In Digital ecosystems and technologies conference, 2007. DEST'07. Inaugural IEEE-IES (pp. 113–117). IEEE.

Van Montagu, M. (2018). Science, ideology and daily life. *Journal of Innovation and Knowledge*, *3*(2), 66–69. doi:10.1016/j.jik.2018.03.005

Weisgram, E. S., & Diekman, A. B. (2017). Making STEM "family friendly": The impact of perceiving science careers as family-compatible. *Social Sciences*, *6*(2), 61. doi:10.3390ocsci6020061

Wojton, M. A. (2009). *A study of a museum-school partnership* (Unpublished doctoral dissertation). The Ohio State University.

Wong, B. (2015). Careers "from" but not "in" science: Why are aspirations to be a scientist challenging for minority ethnic students? *Journal of Research in Science Teaching*, *52*(7), 979–1002. doi:10.1002/tea.21231

Wooley, J. P., McGowan, M. L., Teare, H. J. A., Coathup, V., Fishman, J. R., Settersten, R. A. Jr, Sterckx, S., Kaye, J., & Jeungst, E. T. (2016). Citizen science or scientific citizenship? Disentangling the uses of public engagement rhetoric in national research initiatives. *BMC Medical Ethics*, *6*(17), 33. Advance online publication. doi:10.118612910-016-0117-1

Yoon, S. A., Elinich, K., Wang, J., Van Schooneveld, J. B., & Anderson, E. (2013). Scaffolding informal learning in science museums: How much is too much? *Science Education*, *97*(6), 848–877. doi:10.1002ce.21079

Zimmerman, M. A., Bingenheimer, J. B., & Notaro, P. C. (2002). Natural mentors and adolescent resiliency: A study with urban youth. *American Journal of Community Psychology*, *30*(2), 221–243. doi:10.1023/A:1014632911622 PMID:12002244

ADDITIONAL READING

Cameron, F. (2005). Contentiousness and shifting knowledge paradigms: The roles of history and science museums in contemporary societies. *Museum Management and Curatorship*, *20*(3), 213–233. doi:10.1080/09647770500502003

Davidsson, E., & Jakobsson, A. (Eds.). (2012). *Understanding interactions at science centers and museums*. Springer Science & Business Media. doi:10.1007/978-94-6091-725-7

Falk, J. H., Dierking, L. D., & Foutz, S. (Eds.). (2007). *In principle, in practice: Museums as learning institutions*. AltaMira Press.

Gupta, P., Adams, J., Kisiel, J., & Dewitt, J. (2010). Examining the complexities of school-museum partnerships. *Cultural Studies of Science Education*, *5*(3), 685–699. doi:10.100711422-010-9264-8

Heimlich, J. E. (1993). Nonformal environmental education: Toward a working definition. The Environmental Outlook Columbus, OH: ERIC/CSMEE Informational Bulletin. Columbus, OH: ERIC/CSMEE Informational Bulletin.

Kwiek, N. C., Rooney, K., & Downing, M. (2016). The Generation Rx Laboratory: An Innovative Pharmacology Education Research Hub. *The FASEB Journal*, *30*(1, supplement), lb552–lb552.

Ogodo, J., Irving, K. E., Brosnan, P., & Ding, L. (2019). Rise, Defy, Teach, and Lead: The ENABLE STEM Project. In Recruiting, Preparing, and Retaining STEM Teachers for a Global Generation (pp. 78-99). Brill Sense.

Rennie, L. J., & Johnston, D. J. (2004). The nature of learning and its implications for research on learning from museums. *Science Education*, *88*(S1), S4–S16. doi:10.1002ce.20017

Wagner, L., Speer, S. R., Moore, L. C., McCullough, E. A., Ito, K., Clopper, C. G., & Campbell-Kibler, K. (2015). Linguistics in a science museum: Integrating research, teaching, and outreach at the language sciences research lab. *Language and Linguistics Compass*, *9*(10), 420–431. doi:10.1111/lnc3.12164

KEY TERMS AND DEFINITIONS

ASTC: The Association of Science and Technology Centers – the membership organization for science centers and science museums, but also includes some natural history museums, children's museums, and zoos/aquariums.

Citizen-Science: The involvement of the public in scientific research across a range of phenomena using scientific practices.

COSI: A large science center (using both size and budget definitions) founded in 1964 and located in Columbus, Ohio.

Free-Choice Learning: A phrase offered to the learning community by Falk & Dierking provides a means by which 'learning' is separated from the approach to teaching for learning.

Informal/Nonformal Learning: Terms used to describe institutional settings other than schools; for example, museums are typically referred to as informal settings while a program in a park is often referred to as being in a nonformal setting. Although definitions exist that distinguish these terms based on the structure of learning and the orientation of the learner, they are frequently used as synonyms.

Science Festival: Generally recurring (annual) celebrations dedicated to science, STEM, with many collaborators and stakeholders to provide a wide range of activities, events, and programs for a community over a limited period of time.

Science Museum/Science Center: Early science museums (18th & 19th centuries) as natural history museums with collections on display. In the 20th century, science centers emerged with a greater focus on use of technology (exhibits, experiences included) to educate. Although there is little difference to visitors/publics, the difference in is the centrality of a collection.

ENDNOTE

1 To find a science center near you, ASTC.org has a list of science centers and museums in the U.S. and beyond. Also please note that some children's museums have a very strong science focus. Some of these are members of ASTC as well, but others can be found on the Association for Children's Museums website childrensmuseums.org. For those who are not near a science center, most of the institutions have online resources and/or educational programs designed for classrooms (live and video). In the U.S., many of the science centers have traveling programs designed for schools. All these can be found on search engines using terms science center (or science museum), with one of the following (each will provide different resources): classroom resources, online resources, videos, interactive conference, distance learning, remote programming, traveling programs, etc.

Compilation of References

Achieve. (2013). *Next generation science standards.* Retrieved July 12, 2018, from http://www.nextgenscience.org

Adams, R. (2019, June 12). *N.C. Agriculture: Feeding Families, Growing the Economy*. NC State Extension. https://ipm.ces.ncsu.edu/2019/06/n-c-agriculture-feeding-families-growing-the-economy/

Adedokun, O. A., Hetzel, K., Parker, L. C., Loizzo, J., Burgess, W. D., & Robinson, J. P. (2010). Using virtual field trips to connect students with university scientists: Core elements and evaluation of zipTrips™. *Journal of Science Education and Technology, 21*(5), 607–618. doi:10.100710956-011-9350-z

Adedokun, O., Parker, L., Loizzo, J., Burgess, W., & Robinson, J. (2011). A field trip without buses: Connecting your students to scientists through a virtual visit. *Science Scope, 34*(9), 52–57.

Alegria, S. N., & Branch, E. H. (2015). Causes and consequences of inequality in the STEM: Diversity and its discontents. *International Journal of Gender, Science and Technology, 7*(3), 321–342.

Allaire, F. S., & Kamas, B. (2020). Science That's Out of This World: Connecting Elementary Pre-Service Teacher Science Methods Courses to NASA and the International Space Station. *Connected Science Learning, 2*(2).

Allaire, F.S., King, J.P., & Frenzel, A.C. (in press). *Development and Validation of the Science Teaching Emotion Scales (Sci-TES).* Academic Press.

Alli, R. A. (2018, November 12). *What are the types of autism spectrum disorders?* WebMD. https://www.webmd.com/brain/autism/autism-spectrum-disorders#1

Almieda, L., Azevedo, B., Nunes, T., Vaz-da-Silva, M., & Soares-da-Silva, P. (2007). Why healthy subjects volunteer for phase I studies and how they perceive their participation? *European Journal of Clinical Pharmacology, 63*(11), 1085–1094. doi:10.100700228-007-0368-3 PMID:17891536

Andersen, L., & Ward, T. J. (2014). Expectancy-value models for the STEM persistence plans of ninth-grade, high-ability students: A comparison between Black, Hispanic, and White students. *Science Education, 98*(2), 216–242. doi:10.1002ce.21092

Apple. (2019, October 28). Apple Books *Asset Guide 5.2.11*. https://help.apple.com/itc/booksassetguide/en.lproj/static.html#itcd0c2f7c59

Arctic Research Consortium of the United States (ARCUS). (2015). *In House Report: PolarTREC 2010-2014 Final Evaluation Report.* Prepared by Goldstream Group Inc. (NSF PLR Award #0956825).

Association of Zoos and Aquariums (AZA). (2013). *2013 accreditation standards and related policies.* Retrieved from: https://www.aza.org/accreditation/

Association of Zoos and Aquariums (AZA). (2020). *Research and science*. Retrieved from: https://www.aza.org/research-and-science

ASTC.org. (2020). *Our strategic direction: The work of our members*. www.astc.org/about/strategy/

Audacity. (n.d.). *Free, open source, cross-platform audio software*. https://www.audacityteam.org

Ballantyne, R., Packer, J., Hughes, K., & Dierking, L. (2007). Conservation learning in wildlife tourism settings: Lessons from research in zoos and aquariums. *Environmental Education Research, 13*(3), 367–383. doi:10.1080/13504620701430604

Bandelli, A., & Konijn, E. (2011). An experimental approach to strengthen the role of science centers in the governance of science. In J. C. Marstine (Ed.), *The Routledge Companion to Museum Ethics* (pp. 164–173). Routledge.

Bandura, A. (1997). Self-efficacy: The exercise of control. *The Freeman.*

Banilower, E. R., Smith, S. P., Weiss, I. R., Malzahn, K. A., Campbell, K. M., & Weis, A. M. (2013). *Report of the 2012 national survey of science and mathematics*. Horizon Research, Inc.

Barman, C. (1997). Students' views of scientists and science: Results from a national study. *Science and Children*, (September), 18–24.

Baron, N. (2010). *Escape From the Ivory Tower: A Guide to Making Your Science Matter*. Island Press.

Bauer-Wolf, J. (2019). Early departures. *Inside Higher Ed.* Retrieved from: https://www.insidehighered.com/news/2019/02/26/latinx-black-college-students-leave-stem-majors-more-white-students

Beardsley, D. C., & O'Dowd, D. D. (1961). The college-student image of the scientist. *Science, 122*(3457), 997–1001. doi:10.1126cience.133.3457.997 PMID:17743790

Beauchamp, A.L., Roberts, S.J., Aloisio, J.M., Wasserman, D., Heimlich, J.E., Lewis, J.D., Munshi-South, J., Clark, J.A., & Tingley, K. (in review). *The Impact of Research and Mentoring Experiences on STEM Persistence during the Transition to College.*

Bell, P. (2009). *Learning science in informal environments: People, places, and pursuits*. Committee on Learning Science in Informal Environments.

Bell, P., Lewenstein, B., Shouse, A. W., & Feder, M. F. (Eds.). (2009). *Learning Science in Informal Environments: People, places, and pursuits*. The National Academies Press.

Bell, R. L., Blair, L. M., Crawford, B. A., & Lederman, N. G. (2003). Just do it? Impact of a science apprenticeship program on high school students' understandings of the nature of science and scientific inquiry. *Journal of Research in Science Teaching, 40*(5), 487–509. doi:10.1002/tea.10086

Bencze, J. L., & Bowen, G. M. (2009). A national science fair: Exhibiting support for the knowledge economy. *International Journal of Science Education, 31*(18), 2459–2483. doi:10.1080/09500690802398127

Bentley, P., & Thacker, P. G. (2004). The influence of risk and monetary payment on the research participation decision making process. *Journal of Medical Ethics, 30*(3), 293–298. doi:10.1136/jme.2002.001594 PMID:15173366

Besley, J., & Tanner, A. (2011). What science communication scholars thing about training scientists to communicate. *Science Communication, 33*(2), 239–263. doi:10.1177/1075547010386972

Blanchard, M. R., Collier, K. M., Rajwade, A. M., McCance, K. R., Mcalexander, S. L., Venditti, R. A. (2020). *Analyzing whether teachers' task values influenced their implementation of bioeconomy-focused lessons: A pilot study.* Manuscript in Progress.

Blanchard, M. R. (2010). *STEM teams: Promoting science, technology, engineering, and mathematics (STEM) career interest, skills, and knowledge through strategic teaming*. ITEST Strategies, National Science Foundation.

Blanchard, M. R., LePrevost, C. E., Tolin, A. D., & Gutierrez, K. S. (2016). Investigating technology-enhanced teacher professional development in rural, high poverty middle schools. *Educational Researcher*, *43*(3), 207–220. doi:10.3102/0013189X16644602

Blanchard, M. R., Osborne, J. W., Wallwork, C., & Harris, B. S. (2013). Progress on implementing inquiry in North Carolina: Nearly 1,000 elementary, middle and high school science teachers weigh in. *Science Educator*, *22*(1), 37–49.

Blanchard, M. R., Painter, J., Gray, D., Alsbury, T. L., Denson, C., Williams, B., Ragan, S., & Johnson, J. (2014). *STEM career clubs: Enhancing the potential of underrepresented students in STEM careers through a strategic teaming model*. ITEST Strategies, National Science Foundation.

Blanchard, M. R., & Sampson, V. D. (2018). Fostering impactful research experiences for teachers (RETs). *Eurasia Journal of Mathematics, Science and Technology Education*, *14*(1), 447–465.

Blanchard, M. R., & Sampson, V. D. (2018). Fostering Impactful research experiences for teachers (RETs). *Eurasia Journal of Mathematics, Science and Technology Education*, *14*(1), 447–465.

Blanchard, M. R., Southerland, S. A., & Granger, D. E. (2009). No silver bullet for inquiry: Making sense of teacher change following an inquiry-based research experience for teachers. *Science Education*, *93*(2), 322–360. doi:10.1002ce.20298

Bodzin, A., & Gehringer, M. (2001). Breaking science stereotypes: Can meeting actual scientists change students' perceptions of scientists? *Science and Children*, *38*(4), 36–41.

Bonney, R., Ballard, H., Jordan, R., McCallie, E., Phillips, T., Shirk, J., & Wilderman, C. (2009a). Public participation in scientific research: defining the field and assessing its potential for informal science education. A CAISE Inquiry Group Report. Washington, DC: Center for Advancement of Informal Science Education (CAISE).

Bonney, R., Shirk, J. L., Phillips, T. B., Wiggins, A., Ballard, H. L., Miller-Rushing, A. J., & Parrish, J. K. (2014). Next steps for citizen science. *Science*, *343*(6178), 1436–1437. doi:10.1126cience.1251554 PMID:24675940

Boyd-Kimball, D., & Miller, K. R. (2018). From cookbook to research: Redesigning an advanced biochemistry laboratory. *Journal of Chemical Education*, *95*(1), 62–67. doi:10.1021/acs.jchemed.6b00722

Braund, M., & Reiss, M. (2006). Towards a more authentic science curriculum: The contribution of out-of-school learning. *International Journal of Science Education*, *28*(12), 1373–1388. doi:10.1080/09500690500498419

Brígido, M., Borrachero, A., Bermejo, M., & Mellado, V. (2013). Prospective primary teachers' self-efficacy and emotions in science teaching. *European Journal of Teacher Education*, *36*(2), 200–217. doi:10.1080/02619768.2012.686993

Bronfenbrenner, U. (1977). Toward an experimental ecology of human development. *The American Psychologist*, *32*(7), 513–531. doi:10.1037/0003-066X.32.7.513

Brown, B. A., Henderson, J. B., Gray, S., Donovan, B., Sullivan, S., Patterson, A., & Waggstaff, W. (2016). From description to explanation: An empirical exploration of the African-American pipeline problem in STEM. *Journal of Research in Science Teaching*, *53*(1), 146–177. doi:10.1002/tea.21249

Bruce, M., & Bridgeland, J. (2014). The mentoring effect: Young people's perspectives on the outcomes and availability of mentoring. Washington, DC: Civic Enterprises with Hart Research Associates for MENTOR: The National Mentoring Network.

Bruner, J. (1996). *The culture of education*. Harvard University Press.

Bucholtz, M., Lopez, A., Mojarro, A., Skapoulli, E., VanderStouwe, C., & Warner-Garcia, S. (2014). Sociolinguistic justice in the schools: Student researchers as linguistic experts. *Language and Linguistics Compass*, *8*(4), 144–157. doi:10.1111/lnc3.12070

Buck, G. A., Leslie-Pelecky, D., & Kirby, S. K. (2002). Bringing female scientists into the elementary classroom: Confronting the strength of elementary students' stereotypical images of scientists. *Journal of Elementary Science Education*, *14*(2), 1–9. doi:10.1007/BF03173844

Bullock, O. M., Colón Amill, D., Shulman, H. C., & Dixon, G. N. (2019). Jargon as a barrier to effective science communication: Evidence from metacognition. *Public Understanding of Science (Bristol, England)*, *28*(7), 845–853. doi:10.1177/0963662519865687 PMID:31354058

Burgin, S. R. (2020). A Three-Dimensional Conceptualization of Authentic Inquiry-Based Practices: A Reflective Tool for Science Educators. *International Journal of Science Education*, *42*(9), 1465–1484. Advance online publication. doi:10.1080/09500693.2020.1766152

Burgin, S. R., Oramous, J., Kaminski, M., Stocker, L., & Moradi, M. (2018). High school biology students' use of VMD as an authentic tool for learning about modeling as a professional scientific practice. *Biochemistry and Molecular Biology Education*, *46*(3), 230–236. doi:10.1002/bmb.21113 PMID:29457336

Burgin, S. R., & Sadler, T. D. (2013). Science immersion: Summer programs that expose students to authentic scientific research. *Science Teacher (Normal, Ill.)*, *80*(9), 44–49. doi:10.2505/4/tst13_080_09_44

Burgin, S. R., & Sadler, T. D. (2016). Learning nature of science concepts through a research apprenticeship program: A comparative study of three approaches. *Journal of Research in Science Teaching*, *53*(1), 31–59. doi:10.1002/tea.21296

Burgin, S. R., Sadler, T. D., & Koroly, M. J. (2012). High school student participation in scientific research apprenticeships: Variation in and relationships among student experiences and outcomes. *Research in Science Education*, *42*(3), 439–467. doi:10.100711165-010-9205-2

Burgin, S. R., Sakamaki, Y., Tsuji, M., Watson, O., Heidrick, Z., Childress, M., Chitwood, T., Benamara, M., Martin, E., & Beyzavi, M. H. (2020). Using a Faculty-Developed Documentary-Style Film to Communicate Authentic Chemistry Research to a High School Audience via YouTube. *Journal of Chemical Education*, *97*(8), 2351–2355. Advance online publication. doi:10.1021/acs.jchemed.0c00376

Burrows, A., Dipompeo, M., Myers, A., Hickox, R., Borowczak, M., French, D., & Schwortz, A. (2016). Authentic science experiences: Pre-collegiate science educators' successes and challenges during professional development. *Problems of Education in the 21st Century, 70*, 59-73.

Buxton, C. A. (2006). Creating contextually authentic science in a "low-performing" urban elementary school. *Journal of Research in Science Teaching*, *43*(7), 695–721. doi:10.1002/tea.20105

Buys, N., & Bursnall, S. (2007). Establishing university-community partnerships: Processes and benefits. *Journal of Higher Education Policy and Management*, *29*(1), 73–86. doi:10.1080/13600800601175797

Cafe Scientifique. (2012). *Science for the price of a coffee*. www.Cafescientifique.org

Callanan, M. A. (2012). Conducting cognitive developmental research in museums: Theoretical issues and practical considerations. *Journal of Cognition and Development*, *13*(2), 137–151. doi:10.1080/15248372.2012.666730

Carr, N., & Cohen, S. (2011). The public face of zoos: Images of entertainment, education and conservation. *Anthrozoos: A Multidisciplinary Journal of the Interactions of People & Animals, 24*(2), 175-189.

Cavagnetto, A., Hand, B., & Meier, L. N. (2010). The nature of elementary student science discourse in the context of the science writing heuristic approach. *International Journal of Science Education, 32*(4), 427–449. doi:10.1080/09500690802627277

Chambers, D. W. (1983). Stereotypic images of the scientist: The draw-a-scientist test. *Science Education, 67*(2), 255–265. doi:10.1002ce.3730670213

Chandler, J. D., & Vargo, S. L. (2011). Contextualization and Value-in-context: How Context Frames Exchange. *Marketing Theory, 11*(1), 35–49. doi:10.1177/1470593110393713

Chatterjee, S., Williamson, V. M., McCann, K., & Peck, M. L. (2009). Surveying student attitudes and perceptions towards guided-inquiry and open-inquiry laboratories. *Journal of Chemical Education, 12*(12), 1427–1432. doi:10.1021/ed086p1427

Chen, S., Binning, K. R., Manke, K. J., Brady, S. T., McGreevy, E. M., Betancur, L., Limeri, L. B., & Kaufmann, N. (2020). Am I a science person? A strong science identity bolsters minority students' sense of belonging and performance in college. *Personality and Social Psychology Bulletin.* Advance online publication. doi:10.1177/0146167220936480 PMID:32659167

Chinn, C. A., & Malhotra, B. A. (2002). Epistemologically authentic inquiry in schools: A theoretical framework for evaluating inquiry tasks. *Science Education, 86*(2), 175–218. doi:10.1002ce.10001

Cobb, P., Wood, T., & Yackel, E. (1991). Analogies from the philosophy and sociology of science for understanding classroom life. *Science Education, 75*(1), 23–44. doi:10.1002ce.3730750104

Cohn, J. P. (2008). Citizen science: Can volunteers do real research? *Bioscience, 58*(3), 192–197. doi:10.1641/B580303

Cook-Sather, A., Bovill, C., & Felten, P. (2014). *Engaging students as partners in learning and teaching: A guide for faculty*. Jossey-Bass.

Corin, E. N., Jones, G. M., Andre, T., Hilders, G. M., & Stevens, V. (2017). Science hobbyists: Active users of the science-learning ecosystem. *International Journal of Science Education. Part B, 7*(2), 161–180.

Cossett-Lent, R. (2016). *This is Disciplinary Literacy: Reading, Writing, Thinking and Doing- Content Area by Content Area.* Corwin Literacy.

Crawford, T. (2005). What counts as knowing: Constructing a communicative repertoire for student demonstration of knowledge in science. *Journal of Research in Science Teaching, 42*(2), 139–165. doi:10.1002/tea.20047

Cutucache, C. E., Leas, H. D., Grandgenett, N. F., Nelson, K. L., Rodie, S., Shuster, R., Schaben, C., & Tapprich, W. E. (2017). Genuine faculty-mentored research experiences for In-Service science teachers: Increases in science knowledge, perception, and confidence levels. *Journal of Science Teacher Education, 28*(8), 724–744. doi:10.1080/1046560X.2017.1415615

Dallas, D. (2006). *Café Scientifique— Café Scientifique—déjà vu. Is the café scientifique a fashionable by-product of a comfortable age or an indicator of the changing relationship between science and society?* https://www.cell.com/fulltext/S0092-8674(06)00898-1

Darling-Hammond, L., & Hammond, R. (2015). *Teaching in the flat world: Learning from high-performing systems.* Teachers College Press.

Davis, B. G. (1993). *Tools for teaching*. Jossey-Bass Publishers.

Denham, K., & Lobeck, A. (Eds.). (2010). *Linguistics at school: Language awareness in primary and secondary education.* Cambridge University Press. doi:10.1017/CBO9780511770791

Dennison, J. (2008, March). Understanding the utilization of interactive videoconferencing in the K12 setting. In *Society for Information Technology & Teacher Education International Conference* (pp. 336-339). Association for the Advancement of Computing in Education (AACE).

Doktor, S. (2016, January 7). *5 Ideas to Shift Learning into a Co-created Teacher and Student Partnership.* https://www.pearsoned.com/5-ideas-to-shift-learning-into-a-co-created-teacher-and-student-partnership/

Dolan, E. L., Elliott, S. L., Henderson, C., & Curran-Everett, D. S. (2018). Evaluating discipline-based education research for promotion and tenure. *Innovative Higher Education, 43*, 31–39. doi:10.100710755-017-9406-y

Dorfman, B. S., Issachar, H., & Zion, M. (2020). Yesterday's students in today's world—Open and guided inquiry through the eyes of graduated high school biology students. *Research in Science Education, 50*(1), 123–149. doi:10.100711165-017-9683-6

Douglas, F. (2016). Sociolinguistics in the Museum: Enrichment, Engagement and Education. In R. Lawson & D. Sayers (Eds.), *Sociolinguistic Research: Application and Impact* (pp. 66–86). Routledge.

Dr. Z (2017). *Powers of Observation-year long visiting scientists experiences*. Academic Press.

Enderle, P., Dentzau, M., Roseler, K., Southerland, S., Granger, E., Hughes, R., Golden, B., & Saka, Y. (2014). Examining the influence of RETs on science teacher beliefs and practice. *Science Education, 98*(6), 1077–1108. doi:10.1002ce.21127

Erb, T. O. (1981). Attitudes of early adolescents toward science, women in science, and science careers. *Middle School Research Selected Studies, 6*(1), 108–118. doi:10.1080/08851700.1981.11670215

Estrada, M., Hernandez, P. R., & Schultz, P. W. (2018). A Longitudinal Study of How Quality Mentorship and Research Experience Integrate Underrepresented Minorities into STEM Careers. *CBE Life Sciences Education, 17*(1), ar9. doi:10.1187/cbe.17-04-0066 PMID:29351912

Etzioni, A., & Nunn, C. (1974). The public appreciation of science in contemporary America. *Daedalus, 103*(3), 191–205.

Falk, J. H., & Dierking, L. D. (2010). The 95 percent solution. *American Scientist, 98*(6), 486–493. doi:10.1511/2010.87.486

Falk, J. H., & Dierking, L. D. (2018). Viewing science learning through an ecosystem lens: A story in two parts. In D. Corrigan, C. Buntting, A. Jones, & J. Loughran (Eds.), *Navigating the changing landscape of formal and informal science learning opportunities* (pp. 9–29). Springer. doi:10.1007/978-3-319-89761-5_2

Falk, J. H., & Storksdieck, M. (2010). Science learning in a leisure setting. *Journal of Research in Science Teaching, 47*(2), 194–212.

Farland-Smith, D. & McClusky, S. (manuscript in progress). Why are rabbits not rodents? Examining the role questions across multiple student-scientist-teacher visits in a second grade classroom.

Farland-Smith, D. (2009). Exploring middle school girls' science identities: Examining attitudes and perceptions of scientists when working "side-by-side" with scientists. *School Science and Mathematics, 109*(7), 415–427. doi:10.1111/j.1949-8594.2009.tb17872.x

Farland-Smith, D. (2012). Development and field test of a rubric for assessing the draw-a-scientist test. *School Science and Mathematics, 112*(2), 109–116. doi:10.1111/j.1949-8594.2011.00124.x

Farland-Smith, D. (2017). The evolution of the analysis of the draw-a-scientist test. In P. Katz (Ed.), *Drawing for Science Education* (pp. 169–178). Brill Sense. doi:10.1007/978-94-6300-875-4_15

Farland-Smith, D., & McComas, W. F. (2009). Teaching the human dimension of science. *Science and Children, 46*(9), 32–35.

Farmelo, G. (2004). *The dana centre: Science for adults only, dimensions, biweekly news journal of the association of science-technology centers*. May/June.

Feil, E. G., Baggett, K. M., Davis, B., Sheeber, L., Landry, S., Carta, J. J., & Buzhardt, J. (2008). Expanding the reach of preventive interventions: Development of an internet-based training for parents of infants. *Child Maltreatment, 13*(4), 334–346. doi:10.1177/1077559508322446 PMID:18843143

Feinstein, N. (2011). Salvaging science literacy. *Science Education, 9*(1), 168–185. doi:10.1002ce.20414

Fenichel, M., & Schweingruber, H. A. (2010). *Surrounded by Science: Learning science in informal environments*. The National Academies Press.

Finson, K. D. (2002). *A multicultural comparison of draw-a-scientist test drawings of eighth graders*. Paper presented at the Annual International Conference of the Association of Educators of Teachers of Science, Charlotte, NC.

Finson, K. D., Beaver, J. B., & Cramond, B. L. (1995). Development and field test of a checklist for the Draw-A-Scientist Test. *School Science and Mathematics, 95*(4), 195–205. doi:10.1111/j.1949-8594.1995.tb15762.x

Flick, L. (1990). Scientist in Residence program: Improving children's images of science and scientists. *School Science and Mathematics, 90*(3), 204–214. doi:10.1111/j.1949-8594.1990.tb15536.x

Fort, D. C., & Varney, H. L. (1989). How students see scientists: Mostly male, mostly white, mostly benevolent. *Science and Children, 26*(8), 8–13.

Foutz, S., & Luke, J. (2010). *A youth-directed Café Scientifique summative evaluation (Report prepared for Science Education Solutions)*. Institute for Learning Innovation.

Frenzel, A., Becker-Kurz, B., Pekrun, R., & Goetz, T. (2015). Teaching this class drives me nuts!-Examining the person and context specificity of teacher emotions. *PLoS One, 10*(6), e0129630. Advance online publication. doi:10.1371/journal.pone.0129630 PMID:26053623

Furtak, E. M., & Penuel, W. R. (2019). Coming to terms: Addressing the persistence of "hands-on" and other reform terminology in the era of science as practice. *Science Education, 103*(1), 167–186. doi:10.1002ce.21488

FUSE Network. (n.d.). *Software FUSEe*. https://www.facebook.com/FUSEeOFFICIAL/?ref=py_c

Gallup, J., Perihan, C., Tatsuma, Y., & Ikuta, S. (2021). Creating Inclusive Functional Content Using Dot-Codes: An Exploration of Multistep Recipes for Individuals With Autism in Post-Secondary Settings. In Y. Kats & S. Fabrizio (Eds.), *Education and Technology Support for Children and Young Adults With ASD and Learning Disabilities* (pp. 149–166). IGI Global. doi:10.4018/978-1-7998-7053-1.ch008

Gambone, M., & Connell, J. (2004). *The community action framework for youth*. Academic Press.

Gambone, M., Klem, A., & Connell, J. (2002). *Finding out what matters for youth: Testing key links in a community action framework for youth development*. Youth Development Strategies, Inc., and Institute for Research and Reform in Education.

Gilbert, J. K. (2004). Models and modeling: Routes to more authentic science education. *International Journal of Science and Mathematics Education, 2*(2), 115–130. doi:10.100710763-004-3186-4

Glaser, V. (2013). Meeting current needs and assessing future opportunities to drive the global bioeconomy. *Industrial Biotechnology (New Rochelle, N.Y.), 9*(5), 271–274. doi:10.1089/ind.2013.1594

Glassman, M. (2001). Dewey and Vygotsky: Society, experience, and inquiry in educational practice. *Educational Researcher, 30*(4), 4–14. doi:10.3102/0013189X030004003

Golden, J.S., Handfield, R. B., Daystar, J. S., & McConnell, T. E. (2018). *An economic impact analysis of the U.S. biobased products industry: A Report to the Congress of the United States of America Update.* A Joint Publication of the Duke Center for Sustainability & Commerce and the Supply Chain Resource Cooperative at North Carolina State University.

Gonzales, N., Moll, L., & Amanti, C. (2005). *Funds of Knowledge: Theorizing Practices in Households, Communities and Classrooms.* Lawrence Erlbaum Associates, Publishers.

Goos, M., Galbraith, P., & Renshaw, P. (2002). Establishing a community of practice in a secondary mathematics classroom. In L. Burton (Ed.), *Learning Mathematics: From Hierarchies to Networks* (pp. 36–61). Falmer Press.

Greeno, J. G. (1998). The situativity of knowing, learning, and research. *The American Psychologist, 53*(1), 5–26. doi:10.1037/0003-066X.53.1.5

Gridmark. (n.d.). *Grid Onput.* http://gridmark.co.jp/technologies#section02

Grinell, S. (1988). Science centers come of age. *Issues in Science and Technology, 4*(3), 7–75.

Grossman, B. T., & Farland-Smith, D. (2020). Collaborating with virtual visiting scientists to address students' perceptions of scientists and their work. *Innovations in Science Teacher Education, 5*(3). Retrieved from https://innovations.theaste.org/collaborating- with-virtual-visiting-scientists-to-address-students-perceptions-of-scientists-and-their- work/

Grossman, B. T., & Farland-Smith, D. (2020). Collaborating with virtual visiting scientists to address students' perceptions of scientists and their work. *Innovations in Science Teacher Education, 5*(3). Retrieved from https://innovations.theaste.org/collaborating-with-virtual-visiting-scientists-to-address-students-perceptions-of-scientists-and-their-work/

Gunning, A., & Mensah, F. (2011). Pre-service elementary teachers' development of self-efficacy and confidence to teach science: A case study. *Journal of Science Teacher Education, 22*(2), 171–185. doi:10.100710972-010-9198-8

Habig, B., Gupta, P., Levine, B., & Adams, J. (2020). An Informal Science Education Program's Impact on STEM Major and STEM Career Outcomes. *Research in Science Education, 50*(3), 1051–1074. doi:10.100711165-018-9722-y

Hall, M., Foutz, S., & Mayhew, M. (2010). Design and impacts of a youth directed café scientifique program. *International Journal of Science Education. Part B. Communication and Public Engagement.* Advance online publication. doi:10.1080/21548455.2012.715780

Hattie, J. (2008). *Visual learning: A synthesis of over 800 meta-analyses relating achievement.* Routledge. doi:10.4324/9780203887332

Hattie, J. (2011). *Visual learning for teachers: Maximizing the impact of learning.* Routledge.

Heimlich, J. E., Deedrick, R., & Yocco, V. (2012). *COSI membership study 2011: Final report.* Columbus, OH: COSI Center for Research and Evaluation.

Heimlich, J. E. (2019). Queering STEM learningscapes. In W. Letts & S. Fifield (Eds.), *STEM of desire: Queer theories in science education* (pp. 161–176). Brill.

Heimlich, J. E., & Reid, A. (2017). Environmental learning. In M. A. Peters (Ed.), *Encyclopedia of educational philosophy and theory.* Online. doi:10.1007/978-981-287-588-4_447

Herring, C. (2009). Does diversity pay? Race, gender, and the business case for diversity. *American Sociological Review, 74*(2), 208–224. doi:10.1177/000312240907400203

Hofstein, A., & Lunetta, V. N. (2004). The laboratory in science education: Foundations for the twenty-first century. *Science Education*, *88*(1), 28–54. doi:10.1002ce.10106

Holzberger, D., Philipp, A., & Kunter, M. (2013). How teachers' self-efficacy is related to instructional quality: A longitudinal analysis. *Journal of Educational Psychology*, *105*(3), 774–786. doi:10.1037/a0032198

Honda, M., & O'Neil, W. (1993). Triggering science-forming capacity through linguistic inquiry. In The View from Building 20: Essays in Linguistics in Honor of Sylvain Bromberger. Cambridge, MA: MIT Press.

Houseal, A. K., Abd-El-Khalick, F., & Destefano, L. (2014). Impact of a student-teacher–scientist partnership on students' and teachers' content knowledge, attitudes toward science, and pedagogical practices. *Journal of Research in Science Teaching*, *51*(1), 84–115. doi:10.1002/tea.21126

Hsu, P. L., & Venegas, L. (2018). Activity features of high school students' science learning in an open-inquiry-based internship programme. *International Journal of Science Education*, *40*(12), 1391–1409. doi:10.1080/09500693.2018.1479801

Huinker, D., & Madison, S. (1997). Preparing efficacious elementary teachers in science and mathematics: The influence of methods courses. *Journal of Science Teacher Education*, *8*(2), 107–126. doi:10.1023/A:1009466323843

Ikuta, S., Ishitobi, R., Nemoto, F., Urushihata, C., Yamaguchi, K., & Nakui, H. (2017). Handmade content and school activities for autistic children with expressive language disabilities. In Y. Kats (Ed.), *Supporting the education of children with autism spectrum disorders* (pp. 85–115). IGI Global. doi:10.4018/978-1-5225-0816-8.ch006

Ikuta, S., Morton, D., Kasai, M., Nemoto, F., Ohtaka, M., & Horiuchi, M. (2015). School activities with new dot-code handling multimedia. In L. Lennex & K. Nettleton (Eds.), *Cases on instructional technology in gifted and talented education* (pp. 311–338). IGI Global. doi:10.4018/978-1-4666-6489-0.ch015

Ikuta, S., Nagano, S., Sato, S. E., Kasai, M., Ezoe, T., Mori, K., & Kaneko, C. (2019a). Original teaching materials and school activities with e-books containing media overlays. In S. Ikuta (Ed.), *Handmade teaching materials for students with disabilities* (pp. 76–110). IGI Global. doi:10.4018/978-1-5225-6240-5.ch004

Ikuta, S., Tamai, Y., Odaka, M., Nakazawa, Y., Sato, M., Mitamura, R., Akiyoshi, T., & Abe, S. (2020a). University students–schoolteachers partnership with newly developed technologies. In D. Schmidt-Crawford (Ed.), *Proceedings of Society for Information Technology & Teacher Education International Conference* (pp. 1746–1750). https://www.learntechlib.org/primary/p/215951/

Ikuta, S., Yamashita, S., Higo, H., Tomiyama, J., Saotome, N., Sudo, S., Hoshi, S., Endo, T., Narushima, T., Suzuki, K., & Watanuki, M. (2019b). Original teaching materials and school activities with multimedia-enabled dot codes. In S. Ikuta (Ed.), *Handmade teaching materials for students with disabilities* (pp. 50–75). IGI Global. doi:10.4018/978-1-5225-6240-5.ch003

Ikuta, S., Yoshida, A., Ishitobi, R., Kudo, M., Sekine, M., Yamashita, S., Edagawa, Y., Edagawa, T., & Abe, S. (2020b). Software handling multimedia-enabled dot codes and gifted school activities for students with disabilities. In S. Ikuta (Ed.), *Handbook of research on software for gifted and talented school activities in K-12 classrooms* (pp. 217–242). IGI Global. doi:10.4018/978-1-7998-1400-9.ch010

Independence, E., & Act, S. (2007). *Public Law*, 110–140. https://www.govinfo.gov/content/pkg/BILLS-110hr6enr/pdf/BILLS-110hr6enr.pdf

International Digital Publishing Forum (IDPF). (2011a). *EPUB 3.0.* http://idpf.org/epub/30/

International Digital Publishing Forum (IDPF). (2011b). *EPUB Media Overlays 3.0*. http://www.idpf.org/epub/30/spec/epub30-mediaoverlays.html

International Study Center. (n.d.). *About TIMSS 2019*. https://timssandpirls.bc.edu/timss2019/

Inzer, L. D., & Crawford, C. B. (2005). A review of formal and informal mentoring: Processes, problems, and design. *Journal of Leadership Education*, *4*(1), 31–50. doi:10.12806/V4/I1/TF2

Jennings, L. B., & Mills, H. (2009). Constructing a discourse of inquiry: Findings from a five- year ethnography at one elementary school. *Teachers College Record*, *111*(7), 1583–1618.

Johnson, B., & Kubarek, J. (2019). *Summary of Evaluation Findings: Zoo Crew*. Retrieved from: https://www.PEERAssociates.net

Johnson, D. R. (2011). Women of color in science, technology, engineering, and mathematics (STEM). *New Directions for Institutional Research*, *152*(Winter), 75–85. doi:10.1002/ir.410

Joyce, B. A., & Farenga, S. J. (1999). Intentions of young students to enroll in science courses in the future: An examination of gender differences. *Sci Educ*, *83*(1), 55–75. doi:10.1002/(SICI)1098-237X(199901)83:1<55::AID-SCE3>3.0.CO;2-O

Katz, L. G. (2010). *STEM in the early years*. Collected Papers from the SEED Conference Fall 2010.

Keeble, B. R. (1988). The Brundtland Report: 'Our common future'. *Medicine and War*, *4*(1), 17–25. doi:10.1080/07488008808408783

Kekelis, L., Heber, E., & Countryman, J. A. (2005, May). *Bridge to technology: Designing a program that attracts girls. ASTC Dimensions*.

King, D., Ma, J., Armendariz, A., & Yu, K. (2018). Developing interactive exhibits with scientists: Three example collaborations from the life sciences collection at the Exploratorium. *Integrative and Comparative Biology*, *58*(1), 94–102. doi:10.1093/icb/icy010 PMID:29697800

Kirch, S. A. (2007). Skepticism and open-mindedness for learning, teaching and criticality in science. *Cultural Studies of Science Education*, 2.

Kisling, V. N. Jr. (2001). *Zoo and aquarium history: Ancient animal collections to zoological gardens*. CRC Press.

Koomen, M. H., Rodriguez, E., Hoffman, A., Petersen, C., & Oberhauser, K. (2018). Authentic science with citizen science and student-driven science fair projects. *Science Education*, 102(3), Kosmala, M., Wiggins, A., Swanson, A., & Simmons, B. (2016). Assessing data quality in citizen science. *Frontiers in Ecology and the Environment*, *14*(10), 551–560.

Kuhn, D., Arvidsson, T. S., Lesperance, R., & Corprew, R. (2017). Can engaging in science practices promote deep understanding of them? *Science Education*, *101*(2), 232–250. doi:10.1002ce.21263

Larsen, S., Liston, C., Thiry, H., & Graf, J. (2007). What good is a scientist in the classroom? Participant outcomes and program design features for a short- duration science outreach intervention in K–12 classrooms. *Life Sci Educ*, *6*(1), 49–64. doi:10.1187/cbe.06-05-0165 PMID:17339394

Larson, R., Walker, K., & Pearce, N. (2005). A comparison of youth-driven and adult driven youth programs: Balancing inputs from youth and adults. *Journal of Community Psychology*, *33*(1), 57–74. doi:10.1002/jcop.20035

Lavalley, M. (2018). *Out of the loop: Rural schools are largely left out of research and policy discussions, exacerbating poverty, inequity and isolation*. Center for Public Education.

Lawless, J. G., & Rock, B. N. (1998). Student scientist partnerships and data quality. *Journal of Science Education and Technology*, *7*(1), 5–13. doi:10.1023/A:1022575914118

Lederman. J., Lederman, N., Bartels, S., Jimenez, J., Akubo, M., Aly, S., Bao, C., Blanquet, E., Blonder, R., Bologna Soares de Andrade, M., Buntting, C., Cakir, M., El-Deghaidy, H., ElZorkani, A., Gaigher, E., Guo, S., Hakanen, A., Al-Lal, S. H., Han-Tosunoglu, C., Hattingh, A., … Zhhou, Q. (2019). An international collaborative investigation of beginning seventh grade students' understandings of scientific inquiry: Establishing a baseline. *Journal of Research in Science Teaching*, *56*(10), 486-515.

Lemov, D. (2013). *Teach like a champion: 49 Techniques that put students on the path to college*. Wiley & Sons.

Lent, R. W., Brown, S. D., & Hackett, G. (1994). Toward a unifying social cognitive theory of career and academic interest, choice, and performance. *Journal of Vocational Behavior*, *45*(1), 79–122. doi:10.1006/jvbe.1994.1027

Lent, R. W., Brown, S. D., & Hackett, G. (2000). Contextual supports and barriers to career choice: A social cognitive analysis. *Journal of Counseling Psychology*, *47*(1), 36–49. doi:10.1037/0022-0167.47.1.36

Lewis, C., Perry, R., Hurd, J., & O'Connell, M. P. (2006). Lesson study comes of age in North America. *Phi Delta Kappan*, *88*(4), 273–281. doi:10.1177/003172170608800406

Lewis, C., & Tsuchida, I. (1998). A lesson is like a swiftly flowing river:How research lessons improve Japanese education. *American Educator*, *22*(4), 12–17, 50–52. lessonresearch.net/lesson.pdf

Lidz, J., & Kronrod, Y. (2014). Expanding our reach and theirs: When linguists go to high school. *Language and Linguistics Compass*, *8*(10), 449–463. doi:10.1111/lnc3.12094

Lindahl, B. (2003). Pupils' responses to school science and technology? A longitudinal study of pathways to upper secondary school. *Göteborg Studies in Educational Sciences, 196*.

Lord, T., & Orkwiszewski, T. (2006). Moving from didactic to inquiry-based instruction in a science laboratory. *The American Biology Teacher*, *68*(6), 342–345. doi:10.2307/4452009

Luft, J. A., & Roehrig, G. H. (2007). Capturing science teachers' epistemological beliefs: The development of the teacher beliefs interview. *The Electronic Journal of Science Education*, *11*(2), 39–63.

Lundquist, K., Herndon, C., Harty, T. H., & Gumbartt, J. C. (2016). Accelerating the use of molecular modeling in the high school classroom with VMD lite. *Biochemistry and Molecular Biology Education*, *44*(2), 124–129. doi:10.1002/bmb.20940 PMID:26751137

Maltese, A. V., & Tai, R. H. (2010). Eyeballs in the fridge: Sources of early interest in science. *International Journal of Science Education*, *32*(5), 669–685. doi:10.1080/09500690902792385

Manduca, C. A., Iverson, E. R., Luxenberg, M., Macdonald, R. H., McConnell, D. A., Mogk, D. W., & Tewksbury, B. J. (2017). Improving undergraduate STEM education: The efficacy of discipline-based professional development. *Science Advances*, *3*(2), e1600193. doi:10.1126ciadv.1600193 PMID:28246629

Markham, C. M., Rushing, S. C., Jessen, C., Gorman, G., Torres, J., Lambert, W. E., ... Peskin, M. F. (2016). Internet-based delivery of evidence-based health promotion programs among American Indian and Alaska native youth: A case study. *JMIR Research Protocols*, *5*(4).

Martin, B., Kass, H., & Brouwer, W. (1990). Authentic science: A diversity of meaning. *Science Education*, *74*(5), 541–554. doi:10.1002ce.3730740505

Mason, K., Kahle, J. B., & Gardner, A. L. (1991). Draw-a-scientist-test: Future implications. *School Science and Mathematics, 91*(5), 193–198. doi:10.1111/j.1949-8594.1991.tb12078.x

McAlexander, S. L., Blanchard, M. R., & Venditti, R. (2020a) *Supporting career development and internship experiences of historically underrepresented science and engineering community college to university transfer students.* Manuscript in preparation.

McAlexander, S. L., Blanchard, M. R., & Venditti, R. (2020b). *Investigating internship motivations and outcomes: Undergraduate student and industry perspectives*. Manuscript in preparation.

McAlexander, S. L., McCance, K., Noble, S., Blanchard, M. R., & Venditti, R. (2020). *The development and validation of the bioproduct and bioenergy beliefs (BABB) and career interest (CIBB) surveys.* Manuscript in preparation.

McAlexander, S., Topliceanu, A., Sharma, S., McCance, K., Starkey, H., Blanchard, M. R., & Venditti, R. (2019, October 12). *Building a sustainable future with bioproducts & bioenergy in science curricula* [Conference presentation]. Bridging the Gap 2019: Uniting North Carolina STEM Education, Raleigh, NC.

McCance, K. R., Blanchard, M. R., & Venditti, R. (2020). *Graduate Students' Perceptions of an interdisciplinary collaboration between science, engineering, and education: Developing community and navigating boundaries. Participating in a Science/Engineering and Education Interdisciplinary Collaboration: Developing Community and Navigating Boundaries*. Manuscript in preparation.

McComas, W. F. (2011). A new look at an old tradition. *Science Teacher (Normal, Ill.), 78*(8), 34.

McConnell, T. E., Jeuck, J., Bardon, R., & Hazel, D. (2016). *North Carolina's forests and forest products industry by the numbers, 2013.* North Carolina Cooperative Extension Service Bulletin AG-817. 11

McDonnell, L., Barker, M. K., & Wieman, C. (2016). Concepts first, jargon second improves student articulation of understanding. *Biochemistry and Molecular Biology Education, 44*(1), 12–19. doi:10.1002/bmb.20922 PMID:26537537

McLaughlin, M. W. (2000). *Community counts: How youth organizations matter for youth development*. Public Education Network.

Mead, M., & Metraux, R. (1957). The image of the scientist amongst high school students. In B. Barbar & W. Hirsch (Eds.), *The Sociology of Science* (pp. 38–61). Falmer Press.

Mercer, N., Dawes, L., Wegerif, R., & Sams, C. (2004). Reasoning as a scientist: Ways of helping children to use language to learn science. *British Educational Research Journal, 30*(3), 359–377. doi:10.1080/01411920410001689689

Meredith, D. (2010). *Explaining Research.* Oxford University Press.

Merton, R. K. (1973). *The sociology of science: Theoretical and empirical investigations.* University of Chicago press.

Ministry of Education, Culture, Sports, Science and Technology, Japan. (2003). *Students having significant learning and behavioral difficulties*. https://www.mext.go.jp/b_menu/shingi/chousa/shotou/054/shiryo/attach/1361231.htm

Ministry of Education, Culture, Sports, Science and Technology, Japan. (2019a, September 24). *Report on School Basic Survey.* https://www.mext.go.jp/kaigisiryo/2019/09/__icsFiles/afieldfile/2019/09/24/1421554_3_1.pdf

Ministry of Education, Culture, Sports, Science and Technology, Japan. (2019b, February 22). *Present status on special support services in resource rooms.* https://www.mext.go.jp/component/a_menu/education/micro_detail/__icsFiles/afieldfile/2019/03/06/1414032_09.pdf

Moats, L., & Tolman, C. (2009). *Excerpted from Language Essentials for Teachers of Reading and Spelling (LETRS): The Challenge of Learning to Read (Module 1).* Sopris West.

Mocker, D. W., & Spear, G. E. (1982). *Lifelong learning: Formal nonformal informal and self-directed. Information Series No. 241*. Eric Clearinghouse for Adult and Vocational Education.

Moore, A. J. (1987, October). Native Indian Learning Styles: A review for researchers and teachers. *Journal of American Indian Education, 27*(1). https://jaie.asu.edu/v27/V27S1nat.htm

Moore, D. T. (2010). Forms and issues in experiential learning. *New Directions for Teaching and Learning, 2010*(124), 3–13. doi:10.1002/tl.415

Moss, D. M., Abrams, E. D., & Kull, J. A. (1998). Can we be scientists too? Secondary students' perceptions of scientific research from a project-based classroom. *Journal of Science Education and Technology, 7*(2), 149–161. doi:10.1023/A:1022564507639

Motamedi, V. (2001). A critical look at the use of videoconferencing in United States distance education. *Education, 122*(2), 386–394.

Moulding, B., Bybee, R., & Paulson, N. (2015). *A vision and plan for science teaching and learning*. Essential Teaching and Learning Publications.

Mujtaba, T., Lawrence, M., Oliver, M., & Reiss, M. J. (2018). Learning and engagement through natural history museums. *Studies in Science Education, 54*(1), 41–67. doi:10.1080/03057267.2018.1442820 PMID:29951162

Nagy, W., & Townsend, D. (2012). Words as tools: Learning academic vocabulary as language acquisition. *Reading Research Quarterly, 47*(1), 91–108. doi:10.1002/RRQ.011

National Academies of Science. (2010). *Preparing Teachers: Building Evidence for Sound Policy*. The National Academies Press.

National Education Association. (2012). *Preparing 21st century students for a global society: An educator's guide to "the four Cs."* Washington, DC: Author. Available: https://www.nea.org/assets/docs/A-Guide-to-Four-Cs.pdf

National Institute for Educational Policy Research. (n.d.). *TALIS 2018: Teaching and learning international survey in Japan*. NIER. https://www.nier.go.jp/kenkyukikaku/talis/pdf/talis2018_points.pdf

National Institutes of Health (NIH). (2018, March). *Autism spectrum disorder.* https://www.nimh.nih.gov/health/topics/autism-spectrum-disorders-asd/index.shtml

National Institutes of Health (NIH). (2020, March 5). *What are reading disorders?* https://www.nichd.nih.gov/health/topics/reading/conditioninfo/disorders

National Research Council (NRC). (1996). *National science education standards*. National Academy Press.

National Research Council (NRC). (2008). *Ready, set, Science!: Putting research to work in K–8 science classrooms*. National Academies Press.

National Research Council (NRC). (2012). *A Framework for K-12 science education. Practices, Crosscutting Concepts, and core ideas*. National Academy Press.

National Research Council. (2007). Taking science to school: Learning and teaching in science grades k-8. Committee on science learning, kindergarten through eighth grade. Board of Science Education, Center for Education. Division of Behavioral and Social Sciences and Education. Washington, DC: The National Academies Press.

National Research Council. (2009). *Learning Science in Informal Environments: People, Places, and Pursuits. Committee on Learning Science in Informal Environments. Board on Science Education, Center for Education. Division of Behavioral and Social Sciences and Education*. The National Academies Press.

National Science Foundation, National Center for Science and Engineering Statistics. (2017). *Women, minorities, and persons with disabilities in science and engineering: 2017.* Special Report NSF 17-310. www.nsf.gov/statistics/wmpd/

National Science Teachers Association (NSTA). (2014). *Early Childhood Science Education.* NSTA.

National Science Teachers Association (NSTA). (2018). *NSTA Position Statement: Elementary Science Education.* NSTA.

Neeley, L., Goldman, E., Smith, B., Baron, N., & Sunu, S. (2014). GRADSCICOMM Report and Recommendations: Mapping the Pathways to Integrate Science Communication Training into STEM Graduate Education. In *Conference proceedings.* COMPASS.

NGSS Lead States. (2013). *Next Generation Science Standards: For states, by states.* National Academies Press.

NGSS Lead States. (2013). *Next generation science standards: For states, by states.* The National Academies Press.

Ng, W., & Fergusson, J. (2019). Technology-enhanced science partnership initiative: Impact on secondary science teachers. *Research in Science Education, 49*(1), 219–242. doi:10.100711165-017-9619-1

Niemitz, M., Slough, S., Peart, L., Klaus, A., Leckie, R. M., & John, K. S. (2008). Interactive virtual expeditions as a learning tool: The school of rock expedition case study. *Journal of Educational Multimedia and Hypermedia, 17*(4), 561–580.

Nisbet, M. C., & Mooney, C. (2007a). Framing science. *Science, 316*(5821), 56. doi:10.1126cience.1142030 PMID:17412941

Nisbet, M. C., & Mooney, C. (2007b). The risks and advantages of framing science - Response. *Science, 317*, 1169–1170.

North Carolina Department of Agriculture and Consumer Services. (2015). *North Carolina agricultural statistics. Publication No. 216.* Author.

North Carolina School Report Cards. (n.d.). *School Report Cards.* North Carolina Department of Public Instruction. https://ncreportcards.ondemand.sas.com/src

Nugent, G., Barker, B., Welch, G., Grandgenett, N., Wu, C. R., & Nelson, C. (2015). A model of factors contributing to STEM learning and career orientation. *International Journal of Science Education, 37*(7), 1067–1088. doi:10.1080/09500693.2015.1017863

Office of Energy Efficiency & Renewable Energy. (n.d.). *Bioenergy career map.* https://www.energy.gov/eere/bioenergy/bioenergy-career-map

Olson, J. K., Tippett, C. D., Milford, T. M., Ohana, C., & Clough, M. P. (2015). Science teacher preparation in a North American context. *Journal of Science Teacher Education, 26*(1), 7–28. doi:10.100710972-014-9417-9

Olympus Co. (1999, September 14). *ScanTalk, scan talk reader, & sound print kobo.* https://www.olympus.co.jp/jp/news/1999b/nr990823r300j.jsp

Organisation for Economic Cooperation and Development. (2018). *PISA: Programme for international student assessment.* OECD. http://www.oecd.org/pisa/

Ozgelen, S. (2017). Primary school students' views on science and scientists. In P. Katz (Ed.), *Drawing for Science Education* (pp. 191–203). Brill Sense. doi:10.1007/978-94-6300-875-4_17

Pacific Science Center & Institute for Learning Innovation. (2018). *Portal to the public: Implementation manual & catalog of professional development elements.* Institute for Learning Innovation.

Packer, J. (2006). Learning for fun: The unique contribution of educational leisure experiences. *Curator (New York, N.Y.), 49*(3), 329–344. doi:10.1111/j.2151-6952.2006.tb00227.x

Palmer, D. (2006). Sources of self-efficacy in a science methods course for primary teacher education students. *Research in Science Education*, *36*(4), 337–353. doi:10.100711165-005-9007-0

Parvin, J., & Stephenson, M. (2004). Learning science at industrial sites. *Learning Science Outside the Classroom*, 129-149.

Passerini, K., & Granger, M. J. (2000). A developmental model for distance learning using the Internet. *Computers & Education*, *34*(1), 1–15. doi:10.1016/S0360-1315(99)00024-X

Pekarik, A. J., Doering, Z. D., & Karns, D. A. (1999). Exploring satisfying experiences in museums. *Curator (New York, N.Y.)*, *42*(2), 152–173. doi:10.1111/j.2151-6952.1999.tb01137.x

Peker, D., & Dolan, E. (2012). Helping students make meaning of authentic investigations: Findings from a student-teacher–scientist partnership. *Cultural Studies of Science Education*, *7*(1), 223–244. doi:10.100711422-012-9385-3 PMID:23828722

Penckofer, S., Byrn, M., Mumby, P., & Ferrans, C. E. (2011). Improving Subject recruitment, retention, and participation in research through Peplau's Theory of Interpersonal Relations. *Nursing Science Quarterly*, *24*(2), 146–151. doi:10.1177/0894318411399454 PMID:21471039

Perrin, J. (2014). Features of engaging and empowering experiential learning programs for college students. *Journal of University Teaching & Learning Practice*, *11*(2), 2.

Pew Research Center. (2019). *Trust and mistrust in Americans' views of scientific experts*. Full report available at https://www.pewresearch.org

Prainsack, B. (2014). Understanding participation: the 'citizen science' of genetics. In B. Prainsack, S. Schicktanz, & G. Werner-Felmayer (Eds.), *Genetics as social practice* (pp. 147–164). Ashgate.

Purvis, B., Mao, Y., & Robinson, D. (2019). Three pillars of sustainability: In search of conceptual origins. *Sustainability Science*, *14*(3), 681–695. doi:10.100711625-018-0627-5

Rao, K., & Giuli, C. (2010). Reaching remote learners: Successes and challenges for students in an online graduate degree program in the Pacific Islands. *The International Review of Research in Open and Distributed Learning*, *11*(1), 141–160. doi:10.19173/irrodl.v11i1.785

Rennie, L. J., & Johnston, D. J. (2004). The nature of learning and its implications for research on learning from museums. Science Education, 88(1), S4-S16.

Resnik, D. B. (2001). Research participation and financial inducements. *The American Journal of Bioethics*, *1*(2), 54–56. doi:10.1162/152651601300169112 PMID:11951894

Reveal, H. P. (n.d.). *From print to a whole new reality*. https://www8.hp.com/us/en/printers/reveal.html

Richmond, G. (2017). The power of community partnership in the preparation of teachers. *Journal of Teacher Education*, *68*(1), 6–8. doi:10.1177/0022487116679959

Richmond, G., & Kurth, L. A. (1999). Moving from outside to inside: High school students' use of apprenticeships as vehicles for entering the culture and practice of science. *Journal of Research in Science Teaching*, *36*(6), 677–697. doi:10.1002/(SICI)1098-2736(199908)36:6<677::AID-TEA6>3.0.CO;2-#

Roberts, L. F., & Wassersug, R. J. (2009). Does doing scientific research in high school correlate with students staying in science? A half-century retrospective study. *Research in Science Education*, *39*(2), 251–256. doi:10.100711165-008-9083-z

Roth, W. (1993). Construction sites: Science labs and classrooms. In K. Tobin (Ed.), *The practice of constructivism in science education* (pp. 143–170). Lawrence Erlbaum Associates, Publishers.

Roth, W. M. (1995). *Authentic school science: Knowing and learning in open-inquiry science laboratories*. Kluwer. doi:10.1007/978-94-011-0495-1

Roth, W. M., & Radford, L. (2010). Rethinking the zone of proximal development: (Symmetrically). *Mind, Culture, and Activity*, *17*(4), 299–307. doi:10.1080/10749031003775038

Rushton, E. A., & Reiss, M. J. (2019). From science teacher to 'teacher scientist': Exploring the experiences of research-active science teachers in the UK. *International Journal of Science Education*, *41*(11), 1541–1561. doi:10.1080/09500693.2019.1615656

Ryoko, A. (2018). Trust—What connects science to daily life. *Health Physics*, *115*(5), 581–589. doi:10.1097/HP.0000000000000945 PMID:30260848

Sadler, T. D., Burgin, S., McKinney, L., & Ponjuan, L. (2010). Learning science through research apprenticeships: A critical review of the literature. *Journal of Research in Science Teaching*, *47*, 235–256.

Sandoval, W. A. (2005). Understanding students' practical epistemologies and their influence on learning through inquiry. *Science Education*, *89*(4), 634–656. doi:10.1002ce.20065

Sargent, J. F., Jr. (2017). *The U.S. science and engineering workforce: Recent, current, and projected employment, wages, and unemployment*. https://fas.org/sgp/crs/misc/R43061.pdf

Sargent, L. D., & Waters, L. E. (2004). Careers and academic research collaborations: An inductive process framework for understanding successful collaborations. *Journal of Vocational Behavior*, *64*(2), 308–319. doi:10.1016/j.jvb.2002.11.001

Schleicher, A. (2020). *Teaching and Learning International survey: TALIS 2018 Insights and Interpretations*. OECD. http://www.oecd.org/education/talis/TALIS2018_insights_and_interpretations.pdf

School Digger. (n.d.). https://www.schooldigger.com

Schwan, S., Grajal, A., & Lewalter, D. (2014). Understanding and engagement in places of science experience: Science museums, science centers, zoos, and aquariums. *Educational Psychologist*, *49*(2), 70–85. doi:10.1080/00461520.2014.917588

Science Learning, B. S. C. S. (n.d.). *BSCS 5E instructional model*. https://bscs.org/bscs-5e-instructional-model/

Selvakumar, M., & Storksdieck, M. (2013). Portal to the public: Museum educators collaborating with scientists to engage museum visitors with current science. *Curator (New York, N.Y.)*, *56*(1), 69–78. doi:10.1111/cura.12007

Shanahan, M. C., & Bechtel, R. (2020). "We're taking their brilliant minds": Science teacher expertize, meta-discourse, and the challenges of teacher–scientist collaboration. *Science Education*, *104*(2), 354–387. doi:10.1002ce.21550

Shen, B. S. P. (1975). Science literacy and the public understanding of science. In S. B. Day (Ed.), *Communication of scientific information*. S. Karger and A.G. Basel. doi:10.1159/000398072

Shi, L. (2006). Students as research participants or as learners? *Journal of Academic Ethics*, *4*(4), 205–220. doi:10.100710805-006-9028-y

Shin, J. E. L., Levy, S. R., & London, B. (2016). Effects of role model exposure on STEM and non-STEM student engagement. *Journal of Applied Social Psychology*, *46*(7), 410–427. doi:10.1111/jasp.12371

Shirk, J. L., Ballard, H. L., Wilderman, H. L., Phillips, C. C., Wiggins, T., Jordan, A., Mc, R., Callie, E., Minarchek, M., Lewenstein, B. V., Krasny, M. E., & Bonney, R. (2012). Public participation in scientific research: A framework for deliberate design. *Ecology and Society*, *17*(2), 29. doi:10.5751/ES-04705-170229

Sickler, J. (2018) *Long-term impact of teen science cafés: Results of a pilot study with cafe scientifique New Mexico Evaluation Report*. https://www.informalscience.org/long-term-impact-teen-science-cafes-results-pilot-study-cafe-scientifique-new-mexico

Silver, A., & Rushton, B. S. (2008). Primary-school children's attitudes towards science, engineering and technology and their images of scientists and engineers. *Education 3-13, 36*, 51-67. doi:10.1080/03004270701576786

Silverstein, S. C., Dubner, J., Miller, J., Glied, S., & Loike, J. D. (2009). Teachers' participation in research programs improves their students' achievement in science. *Science*, *326*(5951), 440–442. doi:10.1126cience.1177344 PMID:19833969

Simmel, G. (1950). *The Sociology of Georg Simmel*. Free Press.

Simmons, B. (2010). *Evaluation Report of the NOAA Teacher at Sea Program: 2005-2009*. National Oceanic and Atmospheric Administration - National Marine Fisheries Service - NOAA Teacher at Sea Program. (OMB 0648-0600). Available at https://teacheratsea.noaa.gov/#/Evaluation/

Smith, D. K. (2014). iTube, YouTube, WeTube: Social Media Videos in Chemistry Education and Outreach. *Journal of Chemical Education*, *91*(10), 1594–1599. doi:10.1021/ed400715s

Sneider, C., & Noam, G. G. (2019). The common instrument suite: a means for assessing student attitudes in STEM classrooms and out-of-school environments. *Connected Science Learning, 11*.

Stade, L. (2018, June 2). *The importance of student-teacher relationships: Build the bond*. Linda Stade Education. https://lindastade.com/the-importance-of-student-teacher-relationships/

Stilaloppi, J., & Vargo, S. (2017). Triads: A review and analytical framework. *Marketing Theory*, *17*(4), 395–414. doi:10.1177/1470593117705694

Stylinski, C., Storksdieck, M., Canzoneri, N., Klein, E., & Johnson, A. (2019). Impacts of a comprehensive public engagement training and support program on scientists' outreach attitudes and practices. *International Journal of Science Education, Part B.*, *8*(4), 340–354. doi:10.1080/21548455.2018.1506188

Sullivan, P. & Reed, W. L. (2009). *Connecting youth to high-resource adults: Lessons from effective youth programs*. doi:10.1177/0743558409350505

Swisher, K. (1994). American Indian learning styles survey: An assessment of teachers' knowledge. *Journal of Educational Issues of Language Minority Students*, (13), 59-77. http://www.ncela.gwu.edu/pubs/jeilms/vol13/americ13.htm

Tai, R., Liu, C., Maltese, A., & Fan, X. (2006). Planning early for careers in science. *Science*, *312*(5777), 1143–1144. doi:10.1126cience.1128690 PMID:16728620

Takahashi, A., & Yoshida, M. (2004). How can we start lesson study? Ideas establishing lesson study communities. *Teaching Children Mathematics*, *10*(9), 436–443. doi:10.5951/TCM.10.9.0436

The DAISY Consortium. (n.d.). *Creating the best way to read and publish*. https://daisy.org

Tishler, C. L., & Bartholomae, S. (2003). Repeat participation among normal healthy research volunteers: Professional guinea pigs in clinical trials? *Perspectives in Biology and Medicine*, *46*(4), 508–520. doi:10.1353/pbm.2003.0094 PMID:14593220

Traphagen, K., & Traill, S. (2014). *How cross-sector collaborations are advancing STEM learning:* Working paper.

Trumbull, D. J., Bonney, R., Bascom, D., & Cabral, A. (2000). Thinking scientifically during participation in a citizen-science project. *Science education*, *84*(2), 265–275. doi:10.1002/(SICI)1098-237X(200003)84:2<265::AID-SCE7>3.0.CO;2-5

Tuthill, G., & Klemm, E. (2002). Virtual field trips: Alternatives to actual field trips. *International Journal of Instructional Media*, *29*(4), 453–467.

Uden, L., Wangsa, I. T., & Damiani, E. (2007, February). The future of E-learning: E-learning ecosystem. In Digital ecosystems and technologies conference, 2007. DEST'07. Inaugural IEEE-IES (pp. 113–117). IEEE.

United States Department of Commerce Census Bureau. (2016). *American factfinder*. http://factfinder.census.gov/faces/nav/jsf/pages/index.xhtml

United States Department of Education. (2012, January 10). *Race to the top: North Carolina Report. Year 1 (2010-2011)*. Washington, DC: Author.

Van Montagu, M. (2018). Science, ideology and daily life. *Journal of Innovation and Knowledge*, *3*(2), 66–69. doi:10.1016/j.jik.2018.03.005

Vedel, M., Holma, A.-M., & Havila, V. (2016). Conceptualizing Inter-organizational Triads. *Industrial Marketing Management*, *57*, 139–147. doi:10.1016/j.indmarman.2016.01.005

Venditti, R. & Blanchard, M. R. (2016). *Preparing Diverse and Rural Students and Teachers to Meet the Challenges in the Bioenergy and Bioproducts Industry.* USDA NIFA, $2,750,000/4 years.

Vygotsky, L. S. (1978). *Mind in society: The development of higher psychological processes*. Harvard University Press.

W3C. (n.d.). *HTML & CSS*. https://www.w3.org/standards/webdesign/htmlcss.html

Wagner, L., Speer, S. R., Moore, L. C., McCullough, E. A., Ito, K., Clopper, C. G., & Campbell-Kibler, K. (2015). Linguistics in a science museum: Integrating research, teaching, and outreach in a language sciences research lab. *Language and Linguistics Compass*, *9*(10), 420–431. doi:10.1111/lnc3.12164

Weaver, G. C., Russell, C. B., & Wink, D. J. (2008). Inquiry-based and research-based laboratory pedagogies in undergraduate science. *Nature Chemical Biology*, *4*(10), 577–580. doi:10.1038/nchembio1008-577 PMID:18800041

Weisgram, E. S., & Diekman, A. B. (2017). Making STEM "family friendly": The impact of perceiving science careers as family-compatible. *Social Sciences*, *6*(2), 61. doi:10.3390ocsci6020061

Wenger, E. (2009). *Communities of practice: a brief introduction*. http://www.ewenger.com/theory/

Wenger, E. (2010). Communities of practice and social learning systems: The career of a concept. In C. Blackmore (Ed.), *Social learning systems and communities of practice* (pp. 179–198). Springer. doi:10.1007/978-1-84996-133-2_11

Wheeler, L. B., Maeng, J. L., Chiu, J. L., & Bell, R. L. (2017). Do teaching assistants matter? Investigating relationships between teaching assistants and student outcomes in undergraduate science laboratory classes. *Journal of Research in Science Teaching*, *54*(4), 463–492. doi:10.1002/tea.21373

Wigfield, A., & Eccles, J. S. (2000). Expectancy–value theory of achievement motivation. *Contemporary Educational Psychology*, *25*(1), 68–81. doi:10.1006/ceps.1999.1015 PMID:10620382

Wojton, M. A. (2009). *A study of a museum-school partnership* (Unpublished doctoral dissertation). The Ohio State University.

Wong, B. (2015). Careers "from" but not "in" science: Why are aspirations to be a scientist challenging for minority ethnic students? *Journal of Research in Science Teaching*, *52*(7), 979–1002. doi:10.1002/tea.21231

Wong, S. L., & Hodson, D. (2009). From the horse's mouth: What scientists say about scientific investigation and scientific knowledge. *Science Education*, *93*(1), 109–130. doi:10.1002ce.20290

Wong, S. L., & Hodson, D. (2010). More from the horse's mouth: What scientists say about science as a social practice. *International Journal of Science Education*, *32*(11), 1431–1463. doi:10.1080/09500690903104465

Wooley, J. P., McGowan, M. L., Teare, H. J. A., Coathup, V., Fishman, J. R., Settersten, R. A. Jr, Sterckx, S., Kaye, J., & Jeungst, E. T. (2016). Citizen science or scientific citizenship? Disentangling the uses of public engagement rhetoric in national research initiatives. *BMC Medical Ethics*, *6*(17), 33. Advance online publication. doi:10.118612910-016-0117-1

Wormstead, S. J., Becker, M. L., & Congalton, R. G. (2002). Tools for successful student–teacher–scientist partnerships. *Journal of Science Education and Technology*, *11*(3), 277–287. doi:10.1023/A:1016076603759

Wu, K. (2017). Why can't scientists talk like regular humans. *Scientific American.* Retrieved from https://blogs.scientificamerican.com/guest-blog/effective-communication-better-science/

Wyss, V. L., Heulskamp, D., & Siebert, C. J. (2012). Increasing middle school student interest in STEM careers with videos of scientists. *International Journal of Environmental and Science Education*, *7*(4), 501–522.

Yoon, S. A., Elinich, K., Wang, J., Van Schooneveld, J. B., & Anderson, E. (2013). Scaffolding informal learning in science museums: How much is too much? *Science Education*, *97*(6), 848–877. doi:10.1002ce.21079

Zimmerman, A., Hatchwell, M., Dickie, L., & West, C. (Eds.). (2009). Zoos in the 21st century: Catalysts for conservation? Cambridge University Press.

Zimmerman, M. A., Bingenheimer, J. B., & Notaro, P. C. (2002). Natural mentors and adolescent resiliency: A study with urban youth. *American Journal of Community Psychology*, *30*(2), 221–243. doi:10.1023/A:1014632911622 PMID:12002244

About the Contributors

Shinya Abe is one of the on-staff employees at Gridmark, Inc., Japan that originally developed Grid Onput dot code. He graduated in Department of Digital Media, Faculty of Computer and Information Sciences, Hosei University in 2012. He first had a job with Digital Amuse Inc. that was emerged into Gridmark Inc. He is engaging with creating new software that can handle original Grid Onput dot codes File Linker and GM Authoring Tool that can create a standalone application to reproduce multimedia like movies with a sound pen G-Speak and overlap Grid Onput dot codes on users' designed document, respectively. He hopes the software he developed helps the students tell their feelings and wills and desires to classmates, teachers, and parents. He and his company have been collaborating with Professor Shigeru Ikuta for more than ten years.

Andy Aichele serves as Senior Director of Organizational Learning & Development for COSI. He has a career that includes talent and leadership development for Fortune 500 companies, libraries, and museums. He has a BA in Chemistry from The Ohio State University, and his MBA from Capital University.

Franklin S. Allaire, Ph.D., is an Assistant Professor of Science Education in the Department of Urban Education at the University of Houston-Downtown where he brings twenty years of experience in formal and informal elementary, secondary, and post-secondary science education to undergraduate and graduate courses in elementary and secondary science methods. His research interests focus on issues impacting the success of underrepresented minorities in STEM-related fields and the innovative use of technologies and pedagogies in the professional development of science teachers and teacher candidates.

Frederic Bertley brings his experience as a bench and field immunologist and his passion for education into his leadership of COSI, the award-winning science center in Columbus, Ohio. Dr. Bertley has been widely recognized and honored for his work as a scientist, as an executive in science museums, and as a community leader.

Margaret Blanchard is Professor of Science Education and University Faculty Scholar in the Department of Science, Technology, Engineering, and Mathematics (STEM) Education at North Carolina State University. She also serves as the Director of Graduate Programs and Associate Department Head and received the NCSU Alumni Association Outstanding Extension and Outreach Award. She designs and studies teacher professional development projects, with the goal of enhancing students' STEM learning experiences and stimulating career interest. Her work has focused primarily on teachers, students, leaders, and families in rural North Carolina, supported by grants from the National Science Foundation, the U.S. Department of Education, the U.S. Department of Agriculture and private foundations. She has published widely with her students in both research and practitioner journals.

Stephen Burgin is an assistant professor of secondary science education at the University of Arkansas. Prior to receiving his Ph.D., Dr. Burgin was a high school chemistry teacher. His research interests center on learner involvement in the authentic practices of scientific research.

Kathryn Campbell-Kibler is an associate professor in the Department of Linguistics at The Ohio State University. She received her BA in linguistics and computer science and MA in linguistics from the University of Chicago, and her PhD in linguistics from Stanford University. Her research investigates the social meanings of linguistic variation, asking how speakers and listeners use variation to manage their social business. She focuses in particular on how listeners process individual variables and incorporate them into their social perceptions of speakers. She is also the outreach director for the "the Pod", a working linguistics lab embedded on the floor of the science museum COSI and the director of See Your Speech, an interactive website that gives users visual displays based on acoustic analyses of their own speech.

Karen Collier is a doctoral student in Science Education at North Carolina State University. She has been a high school science teacher in North Carolina for 13 years prior to beginning the Ph.D. program. After completing the program, she hopes to work in the area of teacher preparation for rural schools in eastern North Carolina.

Brandon Grossman is a doctoral student in Education at University of Colorado Boulder. His research focuses on the development of community partnerships in the preparation of prospective science teachers. Currently, he is studying prospective teachers' participation in community makerspaces and how interactions with youth and families in these spaces shape teachers' developing pedagogical frameworks.

Michelle Hall began her career as a geological engineer in industry, then returned to graduate school in geophysics, and ultimately became an academic. The change in careers reminded her how little scientific inquiry was integrated into undergraduate programs and how few skills students were acquiring for work outside of academia. Her interests turned to the challenge of making science and science research more accessible to broad audiences through curricula and research experiences that promote inquiry, data analysis, and scientific thinking. This drove her to develop technology-based, data-rich curricula, explore effective use of data and visualizations for learning, and foster scientist-teacher partnerships to infuse real science in K-12 education. Presently, she is directing free choice learning programs for youth and developing educational games that promote STEM learning. In all of these projects, she has evaluated and conducted research on what works and why.

Joe E. Heimlich, Ph.D., is Senior Director of Research at COSI, and an Academy Professor Emeritus with The Ohio State University. His career has focused on how people learn in non-school settings, and a special attention on adults and environmental learning.With his Ph.D. in Education Psychology, Adult learning, Joe served as a professor in Extension where he was Leader, Environmental Science, and taught in the School of Environment and Natural Resources and the Environmental Science.

Momoka Horiuchi was a student at Ikuta's laboratory, Faculty of Social Information Studies, Otsuma Women's University. She joined Honmachidahigashi elementary school as a voluntary supporter and helped schoolteachers conduct class lessons. She created self-made teaching materials, Our City Machida, with Read-Aloud functionality by using multimedia-enabled dot code, in collaboration with her friend, Nao Suzuki. She is now working at The Ashikaga Bank, Ltd.

Shigeru Ikuta is now a Senior Scholar at Institute of Human Culture Studies, Otsuma Women's University, Japan. He is an education technologist, teacher educator in Science, and special educator with a focus on student learning and development on the basis of communication aids. He completed his graduate work and earned a doctorate in science at Tohoku University in Sendai, Japan. He had been working as a Professor of Computation Chemistry at Tokyo Metropolitan University for twenty-nine years. He moved to University of Tsukuba and has started collaborative works with schoolteachers, affiliated with the University. He has been conducting many school activities in cooperation with the schoolteachers all over the world for more than 15 years using original handmade teaching materials with dot codes, eBooks with Media Overlays, and Augmented Reality in supporting the students' learning both at the special needs and general schools. He is honored to be an Emeritus Professor of Otsuma Women's University and Tokyo Metropolitan University.

Brian Johnson is Co-founder and Partner of Inform Evaluation & Research. He has directed program evaluation and research in a variety of informal and formal learning settings. Previously, Dr. Johnson served as the Director of Educational Research and Evaluation at the Wildlife Conservation Society (WCS), where he established a research and evaluation unit at the WCS zoos and aquarium in New York City. He directed evaluations of informal learning programs for schools and families, as well as exhibit evaluations. He also led a long-term research study focused on STEM learning outcomes for New York City youth. Additionally, Dr. Johnson has managed local environmental education programs at an urban nature center in New York City's Prospect Park and national conservation education programs for Botanic Gardens Conservation International, a worldwide network of public gardens. Dr. Johnson holds a Ph.D. in Environmental Studies, focusing on measuring environmental quality outcomes of environmental education programs.

Becky Kamas is part of NASA Johnson Space Center's Office of STEM Engagement where she serves as the Activity Manager for STEM on Station. STEM on Station focuses on space station-themed resources and opportunities for students and educators such as hands-on lesson plans and in-flight education downlinks where students can talk live with astronauts on the space station. Prior to joining NASA in 2009, Becky was a high school math teacher in rural Central Texas. Becky holds a Bachelor's degree in Mathematics and a Master's degree in Educational Technology from Texas A&M University and is currently pursuing a Ph.D. in STEM Education at Texas Tech University.

Chikako Kaneko is a teacher in the class of special support services in resource rooms for students with weak eyes at the Honmachidahigashi Elementary School, Machida, Tokyo, Japan. She graduated in Faculty of Education, Bunkyo University. She started working as a teacher at MinamiOya Elementary School, Machida, and settled down to the present school via several Special Needs Schools in Hachiouji City and Katsushika-ward for nearly forty years. Personal computers had widely extended the possibility of learning for students with a hearing impairment, but tools and software were so expensive for school use. Now the schoolteachers can use iPad to help students with disabilities extend his/her learning; they can easily find out wonderful applications suitable for each individual. She now hopes that each student with reading and writing disabilities can improve his/her communication skills and gets confidence through various school activities with eBooks with reading aloud and teaching materials with multimedia-enabled dot code.

Larry Krissek is a clastic sedimentologist, with primary interests in the evolution of climates and ocean environments at high latitudes during the past 65 million years. In particular, he and his students have focused on the Late Cenozoic history of ice-rafting in the Northern Hemisphere, and on the mineralogical and geochemical records of Cenozoic sediment sources and weathering patterns in the Ross Sea sector of Antarctica. He also maintains an active interest in earth science education, including: 1) developing activities for classroom use, 2) identifying common misconceptions about earth science topics, and effective strategies for conceptual change, and 3) conducting effective professional development for pre-service and in-service educators. His work is supported by NSF (Polar Programs, Ocean Drilling, and Geoscience Education) and the Ohio Board of Regents. He has won several awards for teaching, and publishes in a variety of peer-reviewed journals.

Joy Kubarek is Co-Founder and Partner of Inform Evaluation & Research, an education research and program evaluation consulting firm focused on science and conservation education. Dr. Kubarek has served in senior leadership positions at informal science education institutions including the Shedd Aquarium in Chicago, Illinois, and she has trained and coached staff in a number of evaluation capacity building efforts. Dr. Kubarek is active in a number of professional organizations including the National Association for Research in Science Teaching and the Association of Zoos and Aquariums. She has presented nationally and internationally on research and evaluation in informal science education.

Michael Mayhew is a Senior Research Scientist, Science Education Solutions. Following early career phases in academia and at NASA, Mike spent many years as a program manager in geophysics, and later geoscience education, at the National Science Foundation. His chief interest is in fostering public understanding of science. He is a Fellow of the American Association for the Advancement of Science.

Shana McAlexander, M.Ed, MBA, is a Doctoral Candidate in the STEM Education program at NC State University. In her educational research she investigates the impacts of community organizations (industry, government, university) on diversifying STEM majors and careers. She formerly served as the Program Manager for a USDA NIFA grant working to increasing interest and awareness of the bioeconomy and related careers across the state of North Carolina. She coordinated the development and implementation of education programs consisting of an online undergraduate certificate, internship program, and an in-service high school teacher professional development effort.

Katherine McCance is a PhD Candidate in Science Education, and Graduate Research Assistant for the USDA grant-funded Sustainable Bioproducts & Bioenergy Program.

Sarah McClusky brings over 15 years of experience in the field of education with a focus on teacher development and the support of best practices for both pre and in-service teachers. Her experiences extend to developing and delivering statewide professional development opportunities to Pre-K through grade 3 teachers in a variety of disciplines such as literacy, science, social studies, social emotional development, assessment and English language learning. Additionally, Dr. McClusky was instrumental in leading a statewide mentoring and coaching initiative for Pre-K to grade 3 teachers across the state of Ohio. She continues to be committed to teacher development as she previously served leadership roles at the Ohio Department of Education in Curriculum and Assessment and Literacy. Dr. McClusky has also presented at state and national conferences on critical topics for practicing teachers such as assessment, state and national standards and the value of reflective practice. Dr. McClusky's research focused on the relationship of teachers' perceptions of quality professional development and teachers' sense of efficacy. Currently, Dr. McClusky is a Visiting Assistant Professor of Education at Ohio Northern University in the Center for Teacher Education teaching an array of coursework in early childhood and middle childhood education.

Leslie C. Moore is Associate Professor in the Department of Teaching and Learning and the Department of Linguistics at The Ohio State University. Her research focuses on language socialization, bi- and multilingualism, informal science education, and linguistic diversity in education. She has published on linguistics outreach, language socialization in religious and secular schooling, and science education for young emergent bilinguals. She is the Education Director of the Language Sciences Research Lab. Her research, outreach, and engagement have been funded by the National Science Foundation, the Ford Foundation, Spencer Foundation, and Fulbright.

Honami Nakano was a student at Ikuta's laboratory, Faculty of Social Information Studies, Otsuma Women's University. She joined Aiwa elementary school as a voluntary supporter and helped schoolteachers conduct class lessons. She created self-made teaching materials of multiplication table for a subject of Math by using multimedia-enabled dot code. She created one more content, eBook of 'Frog and Toad Are friends' with Read-Aloud functionality by using SMIL languages. She is now working at Fujitsu Broad Solution & Consulting Inc.

Yuko Nakazawa is a teacher at Aiwa elementary school in Tama, Tokyo. She has a teacher license for English in addition to elementary school division. She has been welcoming the students coming to the school as voluntary supporters. She has also been collaborating with Shigeru Ikuta to create self-made teaching materials for English classes with multimedia-enabled dot codes developed Gridmark Inc. and to create eBooks with read-aloud.

Mariko Oshima was a teacher at various elementary schools in Tokyo. After the retirement, she joins various schools as a voluntary supporter. Recently, she joins Hikawa elementary school at Okutama in the mountain side, Tokyo, to help students promote reading more books in collaboration with a librarian. She creates self-made reading-aloud books with dot codes developed by Olympus Co. and Gridmark Inc. in collaboration with the assistant language teacher.

Emi Sakurai was a student at Ikuta's laboratory, Faculty of Social Information Studies, Otsuma Women's University. She joined Honmachidahigashi elementary school as a voluntary supporter and helped schoolteachers conduct class lessons. She created self-made teaching materials, eBooks, Our City Machida, with Read-Aloud functionality by using JavaScript, in collaboration with her friend, Misa Takayanagi. She is now working at Mitsubishi UFJ Morgan Stanley Securities Co., Ltd.

Jessica Sickler, M.S. Ed., is Principal of J. Sickler Consulting, an educational research and evaluation firm, with over 15 years of experience studying learning in informal settings. A major focus of her work has been the study and strategic development of public engagement with science initiatives, including understanding scientist motivations, efficacy of training programs, and audience impact, including work with Portal to the Public, The Story Collider, the Teen Science Café Network, and the Science & Entertainment Exchange.

Lindsay Smith has been in the teaching profession for the past nineteen years and is currently a Marine Science, Environmental Earth Science, and AP Environmental Science teacher at Mooresville High School in Mooresville, North Carolina. Prior to Mooresville, she taught for seven years outside the city of Detroit, MI. Additionally, Lindsay was an Albert Einstein Distinguished Educator Fellow for two terms with the NOAA Office of Education in Washington, DC. Her passions as a science teacher are to bring science to all learners, including students with disabilities and lower elementary students. Lindsay invites students of all abilities into her classroom and provides experiences where everyone has the opportunity to participate and be inspired by science. Her other passion is to connect her students to scientists and real-time data to inspire them to be the next generation of scientists. Lindsay participated in the Ocean Exploration Trust Science Communications Fellowship, NOAA Teacher at Sea, and the NSF PolarTREC program. Through these experiences she traveled to Antarctica, the Bering Sea, Galapagos Islands, and the Caribbean Sea to bring real science into her classroom and encourage her students to be scientists. Past awards include North Carolina Science Leadership Fellow, Macomb County Environmental Educator of the Year, South Lake High School Teacher of the Year, Mooresville School District Teacher of the Year, North Carolina Outstanding Earth Science Teacher, and the Southeast Regional Outstanding Earth Science Teacher.

Kana Sugibayashi was a student at Ikuta's laboratory, Faculty of Social Information Studies, Otsuma Women's University. She joined Fuchu Keyakinomori Gakuen Tokyo Metropolitan School for the Physically Disabled and the Intellectually Disabled as a voluntary supporter and helped schoolteachers conduct class lessons. She created a content for a subject of required Physical Education to support the senior high school students who could learn how to practice the basketball by using multimedia-enabled dot code, in collaboration with her friend, Moeka Yoshida. She also created the content for English class lessons. She is now working at Sompo Japan Insurance Inc.

Nao Suzuki was a student at Ikuta's laboratory, Faculty of Social Information Studies, Otsuma Women's University. She joined Honmachidahigashi elementary school as a voluntary supporter and helped schoolteachers conduct class lessons. She created self-made teaching materials, Our City Machida, with Read-Aloud functionality by using multimedia-enabled dot code, in collaboration with her friend, Momoka Horiuchi. She is now working at Tokyo Gas S-net Co.

Misa Takayanagi was a student at Ikuta's laboratory, Faculty of Social Information Studies, Otsuma Women's University. She joined Honmachidahigashi elementary school as a voluntary supporter and helped schoolteachers conduct class lessons. She created self-made teaching materials, eBooks, Our City Machida, with Read-Aloud functionality by using JavaScript, in collaboration with her friend, Emi Sakurai. She is now working at Asahi Kasei Realty & Residence Corp.

Yui Tamai was a student at Ikuta's laboratory, Faculty of Social Information Studies, Otsuma Women's University. She joined Aiwa elementary school as a voluntary supporter and helped schoolteachers conduct class lessons. She created a teaching material for a library that could promote the younger students' reading books, by using multimedia-enabled dot code. She has a teacher license of Science for junior high and senior high school divisions. She now engages in Graduate School of Comprehensive Human Sciences, The University of Tsukuba.

Richard Venditti is the Elis-Signe Olsson Professor in Pulp and Paper Science and Engineering in the Forest Biomaterials Department at NCSU. He has 26 years of experience in research in the areas of renewable materials from plants, the bioeconomy, paper recycling, and environmental LCA. His research and teaching is involved in developing effective systems to transform renewable plant based resources into sustainable products. Venditti uses environmental life cycle analysis to guide and analyze research in bioproducts. He is currently heading a multi-organization research project to understand the fate of microparticles from laundering in the environment. Venditti received a PhD in Chemical Engineering from Princeton University, was named a TAPPI Fellow in 2012, and was named a Fulbright Senior Specialist in Environmental Science in 2009. He has over 180 peer reviewed publications and three patents.

Laura Wagner is a Professor in the Ohio State University department of Psychology. She received a BA in Language Studies and Cognitive Science from Wellesley College, and an MA in Psychology & a PhD in Linguistics from the University of Pennsylvania. Her research focuses on children's language acquisition, particularly their developing understanding of abstract linguistic elements. She is the Director of the Language Sciences Research Lab, a working research lab embedded inside of a science museum. Her most recent grant examines how the engagement activities of the lab impact the students who conduct them as well as the museum visitors.

Satsuki Yamashita is a teacher at Fuchu Keyakinomori Gakuen Tokyo Metropolitan School for the Physically Disabled and the Intellectually Disabled, Japan, and is now a Head of Research Department. He graduated in Faculty of Health and Sport Sciences, the University of Tsukuba, and earned a master' degree at Disabilities Sciences, Graduate School of Comprehensive Human Sciences, University of Tsukuba. After graduation, she worked as a staff of NPO to support disabled persons for more than five years and started as a teacher at the Komei Special Needs School, Tokyo; she moved to the present Special Needs School in 2012. She always tries to find easier-to-understand ways through developing teaching materials and method for each individual. She hopes that students' learning at school will promote them to extend their livings in the society. She has been collaborating with Professor Shigeru Ikuta, Otsuma Women's University to create hand-crafted teaching materials with dot codes and eBooks.

Moeka Yoshida was a student at Ikuta's laboratory, Faculty of Social Information Studies, Otsuma Women's University. She joined Fuchu Keyakinomori Gakuen Tokyo Metropolitan School for the Physically Disabled and the Intellectually Disabled as a voluntary supporter and helped schoolteachers conduct class lessons. She created a content for a required subject of Physical Education to support the senior high school students who could learn how to practice the basketball by using multimedia-enabled dot code, in collaboration with her friend, Kana Sugibayashi. She also created the content for English class lessons. She is now working at Sompo Japan Insurance Inc.

Index

www.ingramcontent.com/pod-product-compliance
Lightning Source LLC
LaVergne TN
LVHW080528300926
849162LV00007B/1562